WASTE AND WASTE MANAGEMENT

AGRICULTURAL WASTES

CHARACTERISTICS, TYPES AND MANAGEMENT

WASTE AND WASTE MANAGEMENT

Additional books in this series can be found on Nova's website
under the Series tab.

Additional e-books in this series can be found on Nova's website
under the e-book tab.

WASTE AND WASTE MANAGEMENT

AGRICULTURAL WASTES:

CHARACTERISTICS, TYPES AND MANAGEMENT

CAMILLE N. FOSTER
EDITOR

nova publishers
New York

Library of Congress Cataloging-in-Publication Data

Agricultural wastes (Foster)
Agricultural wastes : characteristics, types, and management / editor, Camille N. Foster.
pages cm. -- (Waste and waste management)
Includes bibliographical references and index.
ISBN 978-1-63482-359-3 (hardcover)
1. Agricultural wastes. I. Foster, Camille N. II. Title.
TD930.A393 2014
630.2'0186--dc23
2015006789

Published by Nova Science Publishers, Inc. † New York

CONTENTS

PREFACE

Agricultural wastes (AW) can be defined as the residues from the growing and processing of raw agricultural products such as fruits, vegetables, meat, poultry, dairy products and crops. Agricultural wastes can be in the form of solid, liquid or slurries depending on the nature of agricultural activities. Furthermore, agricultural industry residues and wastes constitute a significant proportion of worldwide agricultural productivity. Although the quantity of wastes produced by the agricultural sector is significantly low compared to wastes generated by other industries, the pollution potential of agricultural wastes is high on a long-term basis. This book discusses the characteristics, types and management options for agricultural wastes.

Chapter 1 – Agricultural wastes (AW) can be defined as the residues from the growing and first processing of raw agricultural products such as fruits, vegetables, meat, poultry, dairy products and crops. AW can be in the form of solid, liquid or slurries depending on the nature of agricultural activities. Agricultural industry residues and wastes constitute a significant proportion of worldwide agricultural productivity. Although the quantity of wastes produced by the agricultural sector is significantly low compared to wastes generated by other industries, the pollution potential of agricultural wastes is high on a long-term basis. The opportunity and feasibility for recycling these wastes comes from two directions: the care for environment reflected by new sets of rules and regulation and the potential to add value to these wastes by adding positive elements. Moreover, they can be used as precursors in many other sectors such as membranes, biosorbents or activated carbons for the removal of dyes, organic molecules, heavy metals and fertilizers. Different types of agricultural wastes, i.e., deoiled soya, coconut shell, neem leaves, hyacinth roots, rice husk, rice straw, rice bran, lemon leaf, tea waste, potato plants wastes, tomato wastes, sesame hull, garlic peel, peanut hull, carrot stem, carrot leave, barley straw, banana stalk, olive stones, almond shells, peach stones, apricot stones, cherry stones, grape seeds, *Trapa natans* husk, bamboo, doum-palm seed coat, walnut shells, rose seed, pine sawdust and coir pith are ideal raw materials for different industrial applications due to their low cost, non-toxic content and their abundance. The final products derived from agricultural wastes have shown equal or even better properties compared to conventional products concerning separation, adsorption and fertility. Previous studies and projects dedicated to the development of AW treatment technologies focused mainly on the reduction of the wastes organic load and on the reduction or the recovery of valuable substances and succeeded to develop suitable technologies and methods. However, if land distribution is planned the organic load and the toxic substances of treated wastes should not be the only issues of concern. Specific care should be taken also for inorganic constituents and especially for K, Cl^-, NO_3^-, SO_4^{2-}, P, Mg, Fe, Zn and others, since the very high concentrations disposed on soil change its quality properties drastically, while electrical conductivity and the concentrations of inorganic soil constituents such as K, P, Fe, Cu remain high even after many years from the last disposal. These practices must take into account important specific local conditions, such as waste characteristic, soil type, background levels of nutrients and pollutants for soil, water and plants, the climate, the relevant crops and the local agricultural practices. Emphasis will be given on specific knowledge and technologies developed so far in Mediterranean countries, their impacts, and constraints and knowledge gaps. Furthermore, policy issues for AW use in Europe and especially in the Mediterranean countries at various levels will be considered. Therefore the aim of this study is to examine the properties and uses of new products derived from agricultural wastes and to research and advance agricultural practices with the use of treated agricultural wastes by recycling nutrients and water from treated agricultural wastes.

Chapter 2 – The design of new food products and agricultural practices have generated a wide diversity of co-products and effluents that often contain a high load of organic matter, from which valuable compounds could be isolated. The surplus and concomitant underutilization of these streams establish serious economic and environmental challenges. Whey, a main co-product arising from cheese manufacturing, was previously considered an environmental pollutant but it is now regarded as a source of many valuable compounds. Among current applications, the production of whey protein concentrates and isolates via ultrafiltration represents the major industrial revenue arising from this stream. The recovery of whey proteins generates an enormous volume of another co-product known as whey permeate. This stream has a high organic load being primarily composed of lactose and minerals. However, recent scientific literature demonstrates the presence of other compounds such as oligosaccharides and peptides, possessing unique bioactivities. Because of the worldwide increase in cheese production, the utilization of whey permeate is under strong scrutiny, with many different strategies being developed to add value to this waste stream. The development of feasible industrial processes to transform, isolate and recover valuable compounds is a key step towards the mitigation of environmental and economic problems arising from constant evolution of our food industry. This chapter focuses on the current utilization and research efforts to valorize whey permeate, where specific processing and environmental challenges are addressed along with the state-of-the-art of the processes and utilizations for the naturally occurring compounds in whey permeate, as well as the valuable products that can be generated from this stream.

Chapter 3 – The olive tree is extensively cultivated in countries of the Mediterranean basin; the area currently under cultivation covers roughly 5.5 10^6 ha in the EU and 11.0 10^6 ha in the world. By-products from olive culture and related industries, such as prunings, leaves, olive pomace, and olive stones, are interesting materials for the production of energy, food, fertilizers and other chemicals due to the large available feedstock and their chemical composition. Olive stones are by-products derived from the olive oil extraction industry and from manufacturing of pitted-table olives. Basically, there are two current ways for valorization of olive stones: thermochemical (energy source by combustion, gasification or pyrolysis) and biochemical (ethanol and xylitol production) conversion. Bioconversion of olive stones can also provide other high added-value products such as xylooligosaccharides or natural antioxidants (tyrosol and hydroxytyrosol). Finally, comparison of the different procedures and potential future applications will be discussed as well.

Chapter 4 – Billion metric tons of agricultural residues are generated every year from industry worldwide that may be considered one of the most abundant, cheap and renewable resources on earth. However, they are normally incinerated or dumped causing environmental problems such as air pollution, soil erosion and decreasing soil biological activity. The reuse of these residues not only prevents environmental concerns, but also can provide farmers the opportunity of a second income from plantation. The incorporation of agricultural residues into polymer matrices is currently a trending topic in research due to the relatively high strength, stiffness and low density of natural fibres present in these residues. Nut by-products, such as almonds (brown hulls, shells and seeds coating) or walnuts (shells), among others, have been used as reinforcement in polymeric materials due to their desirable properties: low density and cost, availability, recyclability, environmental friendliness, total degradation without emission of toxic compounds in composting conditions, and good mechanical properties. On the other hand, nut residues (peanuts, almonds, hazelnuts, chestnuts, walnuts, pecan nuts or pistachios) are rich in bioactive compounds which can be extracted and further used as potential natural additives in food packaging materials with antioxidant and/or antimicrobial activity. In this chapter, different strategies for reusing nut by-products in polymer materials obtaining high value-added materials either as reinforcement or as a source of active compounds are reviewed. Finally, the utilization of gums is currently in the spotlight of the chemical industry.

Chapter 5 – Rice (*Oryza sativa L.*) is a very important component of human diet for many people around the globe. Rice world production is approximately 680 million tons year and Asia leads world harvesting. This is an important source of biomass, especially because there is a tendency to rationalize the use of crude oil and derivatives. Enhancing the utilization of biomass may help to avoid climate and environmental problems. The industrial processing of rice generates some byproducts, such as rice bran and broken rice. Both components can be used as nutritional constituents and they will not be discussed in this work. On the other hand, agricultural residues are relevant in the process, especially rice hull, which accounts for about 20% of the rice crop. This work presents some relevant aspects about the utilization of rice hull. There are many possible applications in different areas, including fermentation and production of ethanol, preparation of cellulose, synthesis of inorganic materials, such as pigments, zeolites, cements, composites, fillers, among others.

Chapter 6 – Biorefining has been defined by the International Energy Agency as the sustainable processing of biomass into a spectrum of marketable products and energy. The concept of sustainability, defined by the World Commission on Environment and Development, arose as consequence, among other reasons, of the energy crisis derived from the imminent depletion of the fossil resources. Thus, a global mindset change is required to face the present scenario, through the reconcilement of three important pillars: economical, societal and environmental issues. Since the current energy system results unsustainable because of imbalance concerns that will have environmental, economical and geopolitical implications far into the future, the sustainable development should be achieved by learning how to use/reuse our resources. In this sense, the use of biomass as a source of products and energy is not new, but its use under a sustainable perspective may imply interesting novelties. With this aim, the Biorefinery outlook should be constructed fulfilling some requirements such as the responsible and optimal exploitation of resources, the application of energy efficient processes and the accessibility of the resulting energy and products, i.e., a compliance in terms of a viable, bearable and equitable development. The available source of biomass (the biorefinery feedstock) determines not only the range of products obtained in the Biorefinery, but also the more or less specific technology and the optimal conversion pathway required for its transformation. These three parameters (feedstock, technology and conversion pathway) allow classifying the biorefining processes, and their combination offers a huge range of possibilities for the biomass exploitation. The use of agricultural wastes in a biorefinery concept offers a promising perspective of sustainable development. The agricultural activity generates significant quantities of lignocellulosic residues (over 60% of the total crop) that are usually left on the cropland or incinerated to prevent the spread of pests and uncontrolled fires. Against the high availability of this biorefinery feedstock, some other issues appear concerning the use of agricultural residues as bioproducts source: volume variability, crop seasonality, low density, heterogeneous chemical composition, localized generation ... These factors are negatively considered when agricultural wastes are proposed as biorefinery feedstock. In the present work, several crop residues (woody and non-woody wastes) were chemically characterized for determining their contents of the main lignocellulosic components (cellulose, hemicelluloses and lignin). Other biomass components, such as moisture and ash, were also determined. In addition, hot water and weak soda solubilities were measured in order to establish the treatability of these agricultural wastes in a biorefinery concept. According to the results, and after an exhaustive crop production assessment, some biorefinery scenarios were proposed considering different worldwide agricultural productions.

Chapter 7 – Large amounts of wastes arising from industrial processing of agricultural products constitute alternative renewable bioresources potentially attractive for bioenergy generation and/or for the manufacture of other useful products. Their conversion additionally contributes to reduce environmental pollution. The present chapter examines thermochemical conversion of the wastes generated from industrialization of an agricultural product into biofuels and/or products potentially applicable for environmental remediation. The selected wastes arise from industrial processing of whole branches (leaves and twigs) from a native evergreen tree *Ilex paraguariensis,* belonging to the Aquifoliaceae family, for the manufacture of yerba mate. It is a widespread product massively consumed in Southern Latin America countries to prepare a popular herbal tea–like beverage. The commercial final product generally contains less than ~ 35% twigs, since they provide an unpleasantly bitter taste to the infusion, and therefore huge quantities of unused twigs emerge as a by-product. Kinetics for the pyrolysis of the twigs is characterized by non-isothermal thermogravimetric analysis from room temperature up to 900 ºC to obtain information for the proper design of full–scale pyrolyzers. A deactivation model which assumes an overall first-order process and considers the physicochemical changes taking place in the biomass with the pyrolysis course through variations of the reaction rate constant with the temperature and solid conversion enables a proper representation of the experimental data over the whole temperature range, with estimated energy activation values between 49 and 137 kJ mol^{-1}. Likewise, yield and characteristics of the three kinds of pyrolysis products, comprising bio-char, bio-oil, and gases, are examined from experiments conducted in a bench-scale fixed-bed installation at temperatures in the range 400 – 700 °C. Gas yield increases with increasing temperature, attaining 43% at 700 °C, while the biochar yield decreases from 30% to 20% with temperature rise. Yield of the bio-oil attains a maximum (53%) at 500 °C, likely arising from the competition between primary formation of volatiles, at relatively low temperatures, and secondary degradation of the condensable vapors at the higher temperatures. All the pyrolysis products could be used in energy applications. The obtained biochars with higher heating value (HHV) of 23 – 24 MJ kg^{-1} have potential as environmentally friendly solid biofuel and could be employed for the manufacture of briquettes mainly for domestic use. Accounting for their high stability, as judged from the molar O:C ratio, another possible application could be incorporation of the biochars into the soil for the storage of atmospheric carbon. In turn, the bio-oils show organic fractions with HHV between 28 and 33 MJ kg^{-1}. Density values of the as-produced liquids (~1 kg dm^{-3}) are rather higher than those for conventional hydrocarbon fuels due to their higher contents of oxygen and water. The crude bio-oils could be directly burnt or subjected to further upgrading to attain characteristics similar to those of fuel-oil. Pyrolysis of the twigs yields low to medium heating value-gases (5 – 11 MJ m^{-3}), mostly composed by CO_2, CO, CH_4 and H_2. Gas composition depends on the temperature, even though CO_2 is the major generated species, followed by CO. Proportion of CO_2 decreases with temperature, particularly at 700 ºC, accompanied by enhancements in the HHV of the gaseous mixtures, as a consequence of compositional variations, attaining a maximum value of 11 MJ m^{-3}. They might contribute to the energy sustainability of the process. Besides, phosphoric acid activation of the yerba mate twigs at pre-established moderate conditions leads to good quality activated carbons with well-developed porous structures characterized by textural parameters (BET surface area of ~ 1000 m^2 g^{-1}; total pore volume of 1cm^3 g^{-1}) comparable to those of commercially available samples.

Chapter 8 – Wastewater from many industries such as textile, leather, paper, printing, food, etc. contains large amount of hazardous dyes. Dyes are not biodegradable and photodegradable due to its synthetic origin and complex aromatic nature. Among various physiochemical processes, adsorption techniques are usually widely used to treat dyes laden wastewater. Although commercial activated carbon is the most widely used adsorbent with large success, its use is limited due to high cost and difficulties in regeneration. Therefore there have been explosive growths in research concerning the use of alternative cost effective non-conventional effective adsorbents in the removal of dyes from aqueous solution. In this research direction, agricultural by-product solid wastes which are available in large quantities worldwide with almost through away price are utilized as effective adsorbents in the removal of inorganics and organics from wastewater. The focus of this book chapter is to review extensive literature information about dyes, its classification and toxicity, various treatment methods and finally dye adsorption characteristics by various agricultural by-products solid wastes as adsorbents. The major objective of this chapter is to organize the scattered available information on the adsorptive removal of dyes from its aqueous solution by raw and treated agricultural by-products. Selectively widely used agricultural solid waste adsorbents in the removal of dyes have also been discussed in details here. Finally mechanism, kinetics and adsorptive behaviour of adsorbents under various physicochemical process parameters have been critically analysed and compared. Conclusions have been drawn from the literature reviewed and few suggestions for future research are proposed.

Chapter 9 – One of the most recent trends in environmental technology is the research turn to green chemistry. It is general accepted that one of the most promising techniques for wastewaters treatment is adsorption. In this basis, numerous adsorbent materials have been synthesized up to now. However, there is a novel concept nowadays, which promotes the use of materials with the lowest possible cost. valuation of them for removing of different pollutants (dyes, cations, anions, etc). In the last years, the instant coffee industry has experienced a constant growth as instant coffee has become one of the most popular kinds of coffee drunk by millions of people around the world. As a consequence, large amounts of coffee grounds, which are the solid residues obtained during the processing of coffee powder with hot water or steam to prepare instant coffee, have been generated worldwide (in the order of 6 millions of tons per year). This work investigates the use of coffee wastes or coffee-based materials as adsorbents for the treatment of wastewaters.

Chapter 10 – At the present time, the demand for energy, goods, and materials is surging because of advanced technology and population growth. However, earth's resources are limited. For this reason, the issues concerning using resources effectively and converting them into energy are important. Taiwan creates a vast amount of agricultural waste every year, which is traditionally burned and buried. The authors do not reuse and recycle agricultural waste, and air pollution is increased when wastes are burned. Therefore, it is necessary to create methods for recycling and reusing agricultural wastes and to transform them into an energy source. This chapter is separated into two parts. The first part will convert agricultural waste into sugar. Agricultural waste is replete with wood fiber that can be reduced into sugar by a microbial method. The second part will use the biological hydrogen production capability of *Clostridium acetobutylicum* ATCC824, with sugar being added to the process. Also, this chapter used ultrasonic treatment for the production of biological hydrogen and calculated the natural frequency of ATCC824. The experiment was designed using the Taguchi method for increasing hydrogen production by using an ultrasonic treatment. Our results showed that the best combination is a temperature of 37 °C, 0.5 MHz ultrasonic frequency, 136 mW/cm^2 ultrasonic intensity, 10 s exposure time, pH 7.5, and a bacterial concentration of 20%. The outcome of our research can be applied to the production of biomass energy and the research and development of ripening techniques for accelerating fermentable food with biomechatronics.

In: Agricultural Wastes
Editor: Camille N. Foster
ISBN: 978-1-63482-359-3

Chapter 1

RECYCLING OF AGRICULTURAL WASTES: TREATMENT AND USES

Z. Ioannou[*], V. Kavvadias and C. Karasavvidis
Hellenic Agricultural Organization - DEMETER, Soil Science Institute of Athens, Sof., Lycovrissi, Attiki, Greece

ABSTRACT

Agricultural wastes (AW) can be defined as the residues from the growing and first processing of raw agricultural products such as fruits, vegetables, meat, poultry, dairy products and crops. AW can be in the form of solid, liquid or slurries depending on the nature of agricultural activities. Agricultural industry residues and wastes constitute a significant proportion of worldwide agricultural productivity. Although the quantity of wastes produced by the agricultural sector is significantly low compared to wastes generated by other industries, the pollution potential of agricultural wastes is high on a long-term basis. The opportunity and feasibility for recycling these wastes comes from two directions: the care for environment reflected by new sets of rules and regulation and the potential to add value to these wastes by adding positive elements. Moreover, they can be used as precursors in many other sectors such as membranes, biosorbents or activated carbons for the removal of dyes, organic molecules, heavy metals and fertilizers. Different types of agricultural wastes, i.e., deoiled soya, coconut shell, neem leaves, hyacinth roots, rice husk, rice straw, rice bran, lemon leaf, tea waste, potato plants wastes, tomato wastes, sesame hull, garlic peel, peanut hull, carrot stem, carrot leave, barley straw, banana stalk, olive stones, almond shells, peach stones, apricot stones, cherry stones, grape seeds, *Trapa natans* husk, bamboo, doum-palm seed coat, walnut shells, rose seed, pine sawdust and coir pith are ideal raw materials for different industrial applications due to their low cost, non-toxic content and their abundance. The final products derived from agricultural wastes have shown equal or even better properties compared to conventional products concerning separation, adsorption and fertility. Previous studies and projects dedicated to the development of AW treatment technologies focused mainly on the reduction of the wastes organic load and on the reduction or the recovery of valuable substances and succeeded to develop suitable technologies and methods. However, if land distribution is planned the organic load and the toxic substances of treated wastes should not be the only issues of concern. Specific care

[*] Tel. +30-210-2832031, fax. +30-210-2842129, E-mail: zioan@teemail.gr.

should be taken also for inorganic constituents and especially for K, Cl^-, NO_3^-, SO_4^{2-}, P, Mg, Fe, Zn and others, since the very high concentrations disposed on soil change its quality properties drastically, while electrical conductivity and the concentrations of inorganic soil constituents such as K, P, Fe, Cu remain high even after many years from the last disposal. These practices must take into account important specific local conditions, such as waste characteristic, soil type, background levels of nutrients and pollutants for soil, water and plants, the climate, the relevant crops and the local agricultural practices. Emphasis will be given on specific knowledge and technologies developed so far in Mediterranean countries, their impacts, and constraints and knowledge gaps. Furthermore, policy issues for AW use in Europe and especially in the Mediterranean countries at various levels will be considered. Therefore the aim of this study is to examine the properties and uses of new products derived from agricultural wastes and to research and advance agricultural practices with the use of treated agricultural wastes by recycling nutrients and water from treated agricultural wastes.

Keywords: agricultural wastes, nutrients, inorganic elements, new products

1. Introduction

Agricultural wastes (AW) can be defined as the residues from the growing and first processing of raw agricultural products such as fruits, vegetables, meat, poultry, dairy products and crops. This term includes both natural (organic) and non-natural wastes produced through various farming activities such as dairy farming, horticulture, seed growing, livestock breeding, grazing land, market gardens, nursery plots and even woodlands. AW can be in the form of solid, liquid or slurries depending on the nature of agricultural activities. Agricultural and food industry residues and wastes constitute a significant proportion of worldwide agricultural productivity (estimated at over 30%) (Sarmah, 2009).

This way or another, the term AW relates to all left-overs and residuals of the agriculture production which do not have direct economical value for the farmer and meant for disposal. Special processes and know-how are needed to convert these wastes into valuable product to be used in the next agricultural production. In most cases (and mainly in field crops and vegetables) the economical breakpoint is in doubt considering the costs of removal, transport and processing of these wastes. The opportunity and feasibility for recycling these wastes comes from two directions: the care for environment reflected by new sets of rules and regulation and the potential to add value to these wastes by adding positive elements.

Although the quantity of wastes produced by the agricultural sector is significantly low compared to wastes generated by other industries, the pollution potential of agricultural wastes is high on a long-term basis. For instance, the land spreading of manures and slurries can cause nutrient and organic pollution of soils and waters. Given animal excreta also contains a plethora of organic chemicals and pathogens, the risk for surface- and groundwater contamination can be high (Sarmah, 2009).

Table 1. Characterization of AW depending on the agricultural activity (Loehr, 1978)

Agricultural activity	Types of wastes	Method of disposal
Crop production and harvest	Straw, stover	Land application, burning, plowing
Fruit and vegetable processing	Biological sludges, trimmings, peels, leaves, stems, soil, seeds and pits	Landfilling, animal feed, land application, burning
Sugar processing	Biological sludges, pulp, lime mud	Landfilling, burning, composting, animal feed
Animal production	Blood, bones, feather, litter, manures, liquid effluents	Land application, fertiliser
Dairy product processing	Biological sludges	Landfilling, land spreading
Leather tanning	Fleshings, hair, raw and tanned trimmings, lime and chrome sludge, grease	By-product recovery, landfilling, land spreading
Rice production	Bran, straw, hull	Feeds, mulch/soil conditioner, packaging material for glass and ceramics
Coconut production	Stover, cobs, husk, leaves, coco meal	Feeds, vinegar, activated carbon, coir products

European Commission has funded many projects relative to AW treatment technologies aiming to recover useful by-products and energy, minimize environmental impacts as well as to produce "cleaner" wastes for safe disposal. Also, some technologies to treat AW have been developed by private funding. AW produced in Mediterranean countries depending on the agricultural activity, contain residues from the processing of raw agricultural products such as fruits, vegetables, meat, poultry, dairy products and crops. The AW that are produced in very big quantities in the Mediterranean region include olive oil mill wastewaters (OMW), wine, swine and animal waste and rice straw.

Depending on the agricultural activity, AW can be categorized according to Table 1 (Loehr, 1978).

2. AW Produced in the Mediterranean Region: Qualitative and Quantitative Characteristics

2.1. Olive Oil Mill Wastewaters (OMW)

Generated in huge quantities over a short period every year (November - April), OMW represent a significant environmental problem in Mediterranean countries. One tone of olives produces approximately 0.8 tones of OMW which are characterized as acidic (pH 4-5), with an average chemical oxygen demand (COD) and biochemical oxygen demand (BOD) content of 120 and 60 g/L, respectively, high concentration of suspended solids (7–15 g/L) and phenolic compounds up to 24 g/L. Their disposal may cause adverse effects on soils, surface- and groundwaters. A number of technologies such as physical, physico-chemical, biological and thermal have been developed for OMW treatment but when used individually suffer from drawbacks e.g., low efficiency or high cost. However, combined or advanced alternative methods show encouraging results (Ben Sassi et al., 2006; Niaounakis and Halvadakis, 2006; Mekki et al., 2007; Khoufi et al., 2008; Komnitsas et al., 2011; Camarsa et al., 2010).

2.2. Wine Waste

The wine making process entails the generation of significant amounts of solid waste and wastewater that should be further treated before disposed in the environment. Wine waste can be divided into crush season (August to February) and non-crush season (March to July) waste involving mainly the production of solid waste and wastewater, respectively. In particular, solid waste can cause bad odours and contaminate soil and water resources; wastewater has a high organic content, contains both suspended (SS) and dissolved solids (DS), is usually acidic and high in sulphide compounds which may lead to odour problems and in nitrogen concentration that can cause eutrophication of water sources (Report of LIFE03 ENV/GR/000223 project, 2004).

2.3. Swine Waste

The swine industry produces wastes that exceed the capacity for direct disposal without causing severe environmental impacts such as odour increase, acidification due to emissions of NH_3, SO_2 and NO_x, increase of greenhouse effect due to emissions of CO_2, CH_4 and N_2O, increase of organic and nutrient loading to surface- and groundwater due to the high BOD and nutrient content of piggery effluent, diffuse spreading of heavy metals, etc. Commonly employed waste treatment systems include aerobic and anaerobic lagoons, anaerobic digestion, aerobic biological treatment using continuous flow activated sludge systems or sequencing batch reactors, composting of solid manure, incineration etc. (Deliverable of LIFE03 TCY/CY/000021 project, 2007).

2.4. Animal Waste

Intensive livestock and poultry production results in large volumes of waste that threaten surface- and groundwater quality in the event of waste spills, leakage from waste storage facilities and runoff from fields on which an excessive amount of waste has been applied as fertilizer. Animal waste generally refers to manure but also includes wastewater, urine, bedding, poultry litter and animal carcasses. The most common waste management practices include techniques to (1) limit waste runoff, such as cementing and curbing animal confinement areas or planting grassed buffers around these areas (2) collect and store waste, such as scraping or flushing systems and storage tanks or retention ponds (3) alter or treat waste, such as reformulating feed mixes or composting and (4) utilization of waste, such as organic fertilizer or additive to animal feed (US GAO, 1999).

2.5. Rice Straw

Worldwide rice production is about 600 million t per year resulting in 810 million t of rice straw production, according to Food and Agriculture Organization of the United Nations. The most common treatments of rice straw include mulching in rice fields, on-site burning for producing manure or composting. Burning, which is difficult in most existing combustion systems, affects the quality and the environment mainly due to CO_2 emission (Pütün et al., 2004). Composting is an attractive treatment of rice straw enhancing plant growth and may be suitable for agricultural applications; however rice straw is rich in C and poor in N and its C/N can vary from 50 to 150 limiting the composting process (Campbell et al., 1995).

3. Technologies for the Recycling of Agricultural Wastes in Intensive Crop Production Systems Developed so Far in the Mediterranean Countries

Population growth, an increase in quality of life and changes in behavioral patterns have caused an increase in the quantity of waste produced by modern society. Possible treatment methods include:

1. Landfilling.
2. Waste incineration or combustion to produce energy.
3. Recycling.

The most common waste treatment method used today throughout the world is landfilling. The advantage of this method is that up until the last decade, its cost was low. However recently with an increase in environmental awareness, the desire to move waste disposal sites far away from population centers, the problems of finding new sites and the consideration of external costs, landfilling costs have increased significantly. The second waste treatment alternative – waste incineration or combustion to produce energy – is an effective method, however the costs of establishing and operating incineration plants are very high. In addition, air quality laws and regulations are limiting the incineration of waste due to the serious concern of air pollution. An increase in the quantity of municipal, industrial and agricultural waste simultaneously with a reduction in the volumetric capacity of incineration sites and an increase in incineration costs are encouraging the recycling of the organic component in waste. For example, composting yard waste is a common waste stabilization process in many American states and in Western Europe (Germany) where its products are recycled for use in agriculture and domestic gardening. After years of research in the United States, the EPA determined that the controlled decomposition of organic waste (composting) is one of the best solutions proposed for treating organic waste and is the most effective long-term treatment methods.

3.1. Basic Principles of the Composting Process

The composting process is a controlled biological oxidization process during which solid organic matter is broken down through a prolonged thermophilic process in which the temporary release of phytotoxic materials takes place and solid organic matter is produced (compost) in the end. Compost is defined as a stabilized product of the composting process, which could promote the growth of plants and serve as a substrate for growth in greenhouses and as a soil additive.

The composting process may be divided into four main stages:

1. The mesophilic stage – the initial stage that continues for a few days during which microorganisms, whose source lies in waste, water and air, start to use the organic matter.
2. The thermophilic stage – the stage at which most of the organic matter decomposition takes place using bacteria.
3. The cooling down stage – the stage at which most of the easily decomposed organic matter is used, the metabolism level decreases and cooling begins.
4. The stability and maturity stage – this is the final stage of the composting process.

Of the many indices proposed to estimate compost stability, the C/N ratio is the most reliable for determining the level of availability of decomposing organic matter and therefore serves as the best index for following up after the composting process. The acceptable C/N ratio for solid organic matter is 10-15 in the solid state and 5-6 in the liquid phase. The C/N ratio for extracted material varies according to the material's source and ranges from 20-40. This ratio decreases during the composting process mainly during the thermophile stage when microorganisms utilize the available carbon for energy and for building their bodies.

Stabilization of the C/N ratio indicates the end of the accelerated decomposition of the composting process and stabilization of the compost. The definition of compost maturity is more problematic since this cannot be characterized at this stage by a single index. During the maturity stage, many indices change, and until it is proven otherwise we are assuming that in order to diagnose compost maturity, there is a need to follow up on other indices simultaneously. In addition, great importance is afforded to the designation of the compost when trying to define its maturity: compost intended for use as an isolated substrate must be completely mature; if not, it will cause serious damage to crops due to a lack of nitrogen and oxygen in the substrate. On the other hand, using compost that is not mature enough in soil 6-10 weeks before sowing will not cause problems for plants since decomposition of the organic matter will be completed in the soil.

3.2. Limitations of the Composting Process

The main objective of each composting process is to ensure the regular manufacture of high quality compost. The composting process in principle is not a complex one, however, it is not easy to operate on a large scope especially when the composition of raw materials is not fixed. Therefore, a suitable composting technology should be used in accordance with the composition of the raw materials intended for the process. The raw materials for composting may be divided generally into four components: agricultural waste (cattle, poultry and plant waste), yard waste, municipal solid waste (separated waste and non-separated waste), and sewage sludge. Based on this, it is possible to choose a suitable composting technology.

Agricultural waste – since in general this waste is produced by potential composting consumers and at a reasonable distance from large population centers, there is no need to invest heavily in building a composting plant in a closed building.

The composting process is the biochemical decomposition of organic matter. Therefore, in order to enable the process to take place, ideal conditions must be made available for the microorganisms (bacteria, actinomycytes and fungi). Without providing these conditions, the process is exposed to delays or at times comes to a complete stop. Optimal conditions for composting are: aeration – 2 volumes daily, moisture content – 30-45%, carbon to phosphate ratio (C/P) – 100-150, and particle size – 0.5-2 cm. These conditions could be controlled in different ways according to the raw materials and the composting process.

Aeration of the compost (oxygen supply) is the most important index. Since the process is based on aerobic decomposition, lack of oxygen will cause the creation of anaerobic conditions, slowing down the decomposition and releasing bad odors. On the other hand, over-aeration could cool down the organic matter and decelerate the decomposition processes. Aeration could be carried out using reverse piles or compressed or suctioned air.

Moisture content in compost enables microbial decomposition to take place. A high moisture content will cause the creation of anaerobic conditions and a deceleration of decomposition, whereas too low a moisture content (below 12%) will prevent all biological activity from taking place and the decomposition will stop. Moisture content may be maintained by adding water and sometimes, if required, bulking agents such as shavings to the sewage sludge.

The importance of the C/N ratio is that it ensures the correct dosage of macro-elements essential for their utilization by microorganisms. A ratio exceeding 35 (surplus carbon) will cause a deceleration of the process whereas a low ratio (surplus nitrogen) will cause a loss of nitrogen (as ammonia gas) into the atmosphere and the creation of bad odors. The C/N ratio is maintained by adding nitrogen-rich components such as poultry fertilizer to carbon-rich waste (straw) or vice versa.

Temperature and minimal time of the thermophilic stage are important in order to destroy populations of microorganisms causing disease to humans and plants, as well as bad weed seeds. EPA standards (EPA, 1987) require maintaining the temperature at 55°C for at least three days. Compost temperature may be reduced (if over-heating occurs) by passing compressed air through the organic matter

Particle size of the compost is important in order to ensure the correct air to water ratio. Particles that are too large have a small inner surface area – this causes a reduction in decomposition rate. On the other hand, particles that are too small could create anaerobic conditions. The right particle size is maintained by trimming and screening.

4. Policy Issues for Agricultural Wastes in Europe and Mediterranean Countries

In Europe, the quantity of domestic organic waste including municipal yard waste was about 100 million tons in 2008, even though potential processing and use in soil was still far from exploitation. In all European countries, compost is manufactured and used in agricultural land, and most compost is manufactured from domestic waste separated at the source. The Landfill Directive contributed considerably to this effort, requiring countries that are members of the European Union to reduce the quantity of organic waste buried by 65% by the year 2015.

The goals determined in countries of the European Union required local authorities in each country to take measures to reduce organic waste buried and recycle them into compost. In Holland, for example, it was determined unequivocally according to a strategic waste program that organic waste must be separated at the source. This ruling is anchored in legislation, which requires the local authorities to separate waste at the source.

The European directive dealing with sewage sludge and in particular protecting the soil as a result of the use of sludge in agriculture and additional soil uses, explicitly determines that the use of sewage sludge in agriculture should be encouraged while ensuring controlled use and strict protection of the environment. Some European countries even determined stricter regulations than requirements of the directive regarding the use of sludge in agriculture. This caused countries such as Holland and Germany to develop the alternative of incineration.

Table 2. A comparison of metals concentrations permissible in compost from sewage sludge

Metal	European Union (mg/kg)	United States (mg/kg)
Cadmium	20-40	85
Copper	1,000-1,750	4,300
Nickel	300-400	420
Lead	750-1,200	840
Zinc	2,500-4,000	7,500
Mercury	16-25	57
Chrome	-	3,000

It should be emphasized here that the organic waste treatment policy in Europe relies on two principles. The first involves reducing the environmental effects in general and gaseous emissions in particular. In several research studies that examined domestic organic treatment methods, a clear preference was found for biological treatment (composting and anaerobic digestion) over other alternatives. The second principle affects the concept of organic waste as having the potential to help in rehabilitating soils and stopping their decline. Extensive areas in southern and central Europe suffer from continuous decline, and are lacking organic matter and organic carbon to develop the soil sufficiently.

5. Zeolites in Agricultural Waste Reuse Processes

Zeolites, natural and synthetic, have been widely studied regarding their suitability to be used in many different environmental applications worldwide Zeolites are natural crystalline aluminosilicates. They are among the most common minerals present in sedimentary rocks. Zeolites occur in rocks of diverse age, lithology, and geologic setting. They are generally considered to be low-cost, safe and environment-friendly materials, suitable for a vast variety of uses. The most well-known natural zeolites are clinoptilolite, erionite, chabazite, heulandite, mordenite, stilbite, and phillipsite.

Structurally, zeolites are tectosilicates exhibiting an open three-dimensional structure containing cations needed to balance the electrostatic charge of the framework of silica and alumina tetrahedral units. Pores and voids are the key characteristics of zeolite materials. The pores and interconnected voids are occupied by cations and water molecules. The internal surface area of these channels are reported to reach as much as several hundred square meters per gram of zeolite, making zeolites an extremely effective ion exchangers.

The Si/Al ratio is an important characteristic of zeolites. The charge imbalance due to the presence of aluminum in the zeolite framework determines the ion-exchange property of zeolites and is expected to induce potential acidic sites. The Si/Al ratio is inversely proportional to the cation content, however directly proportional to the thermal stability. Cations can be exchanged by ion exchange and water can be removed reversibly by application of heat.

The unique physical and chemical properties of zeolites, coupled with their abundance in sedimentary deposits and in rocks derived from volcanic parent materials, have made them useful in many agricultural applications. Most of the initial research on the use of zeolites in agriculture took place in the 1960s in Japan. A brief review of the literature has pointed out that Japanese farmers have used zeolite rock over years to control the moisture content and to increase the pH of acidic volcanic soils. Ion-exchange properties of zeolites can be utilized in agriculture because of their large porosity and high cation-exchange capacity. They can be used as both carriers of nutrients and a medium to free nutrients. Zeolites are important materials with very broad applications in agriculture and environmental engineering. Zeolite incorporation in soil was found to increase crop yields and to promote nutrient use efficiency. Other possible uses being investigated include applications as a carrier of slow-release fertilizers, insecticides, fungicides, and herbicides, and as a trap for heavy metals in soils.

5.1. Physical and Chemical Properties of Zeolites

Two major processes have been identified as kinetics of ion-exchange process in zeolites, namely, particle diffusion and film diffusion. Zeolites are one of the greatest cationic interchangers and their cationic interchange capacity is two to three times greater than other types of minerals found in soils. Zeolites are potential adsorbents due to the ability of their microporous structures to adsorb molecules at relatively low pressure. There is a wide variation in the cation-exchange capacity of zeolites because of the differing nature of various zeolite cage structures, natural structural defects, adsorbed ions and their associated minerals. Thus in short, zeolites are natural materials with the ability to exchange ions, absorb gases and vapors, act as molecular-scale sieves and catalyze reactions owing to fixed pore sizes and active sites in the crystal lattice. The size of clinoptilolite channels controls the size of the molecules or ions that can pass through them and therefore a zeolite like clinoptilolite can act as a chemical sieve allowing some ions to pass through while blocking others. Their internal areas mostly fall in the range of 400-850 m^2/g. Zeolites vary widely in their chemical composition, particularly with respect to contents of SiO_2, CaO, K_2O, Al_2O_3, Na_2O, and Fe_2O_3. The techniques for the separation of clinoptilolites from soil are based on the low specific gravity and fine particle-size characteristics of clinoptilolite.

5.2. Clinoptilolite

Clinoptilolite is the most abundant natural zeolite in soils and sediments and the most commonly used one in agricultural practices as a soil amendment and for promoting nitrogen retention in soils. Clinoptilolite is a member of the heulandite group. A temperature-stable heulandite – most commonly referred to as simply clinoptilolite - seems to be the most abundant zeolite in soils over a wide variety of pH conditions, from slightly acidic to strongly alkaline. Extensive deposits of clinoptilolite are found in Western United States, Bulgaria, Hungary, Japan, Australia, and Iran. Clinoptilolite has a high cationic interchange capacity and a great affinity for NH_4^+ ions.

Despite their unique properties, it is evident that zeolites have little been involved in waste reuse processes. A plausible explanation could be the fact that synthetic zeolites are too expensive to be considered as raw materials in large-scale recycling processes and, at the same time, the natural occurring species – whose prices meet the general standards for sustainable technologies – are often found to differ significantly in texture and properties depending on origin and pretreatment so that only a rough approximation of their behavior is possible; Thus, any technology based on natural zeolites should imply experimental research.

5.3. Zeolites for AW Treatment

Zeolites are widely known as catalysts, adsorbents, molecular sieves, substrates for various gas separations and raw materials for soil remediation. Other possible uses being investigated include applications as a carrier of slow-release fertilizers, insecticides, fungicides, and herbicides, and as a trap for heavy metals in soils. Zeolites are reported to have potential applications in many ways in agriculture.

Some of the characteristics of zeolites that make them potentially desirable for improving the properties of soils are a large internal porosity that results in water retention, a uniform particle-size distribution that allows them to be easily incorporated, a and high cation-exchange capacity that enables them to retain nutrients. The addition of zeolite has improved the nutrient status of sand-based root zones, especially in terms of selective retention of NH_4^+ and K^+ ions.

Some specific interactions that can be benefitted by the presence of zeolite materials in soil are: Soil urease adsorption, nitrate leaching, ammonium trapping and rock phosphate dissolution.

A promising alternative option to remove specific contaminants from aqueous solution could be the use of low-cost sorbent materials. Among the different minerals, which possess sorbent properties, zeolites appear to be one of the most promising sorbents for this purpose (Tashauoei et al., 2010). Different kinds of natural zeolites are most frequently suggested as ammonium exchangers for wastewater treatment applications (Hedstrom, 2001). It is well known that aluminosilicate molecular sieves (zeolites) are considered the best sorbents which are used in technological processes of division and deep clearing of liquid and gas mixtures due to their chemical nature and particularities of their porous structure (Nesterenko, 2007). Clinoptilolite is known for its ability to remove ammonium from polluted waters (Rahmani and Mahvi, 2006). Matulova and Klokocnıkova (1994) reported 50% inhibition of algae growth after zeolite addition to water. Haggerty and Results of Erdem et al. (2004) indicate that natural zeolites hold great potential to remove cationic heavy metal species from industrial wastewater.

5.3.1. Zeolites in Composting Processes

Turan (2008) studied the salinity uptake by natural zeolite when used as an ingredient during the composting process. The amounts of 5% and 10% of natural zeolite were applied to poultry litter as volume and compared with the compost made with no amendment. The results clearly showed that the salinity level of poultry litter was too high. It was found that the salinity level in the end compost decreases with increasing the amount of natural zeolite used. Salinity uptake efficiencies were 66.64% and 88.92% for end product containing 5% and 10% natural zeolite, respectively. Significantly, the addition of natural zeolite to poultry litter compost was found to have a beneficial effect on the characteristics of the end product. Similarly, the addition of natural zeolite to municipal solid waste compost was found to have a beneficial effect on the characteristics of the end product.

5.3.2. Zeolites in Piggery Wastes Treatment

A comparison between upflow anaerobic sludge bed reactor (UASB) and anaerobic fixed bed reactor (AFBR) at a similar organic volumetric loading rate of 5 kg DQO/m^3/day was carried out by Sanchez et al. (1995). 60% of the piggery waste COD was removed with the AFBR compared to 40% with the UASB, thus showing a better performance of the AFBR. After 1-h sedimentation secondary process, both anaerobic effluents were treated by ionic exchange with natural zeolite due to their high values of ammoniacal nitrogen (NH~- plus free NH3). The high removal of nutrients reported (90%) shows zeolite to be a good choice as tertiary treatment.

5.3.3. Zeolite Combined with Organic Manure

Gul et al. (2007) conducted a research in two successive seasons to compare the effect of nutrient sources, organic manure and inorganic conventional nutrient solution, in cucumber production performed with different local substrates. In fall, the experiment was designed to test three factors, namely cultivar [(a) Armada, (b) Gordion], nutrient source [(a) inorganic nutrient solution, (b) solid organic manure] and substrate [(a) 3 + 1 perlite + clinoptilolite, (b) 1 + 1 perlite + clinoptilolite, (c) 3 + 1 tuff + clinoptilolite, (d) 1 + 1 tuff + clinoptilolite, v/v]. Results showed that organic manuring decrease the total yield by 22.4% in comparison to inorganic nutrient solution. In organic manure treatment, vigorous variety (Armada) gave higher yield than less vigorous variety (Gordion). In the spring season, the tested factors were decreased to two and tested as nutrient source [(a) inorganic nutrient solution, (b) solid organic manure, (c) organic nutrient solution] and substrate. Armada was the only cultivar. Compared to that of the inorganic nutrient solution, total yield was reduced by 10.9% in the organic nutrient solution system and 31.3% in solid organic manure treatment.

In the case of substrates, yield displayed an increasing trend in perlite-based media. This conclusion can be related to higher water holding capacity of perlite-based media compared to that of tuff-based. The results of this study are in accordance with the findings of Gul et al. (2005). Under the light of previous reports of Harland et al. (1999) and Gul et al. (2005), the hypothesis regarding clinoptilolite additions into perlite and tuff was to improve plant growth and yield due to the increase in the uptake of nutrients especially in organic nutrition. In this study, contrasting to this idea, increasing amount of clinoptilolite in tuff medium fertilized with organic manure resulted in lower yields. Compared to perlite media, the low water holding capacity of clinoptilolite as well as tuff might have decreased the humidity in rooting zone, which is of particular importance in organic nutrition.

6. Production of Adsorbents either Biosorbents or Pyrolized/Activated Carbons from Agricultural Wastes

In a biosorption process, biosorbents, which derived from agricultural wastes, need little processing so as to increase their adsorption ability including rinsing with boiling and cold water, drying and sieving. Biosorbents reduces production cost by using a cheap raw material and eliminate energy costs correlated with thermal treatment as it happens to activated carbons (Salleh et al., 2011). Deoiled soya was washed with deionized water, dried, diluted to H_2O_2 to remove organic impurities and sieved (Gupta et al., 2009). Rice husk was washed with distilled water to remove dust and impurities and then dried in sunlight and in an oven at 60°C, ground and sieved to different sizes (Safa and Bhatti, 2011). Similar procedures followed with tea wastes but before the dry process they were boiled with distilled water to remove caffeine, tannin and other dyes (Uddin et al., 2009). The same procedure applied to sesame hull (Feng et al., 2011), garlic peel (Hameed and Ahmad, 2009) and potato plant wastes (Gupta et al., 2011) leading to the preparation of biosorbents appropriate for dye removal. As far as the preparation of biosorbents for heavy metal and ions removal may concern, raw materials such as tomato wastes (Yargic et al., 2014), peanut hull (Zhu et al., 2009), and rice straw, rice bran, rice husk, hyacinth roots (Singha and Das, 2013) need washing, drying and sieving but there is also an intermediate stage in which they were subjected to low acid or alkali treatment, e.g., 0.1 N NaOH, H_2SO_4, HCL, HNO_3, so as to remove color.

On the contrary, the general process to produce activated carbons is based on the carbonization and activation of the raw materials. There are many raw materials appropriate for the production of activated carbons. Two main categories follows: the first one includes grade low coal such as lignite, waste pulp solution, various agricultural wastes and the second category includes synthetic resins and fibers (Simitzis and Ioannou, 2011, Xuefei et al., 2009). The procedure of carbonization takes place in furnaces with a gradual increase of temperature under a continuous flow of an inert gas such as nitrogen. After the carbonization process of the raw material, an activation process follows so as to ameliorate the pore volume and new porosity will be created. Activation process may be achieved either physically (thermal process) or chemically. Activated carbon presents an extended microporosity increasing their specific surface area and consequently their adsorptive ability. Activated carbons are divided to two basic forms: the granular activated carbon (GAC) and the powdered activated carbons (PAC) (Salleh et al., 2011). Agricultural wastes such as apricot stones, almond shells, cherry stones and grape seeds were carbonized by steam pyrolysis at 600-700°C (Gergova et al., 1994). Other wastes such as pine sawdust, rose seed and cornel seed were dried, crushed and sieved. Wastes were diluted to $ZnCL_2$ solution and stirred for 2h. The mixture was dehydrated in an oven and then pyrolyzed up to 800 ^{o}C under nitrogen flow. Activated carbon was washed with 3M HCL solution by heating at 90°C, then filtered, rinsed by distilled water and dried (Açikyildiz et al., 2014). Similar procedure followed with wild olive stones (Boumaza et al., 2012) and walnut shells (Martínez et al., 2006).

6.1. Organic Compounds (Dyes, Pesticides, Pharmaceuticals, Industrial Solvents) Removal

The removal of dyes from wastewaters is considered an environmental challenge and an effective removal process is their adsorption onto agricultural wastes. Dyes can be classified to anionic, cationic and non-ionic dyes. Cationic dyes are basic dyes while anionic dyes are direct, acid and reactive dyes (Salleh et al., 2011). Dye adsorbents derived from agricultural wastes can be classified either to biosorbents or pyrolyzed/activated carbons.

An agricultural waste (deoiled soya) was used for the removal of a synthetic azo dye, carmoisine A, from aqueous solutions. Carmoisine A was the primary red coloring material for foods such as jams and preservatives for many years. Batch experiments, kinetic studies and column operations took place so as to understand dye adsorption onto deoiled soya (Gupta et al., 2009). The biosorption of textile dyes such as Everdirect Orange-3GL and Direct Blue-67 onto rice husk was also studied. A factorial experimental design technique was applied to study the effect of initial dye concentration, biosorbent dose and pH at five levels. pH was found to be the most significant parameter for both dyes (Safa and Bhatti, 2011). Other studies examined the adsorption of methylene blue (MB) on garlic peel (Hameed and Ahmad, 2009) or sesame hull (Feng et al., 2011) or tea waste (Uddin et al., 2009). A batch process was carried out, examining the contact time, MB initial concentration, optimal pH and temperature for the three adsorbents. Adsorption isotherms were modelled with the Langmuir, Freundlich and Temkin equations. The equilibrium data were well represented by the Langmuir isotherm for the removal of MB dye from aqueous solutions onto sesame hull and tea waste and by the Freundlich isotherm for the removal of MB dye from aqueous solutions onto garlic peel. The kinetic data were analyzed using pseudo-first and pseudo-second order equations. According to the results, pseudo-second order model described well the kinetic adsorption data of MB removal onto all adsorbents. Similar studies examined the removal of methylene blue and malachine green dyes from aqueous solutions onto potato (*Solanum tuberosum*) plant wastes (Gupta et al., 2011) and waste materials of Daucus carota plant, i.e., carrot stem powder and carrot leaves powder (Kushwaha et al., 2014). The effect of physicochemical parameters was investigated. The adsorbents were characterized through Fourier transform infrared spectroscopy (FTIR) and scanning electron microscopy (SEM). Adsorption kinetics and isotherm studies were also studied. Direct and acid dyes such as C.I.Direct red 80 (DR80), C.I.Direct red 81 (DR81), C.I.Acid blue 92 (AB92), C.I.Acid red 14 (AR14), removed from aqueous solutions by the application of Soy Meal Hull (SMH) (Arami et al., 2006). The material was characterized by Fourier transform infrared spectroscopy (FTIR) and scanning electron microscopy (SEM). The results have shown that SMH presents functional groups, i.e., hydroxyl, amine and carbonyl groups and adsorption capacities equal to 178.57, 120.48, 114.94 and 109.89 mg/g for DR80, DR81, AB92 and AR14, respectively. Consequently, SMH can be characterized as a low-cost, natural adsorbent for dye removal from wastewaters.

Pyrolyzed or activated carbons can be prepared from a variety of raw materials especially agricultural solid wastes. The main characteristics of a carbon material so as to be used in adsorption applications are high specific surface area, adequate pore size distribution and activated sites, i.e., chemical groups containing oxygen or other heteroatoms. Coconut tree flower carbon (CFC) and Jute fibre carbon (JFC) were studied for their adsorption abilities on Reactive red dye (RR) (Senthilkumaar et al., 2006). Kinetic studies have shown that pseudo-second order equation describes well the dye adsorption while isotherm data were fitted well to Langmuir equation indicating that the adsorption capacities were equal to 181.9 and 200 mg/g for CFC and JFC, respectively. The overall rate of RR adsorption appeared to be controlled by chemisorption. Activated carbons, which derived from apricot and cherry stones, almond shells and grape seeds through steam pyrolysis (Gergova et al., 1994), or pine sawdust (PS), rose seed (RS), and cornel seed (CS) through activation (Açikyildiz et al., 2014) were characterized by N_2 adsorption and methylene blue adsorption. The results have shown that the iodine and methylene blue numbers increased with the increase of treatment temperature and soak time for activated carbons derived from apricot stones. The final pore structure of activated carbons depends on the initial raw materials, resulting to microporous structure of activated carbons derived from apricot stones and to meso- and macroporous structure of activated carbons derived from cherry stones and grape seeds while activated carbons derived from almond shells presented an intermediate structure. As far as activated carbons derived from pine sawdust, rose seed, and cornel seed may concern, they presented high specific surface areas equal to 1825, 1265 and 1355 m^2/g and methylene blue indexes were found around 300, 297 and 299 mg/g for PS, RS and CS, respectively. Similar studies concerning the adsorption of basic red 46 dye on activated carbons derived from wild olive stone (Boumaza et al., 2012) led to high adsorption yields.

There are several procedures available for pesticides removal from water and wastewaters such as photocatalytic degradation, ultrasound combined with photo-Fenton treatment, advanced oxidation processes, aerobic degradation, electrodialysis membranes, ozonation and adsorption through carbons. Activated carbon was prepared from banana stalk by KOH and CO_2 activation as adsorbent for the removal of pesticides, i.e., 2,4-dichlorophenoxyacetic acid (2,4-D) and bentazon (Salman et al., 2011). Kinetic adsorption data were better described by pseudo-second order model while equilibrium adsorption data were fitted well to Freundlich model. Higher adsorption capacity was presented to 2,4-D (196.33 mg/g) than bentazon (115.07 mg/g).

Pharmaceuticals have been detected at high concentration levels in aquatic environments. Clofibric acid (CA) and carbamazerine (CBZ) are drugs that used widely in medicine and were found in wastewaters and groundwater. A new biosorbent from waste rice straw was employed to remove both pharmaceuticals from aqueous solutions (Liu et al., 2013). The adsorption of both pharmaceuticals followed pseudo-second order equation. Intraparticle diffusion was the rate limiting step. Equilibrium adsorption data were fitted well to non-linear Freundlich isotherm. Maximum adsorption capacities were equal to 126.3 and 40 mg/g for CA and CBZ, respectively.

Industrial solvents have a wide use during the production of adhesives, resins, corrosion inhibitors, textile water-repellents. Pyridine is an organic liquid with high solubility. Agricultural waste, e.g coconut coir, was activated by H_3PO_4 for the removal of pyridine (Ahmed et al., 2014). The adsorption process took place in a batch/column system examining kinetics, isotherm modelling, error and thermodynamic analysis. The adsorbent presented acidic and basic functional groups, i.e., hydroxyl, carboxylic acid and bounded water molecules. The specific surface area was equal to 1254.67 m^2/g. The adsorptive capacity was 107.18 and 140.94 mg/g in batch and in column system, respectively. Adsorption kinetic and equilibrium data described well with the pseudo-second-order equation and Langmuir model, respectively. Thermodynamic parameters have shown the endothermic and spontaneous nature of the process. Another study examined the adsorption of trichloroethylene, which is an industrial degreaser, on zerovalent iron/activated carbon adsorbent (Su et al., 2013). The adsorbent derived from coir pith through pyrolysis. Such adsorbents can effectively dehalogenate the chlorinated compounds in water.

Other organic compounds such as phenols and substituted phenols were also removed using activated carbons derived from coconut shell (Singh et al., 2008), olive stones (Ioannou and Simitzis, 2009) and apricot stone shells (Daiffullah and Girgis, 1998). The results indicated that the activated carbons derived from agricultural solid wastes can be used as potential adsorbent for phenols in wastewaters. The removal of oil from water can be achieved by a chemically modified agricultural by-product such as barley straw (Ibrahim et al., 2010). The adsorption capacity was equal to 576 mg/g at 25 ^{o}C. Several factors influenced the adsorption process, i.e., temperature, solution pH, particle size, adsorbent.

6.2. Heavy Metals and Other Ions Removal

Various methods have been developed for the remediation of contaminated soils and waters from heavy metals, i.e., thermal, biological, physical and chemical treatments. Among all the other methods, adsorption is highly selective, more efficient, easy to operate and low cost method (Zhu et al., 2009).

One of the most dangerous water pollutants is hexavalent chromium, Cr (VI), which is widespread in industrial wastewaters. A low cost aquatic plant residue, *Trapa natans L.*, was used as raw material for the preparation of Fe-modified activated carbon (THAC-Fe) (Liu et al., 2010). Activated carbon was tested for its ability to adsorb Cr(VI) from aqueous solutions. Iron addition to activated carbons increased the adsorption ability with the maximum adsorption capacity to reach 11.83 mg/g Cr(VI). Kinetic data fitted well to pseudo-second order equation while equilibrium data followed the Temkin and Freundlich models. The entire adsorption process was controlled by external mass transfer and intraparticle diffusion. Similar studies examined the removal of nickel from aqueous solutions by activated carbons derived from doum-palm seed coat (El-Sadaawy and Abdelwahab, 2014) and zinc, cadmium and copper by activated carbons derived from almond shells, olive stones and peach stones (Ferro-Garcia et al., 1988). Batch experiments were conducted investigating the role of solution pH, initial nickel concentration, adsorbent dose and contact time. Kinetic data fitted well to pseudo-second order equation while equilibrium data followed the Freundlich model. The maximum nickel adsorption capacity was 13.51 mg/g (El-Sadaawy and Abdelwahab, 2014). According to Ferro-Garcia et al., different parameters were examined such as solution pH and extent of heavy metal adsorption in the presence of CL^-, CN^-, SCN^-, EDTA. Chemical structure and porous texture are the main parameters which influenced the adsorption of zinc, cadmium and copper by activated carbons. Other studies (Baccar et al., 2009) concerning the uptake of Cu(II) ions from activated carbons derived from olive waste cakes and a chemical modification using $KMnO_4$ as oxidant ($KMnO_4$-modified carbons) and unmodified activated carbons showed high adsorption capacity equal to 35.3 and 12 mg/g for modified and unmodified carbons, respectively.

Although the production of activated carbons consists of pyrolysis at high temperatures and physical or chemical activation, biosorbents, which derived from agricultural wastes, need little processing so as to increase their adsorption ability. Removal of copper(II) from aqueous solutions by chemically – treated tomato waste (*Solanum lycopersicum*) (Yargiç et al., 2014) or by peanut hull (Zhu et al., 2009) or by coconut shell, neem leaves, hyacinth roots, rice straw, rice bran and rice husk (Singha and Das, 2013) were investigated. The Cu(II) removal was pH dependent to all cases indicating that the optimum pH for adsorption was equal to 8 (Yargiç et al., 2014), 5.5 (Zhu et al., 2009) and 6 (Singha and Das, 2013), respectively. Kinetic data were best described by pseudo-second order equation to all cases. The maximum biosorbent capacity was equal to 22.37 mg/g for 125 mg/L copper solution (Yargiç et al., 2014), 21.25 mg/g (Zhu et al., 2009) and 19.89, 17.49, 21.80, 18.35, 20.98, 17.87 mg/g for coconut shell, neem leaves, hyacinth roots, rice straw, rice bran and rice husk, respectively (Singha and Das, 2013). Another study (Brown et al., 2000) assessed the removal of copper(II), cadmium(II), zinc(II) and molybdenum(II) from wastewaters by peanut hulls and hull pellets indicating that almost 90% of metals removal occurred at the first 20 min of contact.

The removal of other ions except of heavy metals by adsorbents, which were prepared from agricultural wastes, were also reported in literature. Dried orange juice residue (DOJR) (Paudyal et al., 2013) or *Citrus limonum* (lemon) leaf (Tomar et al., 2014) was converted to a promising adsorbent for fluoride ions from water. Different metal loaded DOJR were produced for the adsorption of fluoride ion traces from wastewaters leading to maximum adsorption capacities with values between 0.67 and 1.43 mg/g depending on the metal, i.e., Zr(II), Ce(IV), Al(III) loaded DOJR. Adsorption of fluoride ion from aqueous solutions on *Citrus limonum* (lemon) leaf led to 70% maximum defluoridation capacity with optimal pH equal to 2.

7. Production of Membranes from Agricultural Wastes

Research studies have focused on the adsorptive materials made from agricultural wastes due to their low cost, abundance and ecological friendly nature. Porous mixed matrix membranes (MMMs) were prepared by particles of banana peel, tea waste and shaddock peel as fillers in polyethersulfone (PES) (Lin et al., 2014). The prepared MMMs were applied in the adsorption of methylene blue (MB) and methyl violet 2B (MV) dyes using either a batch or flow process. Moreover, desorption took place so as to regenerate membranes. According to the results, the saturated dye adsorption capacities for MMMs were equal to 294-340 mg/g for MB and 308-370 mg/g for MV in a batch process. Desorption reached 95% leading to membrane regeneration. In flow type operation, high dye removal and recovery efficiency could be retained after three adsorption/desorption cycles.

CONCLUSION

Agricultural wastes can be used as raw materials for the production of fertilizers, membranes, biosorbents or activated carbons for the removal of dyes, organic molecules and heavy metals. Different types of agricultural wastes, i.e., deoiled soya, coconut shell, neem leaves, hyacinth roots, rice husk, rice straw, rice bran, lemon leaf, tea waste, potato plants wastes, tomato wastes, sesame hull, garlic peel, peanut hull, carrot stem, carrot leave, barley straw, banana stalk, olive stones, almond shells, peach stones, apricot stones, cherry stones, grape seeds, *Trapa natans* husk, bamboo, doum-palm seed coat, walnut shells, rose seed, pine sawdust and coir pith are ideal raw materials for different industrial applications due to their low cost, non-toxic content and their abundance. The final products derived from agricultural wastes have shown equal or even better properties compared to conventional products concerning separation, adsorption and fertility.

Most of the projects related to sustainable use of AW are focused on the development of innovative technologies of wastes treatment as well as, of innovative technologies for the improvement of production processes which further produce "cleaner" wastes. Treated wastewaters or composted sludges produced by these technologies could potentially be used for irrigation and/or fertilization of crops after evaluation and definition of specific terms and conditions regarding their suitability to support plant growth, without causing phytotoxicity and environmental problems, in general. In addition, relevant studies have proven that natural zeolites can be very effectively used as a slow release fertilizer and cultivation yields can be highly increased thanks to its capacity to hold water and equilibrate the root environment particularly when they used with AW. Successful application cases have been reported in many countries but not in European member states. Despite unique properties of zeolites, it is evident that zeolites have little been involved in waste reuse processes. Thus, any technology based on natural zeolites should imply experimental research.

REFERENCES

Açikyildiz, M., Gürses, A., Karaca, S., 2014. Preparation and characterization of activated carbon from plant wastes with chemical activation, *Microporous and Mesoporous Materials* 198, 45 – 49.

Ahmed, M.J.K., Ahmaruzzaman, M., Reza, R.A., 2014. Lingocellulosic – derived modified agricultural waste: Development, characterization and implementation in sequestering pyridine from aqueous solutions, *Journal of Colloid and Interface Science* 428, 222 – 234.

Arami, M., Limaee, N.Y., Mahmoodi, N.M., Tabrizi, N.S., 2006. Equilibrium and kinetics studies for the adsorption of direct and acid dyes from aqueous solution by soy meal hull, *Journal of Hazardous Materials* 135, 171-179.

Baccar, R., Bouzid, J., Feki, M., Montiel, A., 2009. Preparation of activated carbon from Tunisian olive – waste cakes and its application for adsorption of heavy metal ions, *Journal of Hazardous Materials* 162, 1522 – 1529.

Ben Sassi, B., Boularbah, A., Jaouad, A., Walker, G. and Boussaid, A. 2006. A comparison of Olive oil Mill Wastewaters (OMW) from three different processes in Morocco, *Process Biochemistry,* 41, 74–78.

Boumaza, S., Kaouah, F., Berrama, T., Trati, M., Bendjama, Z., 2012. Optimization of preparation conditions for activated carbon from waste materials of agricultural origin for the removal of basic red 46, *Chemometrics and Intelligent Laboratory Systems* 118, 311 – 316.

Brown, P., Jefcoat, I. A., Parrish, D., Gill, S., Graham, E., 2000. Evaluation of the adsorptive capacity of peanut hull pellets for heavy metals in solution, *Advances in Environmental Research* 4, 19 – 29.

Camarsa, G., Gardner, S., Jones, W., Eldridge, J., Hudson, T., Thorpe, E., O'Hara, E., 2010. LIFE among the olives: Good practice in improving environmental performance in the olive oil sector. *European Commission. Environment Directorate-General.* Luxembourg: Office for Official Publications of the European Union, ISBN 978-92-79-14154-6.

Campbell, L.S., Davies, B.E., 1997. Experimental investigation of plant uptake of caesium from soils amended with clinoptilolite and calcium carbonate. *Plant and Soil,* 189, 65-74.

Daifullah, A. A. M., Girgis, B. S., 1998. Removal of some substituted phenols by activated carbon obtained from agricultural waste, *Water Research* 32, 1169 – 1177.

El – Sadaawy, M., Abdelwahab, O., 2014. Adsorptive removal of nickel from aqueous solutions by activated carbons from doum seed (*Hyphaenethebaica*) coat, *Alexandria Engineering Journal* 53, 399 – 408.

Environment Protection Authority, 2004. *EPA Guidelines for Wineries and Distilleries.*

Erdem, E., Karapinar, N., & Donat, R., 2004. The removal of heavy metal cations by natural zeolites. *Journal of Colloid and Interface Science,* 280, 309-314.

Feng, Y., Yang, F., Wang, Y., Ma, L., Wu, Y., Kerr, P.G., Yang, L., 2011. Basic dye adsorption onto an agro – based waste material – Sesame hull (*Sesamum indicum L.*), *Bioresource Technology* 102, 10280 – 10285.

Ferro – Garcia, M.A., Rivera – Utrilla, J., Rodriguez – Gordillo, J., Bautista – Toledo, I., 1988. *Adsorption of zinc, cadmium and copper on activated carbons obtained from agricultural by – products, Carbon* 26, 363 – 373.

Gergova, K., Petrov, N., Eser, S., 1994. Adsorption properties and microstructure of activated carbons produced from agricultural by-products by steam pyrolysis, *Carbon* 32(4), 693-702.

Gul, A., Erogul, D., Ongun, A.R., 2005. Comparison of the use of zeolite and perlite as substrate for crisp-head lettuce. *Scientia Horticulturae* 106, 464–471.

Gupta, V. K., Mittal, A., Malviya, A., Mittal, J., 2009. Adsorption of carmoisine A from wastewater using waste materials – Bottom ash and deoiled soya, *Journal of Colloid and Interface Science* 335, 24 – 33.

Gupta, N., Kushwaha, A.K., Chattopadhyaya, M.C., 2011. Application of potato (*Solanum tuberosum*) plant wastes for the removal of methylene blue and malachite green dye from aqueous solution, *Arabian Journal of Chemistry*, In Press

Hameed, B. H., Ahmad, A. A., 2009. Batch adsorption of methylene blue from aqueous solution by garlic peel, an agricultural waste biomass, *Journal of Hazardous* Materials 164, 870 – 875.

Harland, J., Lane, S., Price, D., 1999. Further experiences with recycled zeolite as a substrate for the sweet pepper crop. *Acta Horticulturae* 481, 187–194.

Hedstorm, A., 2001. Ion exchange of ammonium zeolites: A literature review. *Journal of Environmental Engineering,* August, 673.

Hedstorm, A., 2001. Ion exchange of ammonium zeolites: A literature review. *Journal of Environmental Engineering,* August, 673.

Ibrahim, S., Wang, S., Ang, H.M., 2010. Removal of emulsified oil from oily wastewater using agricultural waste barley straw, *Biochemical Engineering Journal* 49, 78 – 83.

Ioannou, Z., Simitzis, J., 2009. Adsorption kinetics of phenol and 3-nitrophenol fromaqueous solutions on conventional and novel carbons, *Journal of Hazardous Materials,* 171, 954-964.

Khoui, S., Aloui, F., Sayadi S., 2008. Extraction of antioxidants from olive mill wastewater and electrocoagulation of exhausted fraction to reduce its toxicity on anaerobic digestion. *Journal of Hazardous Materials,* 151: 531-539.

Komnitsas K., Zaharaki D., Doula M., Kavvadias V., 2011. Origin of Recalcitrant Heavy Metals Present in Olive Mill Wastewater Evaporation Ponds and Nearby Agricultural Soils, *Environmental Forensics* 12, 1–8.

Kushwaha, A. K., Gupta, N., Chattopadhyaya, M. C., 2014. Removal of cationic methylene blue and malachite green dyes from aqueous solution by waste materials of *Daucus carota*, *Journal of Saudi Chemical Society* 18, 200 – 207.

LIFE03 ENV/GR/000223 (2004). *DIONYSOS: Development of an economically viable process for the integrated management via utilization of winemaking industry waste;* Production of high added value natural products and organic fertilizer

LIFE03 TCY/CY/00002 (2007). PIGWASTEMAN: Guidelines to the Cyprus competent authorities for policy formulation for sustainable management of pig farming wastes in compliance with EU practice.

Lin, C.H., Gung, C.H., Sun, J. J., Suen, S.Y., 2014. Preparation of polyethersulfone / plant – waste – particles mixed matrix membranes for adsorptive removal of cationic dyes from water, *Journal of Membrane Science* 471, 285 – 298.

Liu, W., Zhang, J., Zhang, C., Wang, Y., Li, Y., 2010. Adsorptive of Cr (VI) by Fe – modified activated carbon prepared from *Trapa natans* husk, *Chemical Engineering Journal* 162, 677 – 684.

Liu, Z., Zhou, X., Chen, X., Dai, C., Zhang, J., Zhang, Y., 2013. Biosorption of clofibric acid and carbamazepine in aqueous solution by agricultural waste rice straw, *Journal of Environmental Sciences* 25(12), 2384 – 2395.

Loehr, 1978. *Best management practices for agriculture and silviculture Proceedings of the Cornell Agricultural Waste Management Conference.* Ann Arbor Science Publishers

Martínez, M. L., Torres, M. M., Guzmán, C. A., Maestri, D. M., 2006. Preparation and characteristics of activated carbon from olive stones and walnut shells, *Industrial Crops and Products* 23, 23 – 28.

Mekki, A., Dhouib, A., Sayadi, S., 2007. Polyphenols dynamics and phytotoxicity in a soil amended by olive mill wastewaters. *Journal of Environmental Management* 84, 134–140.

Muhlbachova, G., Simon, T., 2003. Effects of zeolite amendment on microbial biomass and respiratory activity in heavy metal contaminated soils. *Plant Soil and Environment,* 49, 536-541.

Nesterenko, V. P., 2007. Use of ion-exchange composites based on natural zeolites for cleaning of water solutions with purpose to create environmentally safe technologies. In: *Proceedings of the European Congress of Chemical Engineering* (ECCE-6), Copenhagen, 16-20 September.

Niaounakis, M., Halvadakis, C.P., 2006. *Olive processing waste management, Literature Review and Patent Survey,* Second Edition, Elsevier, Amsterdam.

Paudyal, H., Pangeni, B., Inoue, K., Kawakita, H., Ohto, K., Ghimire, K. N., Harada H., Alam S. 2013. Adsorptive removal of trace concentration of fluoride ion from water by using dried orange juice residue, *Chemical Engineering Journal* 223, 844 – 853.

Putun, E., Apaydin, E., Putun, E., 2004. Rice straw as a bio-oil source via pyrolysis and steam pyrolysis. *Energy* 29, 2171 - 2180.

Rahmani, A. R., Mahvi, A. R., 2006. Use of ion exchange for removal of ammonium: A biological regeneration of zeolite. *Global NEST Journal,* 8(2), 146-150.

Safa, Y., Bhatti, H.N., 2011. Adsorptive removal of direct textile dyes by low cost agricultural waste: Application of factorial design analysis, *Chemical Engineering Journal* 167, 35 – 41.

Salleh, M.A.M., Mahmoud, D.K., Karim, W.A.W.A., Idris, A., 2011. Cationic and anionic dye adsorption by agricultural solid wastes: *A comprehensive review, Desalination* 280, 1 – 13.

Salman, J. M., Njoku, V. O., Hameed, B. H., 2011. Adsorption of pesticides from aqueous solution onto banana stalk activated carbon, *Chemical Engineering Journal* 174, 41 – 48.

Sarmah, A.K., 2009. Agricultural Wastes, Chapter 1. *Potential risk and environmental benefits of waste derived from animal agriculture;* Editors: G. S. Ashworth and P. Azevedo Nova Publishers, p.p. 1-17.

Senthilkumaar, S., Kalaamani, P., Porkodi, K., Varadarajan, P. R., Subburaam, C. V., 2006. Adsorption of dissolved Reactive red dye from aqueous phase onto activated carbon prepared from agricultural waste, *Bioresource Technology* 97, 1618 – 1625.

Simitzis, J., Ioannou, Z., 2011. Activated carbonaceous materials based on thermosetting binder precursors in: Activated carbon: Classification, Properties and Applications, Chap. 12, J.F. Kwiatkowski (editor), *Nova Publishers,* p.p. 377-392.

Singh, K. P., Malik, A., Sinha, S., Ojha, P. 2008. Liquid – phase adsorption of phenols using activated carbons derived from agricultural waste material, *Journal of Hazardous Materials* 150, 626 – 641.

Singha, B., Das S. K., 2013. *Adsorptive removal of Cu(II) from aqueous solution and industrial effluent using natural/agricultural wastes,* Colloids and Surfaces B: Biointerfaces 107, 97-106.

Su, Y.F., Cheng, Y.L., Shih, Y.H., 2013. Removal of trichloroethylene by zerovalent iron/activated carbon derived from agricultural wastes, *Journal of Environmental Management* 129, 361 – 366.

Tashauoei, H. R., Movahedian Attar, H., Amin, M. M., Kamali, M., Nikaeen, M., Dastjerdi, V. M., 2010. Removal of cadmium and humic acid from aqueous solutions using surface modified nanozeolite A. *International Journal of Environmental Science and Technology, 7*(3), 497-508.

Tomar V., Prasad S., Kumar D., 2014. Adsorptive removal of fluoride from aqueous media using *Citrus limonum* (lemon) leaf, *Microchemical Journal* 112, 97 – 103.

Turan, N.G., 2008. The effects of natural zeolite on salinity level of poultry litter compost. *Bioresource Technology* 99, 2097–2101.

Uddin, M.T., Islam, M.A., Mahmud, S., Rukanuzzaman, M., 2009. Adsorptive removal of methylene blue by tea waste, *Journal of Hazardous Materials* 164, 53 – 60.

US GAO 1999. *Animal Agriculture: Waste Management Practices.* Report to the Honorable Tom Harkin, Ranking Minority Member, Committee on Agriculture, Nutrition, and Forestry, US Senate, US General Accounting Office, 1999.

Xuefei, Z., Shiquan, L., Hongzhao, L., Lijuan, G., 2009. Preparation and characterization of activated carbon foam from phenolic resin, *Journal of Environmental Sciences Supplement,* S121-S123.

Yargic, A.Ş., Şahin, R. Z. Y., Özbay, N., Önal, E., 2014. Assessment of toxic copper (II) biosorption from aqueous solution by chemically – treated tomato waste (*Solanum lycopersicum), Journal of Cleaner Production* 88, 152-159.

Zhu, C.S., Wang, L.P., Chen, W.B., 2009. Removal of Cu (II) from aqueous solution by agricultural by – product: Peanut hull, *Journal of Hazardous Materials* 168, 739 – 746.

In: Agricultural Wastes
Editor: Camille N. Foster

ISBN: 978-1-63482-359-3

Chapter 2

CURRENT UTILIZATION OF DAIRY INDUSTRY CO-PRODUCTS

Joshua L. Cohen[1], Mareen Geissler[1], Daniela Barile[1,2], Christopher W. Simmons[1] and Juliana Maria Leite Nóbrega de Moura Bell[1,2]

[1]Department of Food Science& Technology, University of California Davis, One Shields Avenue, Davis, CA, US

[2]Foods for Health Institute, University of California Davis, One Shields Avenue, Davis, CA, US

The design of new food products and agricultural practices have generated a wide diversity of co-products and effluents that often contain a high load of organic matter, from which valuable compounds could be isolated. The surplus and concomitant underutilization of these streams establish serious economic and environmental challenges. Whey, a main co-product arising from cheese manufacturing, was previously considered an environmental pollutant but it is now regarded as a source of many valuable compounds. Among current applications, the production of whey protein concentrates and isolates via ultrafiltration represents the major industrial revenue arising from this stream. The recovery of whey proteins generates an enormous volume of another co-product known as whey permeate. This stream has a high organic load being primarily composed of lactose and minerals. However, recent scientific literature demonstrates the presence of other compounds such as oligosaccharides and peptides, possessing unique bioactivities. Because of the worldwide increase in cheese production, the utilization of whey permeate is under strong scrutiny, with many different strategies being developed to add value to this waste stream. The development of feasible industrial processes to transform, isolate and recover valuable compounds is a key step towards the mitigation of environmental and economic problems arising from constant evolution of our food industry. This chapter focuses on the current utilization and research efforts to valorize whey permeate, where specific processing and environmental challenges are addressed along with the state-of-the-art of the processes and utilizations for the naturally occurring compounds in whey permeate, as well as the valuable products that can be generated from this stream.

1. Current Status of Major Dairy Streams Production and Their Utilization

Whey is the liquid resulting from the coagulation of milk proteins (caseins) during cheese manufacturing and accounts for approximately 90% of the initial weight of milk used. Humans have been interacting with whey from cheese manufacturing for thousands of years for various purposes. Historically, the most common outlets for this co-product have been human consumption either directly or through coagulation of whey proteins into a new food product known as Ricotta. Additional applications of whey include its use as animal feed, especially for pigs, and in land spreading [1].

Typical whey composition is based upon the type of cheese produced and the technique (i.e., enzymatic vs. acid coagulation). Lactose, protein and mineral concentrations usually range from 44-52 g/L, 6-10 g/L, and 2.5-7.2 g/L respectively [2]. High amounts of lactose and protein cause a high biochemical oxygen demand (BOD) and high chemical oxygen demand (COD) of >35,000 ppm and >60,000 ppm, respectively [3]. Unfortunately, whey disposal leads to serious environmental problems in water as well as soil. In fact, it endangers both the physical and chemical structure of soil, ultimately decreasing crop yields. The subsequent safe disposal and recycling of the products, in water bodies or on land, must be within the federal environmental limits. Because transport of liquid whey over any distance is both difficult and uneconomical, many concentration and separation techniques have been devised. Milk proteins have a high nutritional value compared to other proteins because of their relatively high content of essential amino acids and their good digestibility. Whey, in particular, has the highest content of branched-chain amino acids (BCAAs), especially leucine. It also contains the biggest amount of cysteine among non-meat products. Whey protein has the highest biological value of all food proteins, meaning it is used efficiently by the human body [4]. To obtain whey proteins in a pure form, whey is filtered through an ultrafiltration membrane which, because of the structure of its pores, filters out the macromolecules while allowing the liquid phase containing the soluble molecules to permeate. This process allows whey proteins to be separated from lactose, minerals and other components. Two major fractions are generated by the ultrafiltration of whey: the retentate—composed of retained macromolecules by the membrane—and the permeate—the liquid phase that permeates the membrane. Figure 1 gives a general overview about the cheese whey production and the main processing steps involved in the recovery of whey proteins as well as the side-streams generated by this process. Casein coagulation (enzymatic or acid) during cheese manufacturing generates a curd fraction (coagulated protein) that represents approximately 10% of the initial weight of milk and a liquid fraction (whey) that accounts for as much as 90% of the initial weight of milk used. Whey protein concentrates (WPC) and whey protein isolates (WPI) are the two major industrial outlets for cheese whey processing [5]. The protein concentration of those fractions vary according to the recovery processing used (membrane filtration, precipitation or complexation with agents, physical and chromatographic separations). Usually WPC and WPI have protein contents ranging from 35-

80% and 80-95%, respectively [5, 6]. Until the 1970s, most of the whey protein was available in a heat denatured form which resulted in a final product with poor functionality and limited application [7]. Those drawbacks have hindered process economics and have opened up, the opportunity to use membrane technology as a less harsh processing method to recover whey proteins with improved functionalities [1, 8]. The use of membrane filtration, alone or in combination with additional purification techniques such as diafiltration and ion exchange chromatography, enables the concentration of whey proteins to different degrees of concentration and purity [1].

However, the recovery of whey proteins by ultrafiltration (UF) generates another side-stream known as whey permeate (Fig 1). Approximately 93% of whey permeate is water, with 65 – 85% of the solids being lactose, 3 – 5% being protein and 8 - 20% being ash/minerals [9]. Considering the differing degrees of protein purification for different applications, a wider variety of whey permeate composition may be observed.

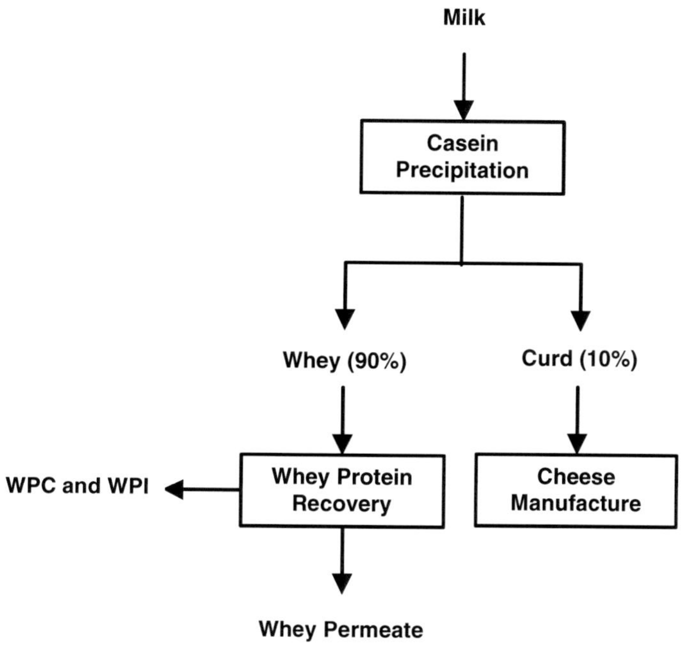

Figure 1. Recovery of whey proteins and their respective side-streams.

The generation of this side-stream begets the need to develop new techniques to recover compounds in whey permeate to add value to this stream. At present, the disposal of whey permeate is just as problematic as the disposal of whey due to their similar lactose content. The disposal of untreated whey was prevented in the late 20th century throughout the world and the same should be expected to occur with whey permeate disposal [3].

While whey is a well-researched and characterized stream with a large market, whey permeate is greatly underutilized and represents serious environmental and economic issues. Hence, it is crucial to develop processing techniques to recover and to valorize whey permeate compounds. Figure 2 shows the trend in number of publications, normalized for total annual publications, on whey, whey proteins, and whey permeate from 1950 to 2011, which demonstrates the dearth of knowledge and interest regarding whey permeate compared to its well-established counterparts.

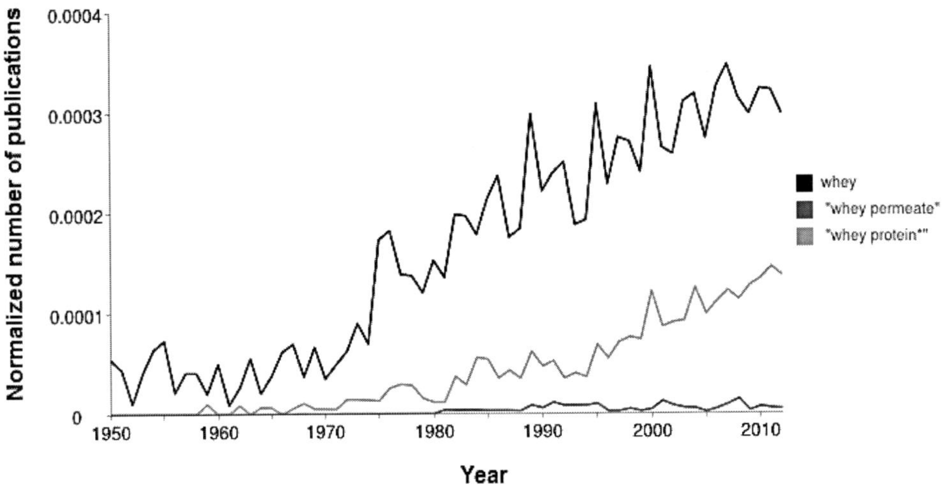

Figure 2. Number of publications normalized for annual publications in MEDLINE database with terms in titles and abstracts matching the queries “whey”, “whey permeate” and “whey protein*” (10). Note: maximum value of 0.0004 refers to 0.04% of all articles in MEDLINE database for that year matching query.

Figure 3 illustrates possible pathways for whey permeate utilization, with major research efforts focusing on the areas of food, animal feed, agricultural applications, and as an energy source. This chapter focuses on the current uses of whey permeate and research efforts to valorize future whey permeate utilization by generating and/or isolating high-value products, where specific processing and environmental challenges are addressed.

2. Major Applications of Whey Permeate

The most common use of whey permeate is for the recovery of lactose by crystallization. Lactose is widely utilized in the food industry and in the pharmaceutical industry. Furthermore, lactose is a promising source of value-added derivatives that, amongst others, show prebiotic effects or serve as sugar replacers.

Beyond these applications, the economic value of whey permeate utilization could be maximized with the development of processing techniques able to isolate the bioactive compounds present in this stream. Whey permeate has been shown to contain bioactive oligosaccharides and peptides that could be used in the development of health promoting foods/supplements [11, 12].

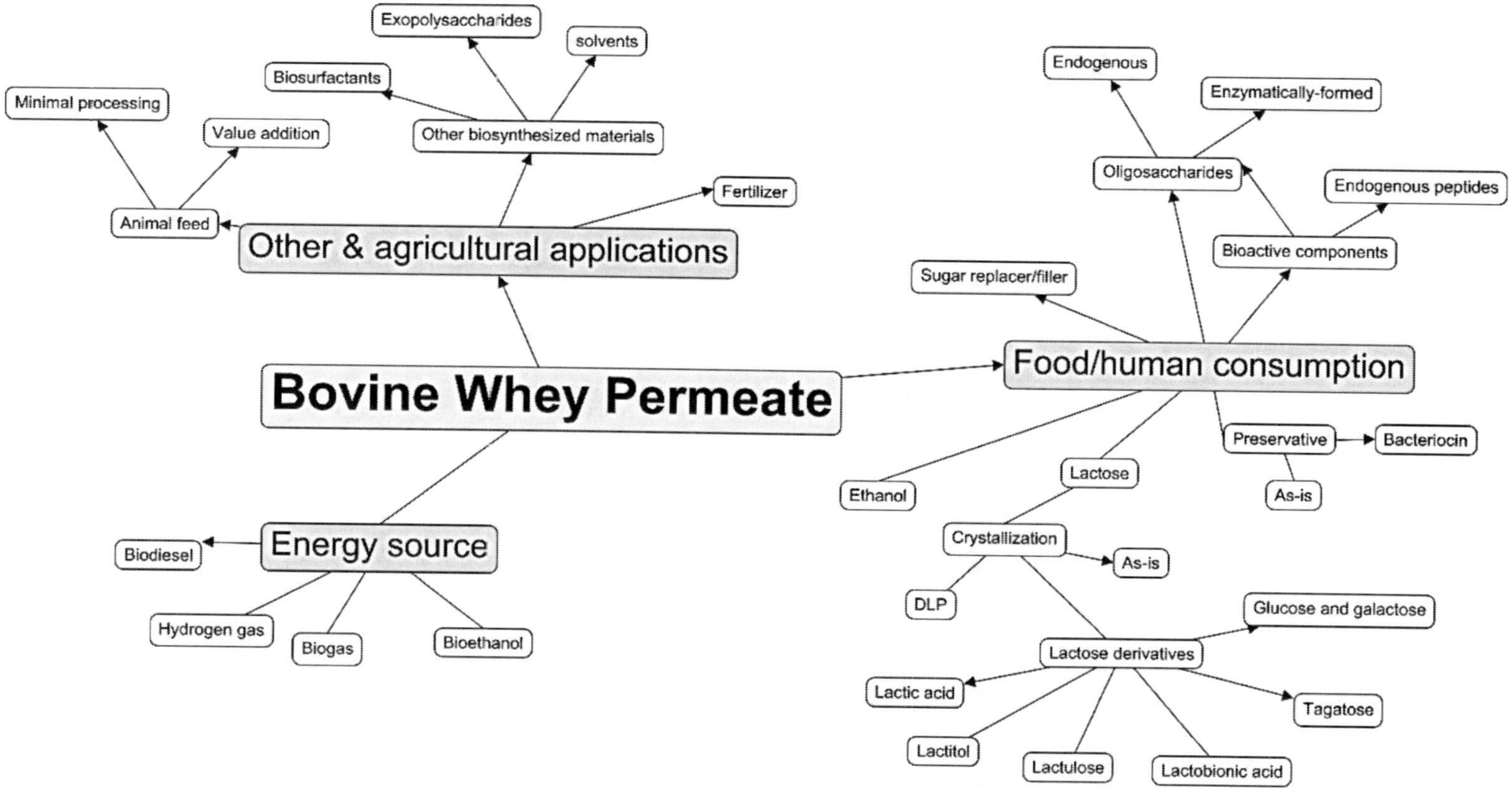

Figure 3. Concept map of whey permeate utilization in this book chapter generated using CmapTools (Pensacola, FL).

Considering the increasing annual production of whey permeate, the development of new processing techniques to recover and produce the aforementioned compounds and the use of high throughput analytical techniques will likely enable the production of large quantities of high-value added compounds that will ameliorate both environmental problems and poor financial viability of dairy processors.

2.1. Agricultural Applications of Whey Permeate

At its most basic level, whey permeate is an agricultural co-product, and as such, has direct applications to the agriculture industry. Although frowned upon nowadays, land spreading on crop fields was once a common outlet for whey permeate as well as for other waste streams. Alternatively, whey permeate can be used as-is or dried to a powder for swine feed, with reduced use as animal feed for ruminants. While this process is certainly able to absorb a good portion of the waste stream and mitigate its environmental impacts, it does not valorize the product nor the cheese manufacturing industry. New applications still need to be developed to better utilize the forecasted production of whey permeate for the next century.

2.1.1. Land Spreading

Many research articles cite the land spreading of whey and whey permeate as a common method of disposal. The results, which date back as far as 1923, are quite varied [13]. While most of the research conducted has focused exclusively on the use of whey, this dearth of knowledge highlights the need of examining the effects of land spreading of whey permeate.

Whey permeate has several nutrients that facilitate plant growth. Some of them are present in relatively large amounts, such as phosphorus and potassium [14]. Previous results have shown that when whey is applied to cropland the soil can act as a sink and the whey penetrates into the depths of the soil. Whey nutrients absorbed by the soil have been shown to persist for future crops [2]. However, dosing of whey in excess of 4 acre-inches on cornfields has been shown to inhibit plant growth [15]. Various spreading practices can be employed based on the type of soil, the type of crop being grown, the season, and the region [16]. A more recent study has confirmed that whey may be beneficial, but the excess of nitrite and nitrate formation from the whey application could lead to leaching into drinking water and therefore result in health complications for humans and animals within an ecosystem [17].

2.1.2. Animal Feed

Waste streams from food processors have long been a source of nutrition for livestock. Brewer's and distiller's grains have become a staple in dairy cow diets, while various forms of (literally) garbage have been used as a source of feed for swine recorded as early as 1914. While the latter has since been phased out, feeding by-products, co-products, and waste streams from the food industry to livestock remains an effective and attractive outlet for these products [18].

Whey permeate is often used as a carbohydrate supplement for swine nutrition, due to its high lactose content as well as its nutritionally valuable micronutrient content. Feeding dry whey permeate to weaning pigs as a supplement to a traditional diet increases the average daily gain of the pigs, while also increasing their average daily feed intake. These, along with other metrics, demonstrate that whey permeate effectively improves growth performance in some pigs, with good digestibility as well [19, 20]. Additionally, whey permeate is often added to the total mixed rations (TMR) of lactating cows. At a final concentration of 5% sugar in the ration, the addition of whey permeate improves milk production and is economical [21].

However, there are massive energy implications associated with drying whey permeate. Because diafiltration (the use of a solvent, typically water, to wash out undesirable components from the retentate) is applied to increase the degree of purity of proteins recovered by membrane filtration, the solids content of whey permeate is reduced to typically 3-5% solids [22, 23]. Although the process of spray drying is technically feasible and currently used, the use of permeate powders is limited to some baking and confectionery applications, as well as animal feed as discussed above [24].

While the simple and direct use of whey permeate powder, or even isolated lactose, has become the standard outlet for whey permeate-derived animal feed, much more attractive is the prospect of a high-value product for animal and human nutrition [25]. Alternatively, single cell proteins can be added into whey permeate as a protein supplementation strategy. Single cell proteins refer to the mixture of proteins extracted from microorganisms such as bacteria, yeasts, fungi, or algae. This product can be substituted up to 20% of the protein fraction of swine feed, which represents a potential market for a single cell protein extract from agricultural waste streams [26].

Fermentation of whey permeate to produce single cell protein as well as single cell oil was already suggested nearly 40 years ago [27]. Beyond recovering up to 3.2 g protein/liter of permeate, a 95% reduction in the chemical oxygen demand (COD) was observed. The most common microorganism for single cell protein production on whey permeate is the dairy yeast *Kluyveromyces marxianus*. More recently, industrially attractive single cell protein from both sour and sweet whey permeates has illustrated that, with optimized processing conditions, a commercially viable process could be developed. The optimal results on concentrated sweet whey permeate (120 g/L lactose) were a biomass accumulation of up to 50 g/L and an increase in many essential amino acids compared to the original whey permeate feedstock [28]. Although animal feed is a simple outlet for whey permeate, this application represents a very low-value opportunity for the use of this stream. The development of alternative bioprocessing techniques could lead to the production of higher value products thus contributing to the valorization of whey permeate.

2.1.3. Chemical and Cosmetics Applications of Whey Permeate

Industrial biotechnology requires high availabilities of low-cost and easily-accessible carbon as feedstock for the production of bio-based chemicals and components. In order to be commercially viable and truly adoptable in light of heavy petrochemical influence, the fermentable feedstock must be plentiful and inexpensive. In fact, the cost of feedstock is a main determining factor for the production and impact of biotechnologically-produced chemicals and products [29]. Biobased chemicals can be used in several key industries such as the chemical, cosmetics, and biofuels industries. In pilot and lab scales, whey permeate has been shown to meet the needs of the industrial biotechnology sector as a potential feedstock.

2.1.3.1. Solvents

One method of reducing the environmental impact of whey permeate while producing a useful end product is by solventogenic fermentation. Most commonly, *Clostridium acetobutylicum* is used to produce acetone, butanol, and ethanol (ABE). Grain or molasses, depending on the region, are the most common industrial feedstock for ABE fermentation. Until the end of World War II, most of the butanol and some of the acetone produced was via fermentation, but this application declined rapidly in the following decades due to a boom in the oil industry [30]. Due to inherent political instability surrounding the oil industry, plants producing ABE via fermentation located in China and the United States have faced pressure from the petrochemical industry and have shut down and resumed production several times. As of the early 2000s, bio-based solvents were projected to have captured 12.5% market share for the following several years [29]. Both acetone and butanol are important solvents in the chemical industry, while the latter has shown promise as a biofuel [31].

Solventogenic fermentation was first adapted for whey permeate in 1985, due in large part to its high content of low-cost lactose [32]. While in the early investigations of solventogenic fermentations of whey permeate the yield was low and lactose consumption was incomplete, processes such as membrane perstraction to reduce the effects of product inhibition have dramatically improved yields of solvents on a small scale. For example, only 60% of lactose was utilized in a solventogenic fermentation of whey permeate due to toxicity of the solvents in the culture broth. By removing the solvent using membrane perstraction with oleyl alcohol, the yield was increased from 9.34 g/L to nearly 100g/L [33]. A major challenge associated with biobutanol and acetone production via fermentation is the substrate cost, which is addressed by the readily available, cheap and fermentable lactose in whey permeate. Downstream processing, especially in a dilute solution such as whey permeate, poses a large problem for future commercialization of solventogenesis with whey permeate [34].Upstream concentration of whey permeate by membrane filtration prior to solventogenic fermentation may be necessary to reduce the initial volume of diluted whey permeate.

2.1.3.2. Exopolysaccharides

Exopolysaccharides are the secreted polysaccharides from microorganisms. These complex sugars play a large role in the formation of biofilms and modulation of texture in fermented foods [35, 36]. While these polysaccharides are commonly produced by lactic acid bacteria (LAB), they can also be produced by fungi such as *Aureobasidium pullulans*, which has been shown to produce pullulans on deproteinized whey permeate [37]. More recently, xanthan gum, a ubiquitous stabilizer/thickener in the food industry, has been produced from whey permeate. While the medium must first undergo lactose hydrolysis with β-galactosidase to release glucose and galactose for optimal polysaccharide production, this study illustrates an outlet for adding value to whey permeate [38].

2.1.3.3. Biosurfactants

Beyond modifying texture of food products, whey permeate can be transformed into products which modify the rheological properties of cosmetics or act as common household or industrial detergents. Biosurfactants are recognized for their biodegradability and low environmental impact, and have gained popularity over the past several decades. Beyond their cosmetic or typical cleaning applications, they have also been suggested as a remediation tool due to their ability to facilitate adsorption by microorganisms which degrade the hydrocarbons [39]. One of the most common types of biosurfactants is glycolipids, which have a hydrophilic sugar-containing head group and a hydrophobic tail. In order to biosynthesize these surfactants, the fermentation broth must contain a sugar source as well as a fatty acid source, in the form of a free fatty acid, fatty acid methyl or ethyl ester, or even a triglyceride. A popular example of a biosurfactant is the sophorolipid, which is produced in great quantities (over 400 g/L) by the fungus *Candida bombicola* [40]. It has been produced by combining whey permeate with glucose and oleic acid, which are collectively metabolized into sophorolipids [41]. Due to its cheap and plentiful source of sugars, whey permeate could be an attractive feedstock for biosurfactant production.

2.2. Major Food Applications of Whey Permeate

Unlike whey protein concentrates as a food ingredient, limited research has evaluated the addition of liquid whey permeate (WP) and whey permeate powder (WPP) as ingredients in food products such as beverages, chocolate or jam [42, 43]. Most of the research conducted has focused on the use of WP as a source of lactose for sugar replacement in lactose-derived applications, with other minor compounds remaining unnoticed. Owing to its lower relative sweetness compared with sucrose, higher amounts of lactose are required to reach the same sweetness intensity [44]. Hence, WP and WPP find more promising direct applications in the beverage industry, in particular for sport beverages. For this application, an enrichment of electrolytes is desired, and a reduction of the mineral content of whey permeate is not necessary. The effects of WP addition into beverages with sensory studies have been reported, demonstrating that WP can replace up to 50% of water in beverages and score consumer acceptance values comparable to commercially available sport drinks [42]. Furthermore, acceptance of unhydrolyzed whey permeate was similar to hydrolyzed WP. As neither hydrolysis nor the removal of minerals is needed, whey permeate could be incorporated as-is in certain beverage formulas. Further research must be conducted regarding incorporation of whey permeate as-is into other food products; in particular where saltiness and its other flavors can be masked.

2.2.1. Lactose Characteristics and Applications

After water, the major component of whey permeate is lactose, which is typically about 4-5% in liquid whey permeate [45]. As such, the isolation and recovery of lactose is of great interest for this stream, where it can be used either as an ingredient or it can be converted into a variety of derivatives such as lactic acid, lactulose and galactooligosaccharides, amongst others.

While the major use of lactose in the US market (66%) is as ingredient in infant formula, a good portion (16%) is used in the confectionery industry. In the EU, it is used mainly in the pharmaceutical industry, in infant formula, and in various processed foods including meats [46].

The addition of lactose in infant formula aims at enriching the amount of lactose in bovine milk (4.6% lactose) derived formula to match the higher amount present in human milk (7% lactose) [46, 47]. Lactose is an attractive ingredient for the food industry due to its low hygroscopic properties and its flavorless characteristics, the latter of which allows it to be used as a filler and flavor enhancer. Additionally, its reducing end enables its participation in the Maillard reaction where important flavors and color compounds are formed [46]. On the other hand, its use is restricted in food products due to its limited solubility and lower relative sweetness [48].

2.2.2. Lactose Recovery Processes

Crystallization is the most widely used process for lactose recovery. Commercial lactose recovery via crystallization from ultrafiltrate or concentrated whey yields 400,000 tons of crystalline lactose annually [49].

The lactose manufacturing process aims at maximizing the yield of crystal mass in a minimum time, and to produce crystals that can undergo the washing step with a minimum loss. In order to increase the yield and reduce the reaction time, novel methods such as anti-solvent crystallization, ultrasound crystallization and combined anti-solvent ultrasound crystallization have been examined in the last several years [50–52]. Food-grade lactose recovery in industrial scale is stated to be approximately 65% due to scaling of minerals in the process as well as incomplete washing steps [46]. Due to its higher lactose/solids ratio after protein removal, whey permeate has become an attractive source of lactose. The crystallization process involves several steps including evaporation, crystal formation, centrifugation, washing, and spray drying [46, 50, 53].

In the first step, the initial amount of solids is increased to 50-70% by evaporation. It is advantageous to remove the minerals by electrodialysis or ion exchange resins to avoid fouling on the heat exchanger surface during evaporation [53]. The evaporation step is followed by the initiation of the crystals growth, where the concentrated permeate is fed into a tank and the crystals form either spontaneously or by seeding during a gradual and controlled cooling process. After reaching the desired level of growth, crystals are removed via centrifugation to produce crude lactose slurry. Crude lactose is further washed and centrifuged to increase the purity of the crystals, a process in which it is difficult to control the crystal shape and size as well as the final purity of the crystals.

The presence of proteins and minerals has been associated with the formation of small lactose crystals and lower recovery yields. Partial removal of whey permeate salts by nanofiltration has led to an increase in lactose recovery yield of 6-10% [53, 54].

High evaporation costs and long crystallization times have prompted the search for alternative methods for lactose recovery. Several investigations have described processes to increase the lactose yield, speed up the crystallization and control desired parameters. Some of the proposed techniques include membrane filtration, anti-solvent crystallization, sonocrystallization and anti-solvent sonocrystallization [51, 52, 55, 56].

The use of membranes with different molecular weight cut-offs enables the partial removal of whey permeate salts which affects both lactose solubility and lactose recovery yield. Sequential filtration steps can be used to achieve desired yield and degree of purity; most commonly ultrafiltration (UF) and nanofiltration (NF). Atra et al. [55] examined the performance of nanofiltration of whey permeate. A lactose yield higher than 90% was achieved at 30°C and concentration factor 5. With respect to lactose recovery, membrane processes are most effective compared to crystallization, but less profitable for small and medium scale recoveries due to high capital costs, recurring costs and limited membrane life associated with any membrane filtration system [56].

The addition of anti-solvents leads to a decrease in lactose solubility without creating an additional liquid phase. Due to the reduced lactose solubility, supersaturation and hence crystallization occur faster. Patel and Murthy [51] demonstrated that the recovery of lactose could reach almost 90% when using 85% v/v acetone in conjunction with appropriate process conditions such as stirring time and speed. Furthermore, seeding also decreased the total required crystallization time, and a lactose recovery greater than 90% of lactose was reported.

Ultrasound-assisted crystallization (sonocrystallization) is a new technology that relies upon the use of ultrasound to alter induction periods and saturation. It is typically combined with anti-solvent crystallization (usually acetone or ethanol) to develop anti-solvent sonocrystallization. This process can further decrease the crystallization time to several minutes by reducing the solubility of lactose in water and by inducing a more rapid nucleation event. With this technology, a five-fold decrease in the crystallization time of lactose has been achieved [56].

The removal of lactose via crystallization yields a new stream known as delactosed permeate (DLP). Until recent years, little effort has been made to investigate uses of this novel waste stream. While the lactose content is still quite high on a solids basis (nearly 60%) the mineral content, often measured as ash, increases from 8% to nearly 30% after lactose crystallization. This vast increase in minerals, including an increase to over 3.5% calcium, has led to its use as a dried supplement in animal feed. As with whey permeate, DLP is also spread on land for use as a fertilizer, with fewer deleterious effects due to a lower BOD [57]. Additionally, DLP has been used as a salt replacement in food products [14]. Before drying, a typical moisture content of DLP is 60-70%. The high moisture content of DLP [58] represents a challenge for the development of an economically feasible process for this new stream. Because lactose is commonly crystallized from whey permeate in large cheese manufacturing plants, DLP could pose unique issues to the dairy industry as a new waste stream with which to contend.

2.2.3. Lactose Derivatives

The presence of reactive functional groups makes lactose an interesting substrate for conversion into other compounds. Many of these derivatives exhibit prebiotic effects or find other applications in the food and pharmaceutical industry. The production of these derivatives, which are amongst others lactic acid, lactosucrose and galacto-oligasaccharides, shows an annual increasing growth rate of 5 to 20% [59]. Increasing research of new feasible techniques for those derivatives is necessary to meet the demand of an important emerging market. Much research has been conducted regarding the production of those derivatives from pure lactose solutions and, to a lesser extent, from whey and whey permeate. Although many researchers propose whey permeate as a promising substrate for the production of lactose derivatives [60, 61], very little research has actually been conducted using this stream as a starting material.

Compared to whey, the lactose amount in whey permeate is much higher on a solids basis, and as such this waste stream could be considered a source for the production of lactose derivatives.

2.2.3.1. Lactic Acid

Annual world lactic acid production in 2007 was 150,000 metric tons and the forecasted production for 2017 is expected to reach 367,300 metric tons [64]. Approximately 70% of the lactic acid produced is used in the food industry with the remainder being used in the cosmetic and pharmaceutical industries. A novel outlet is in the production of biodegradable plastics where lactic acid serves as precursor for polymers [60, 62].

Table 1. Summary of the types of lactose derivatives discussed in this chapter along with their structures and applications

Name	Structure	Application	Ref
Lactic acid	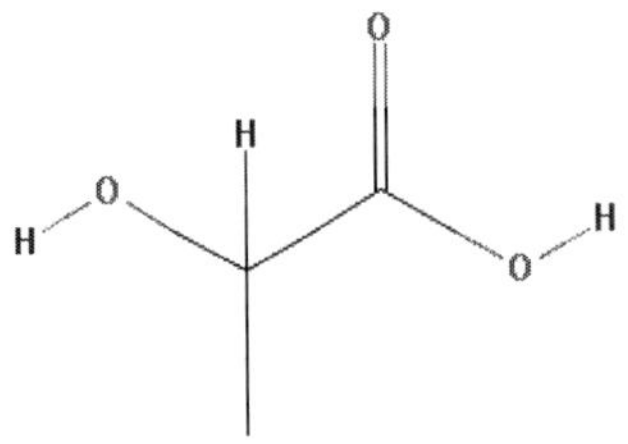	Dairy acid, preservative, flavor ingredient, biodegradable packaging, precursor for polymers and small chemicals	[62]
Lacto-bionic acid	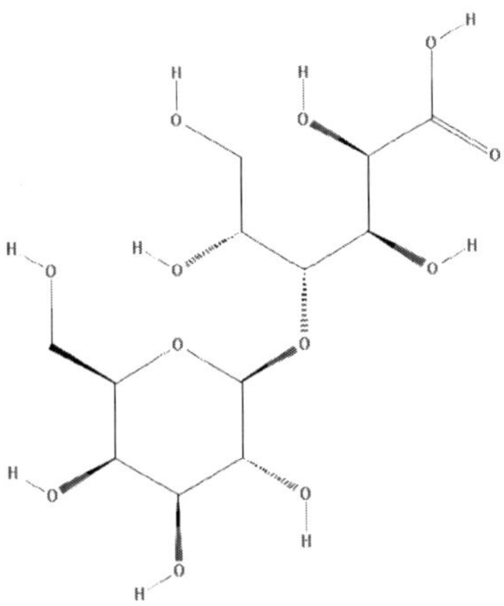	Calcium fortification, cold storage organ transport	[59]
Lactulose	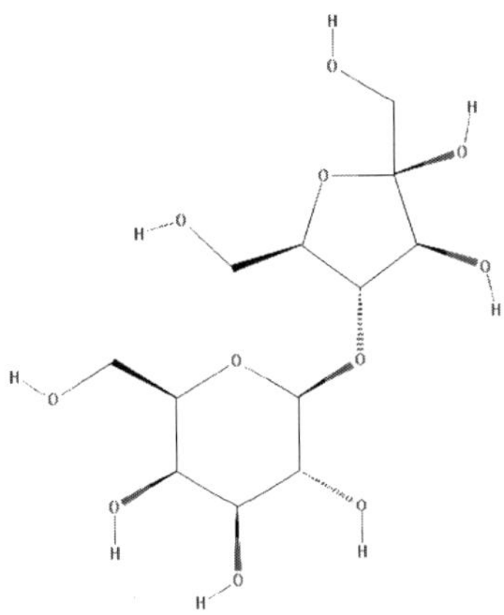	Prebiotic, laxative, alleviates chronic hepatic encephalopathy	[59]

Name	Structure	Application	Ref
Lacto-sucrose	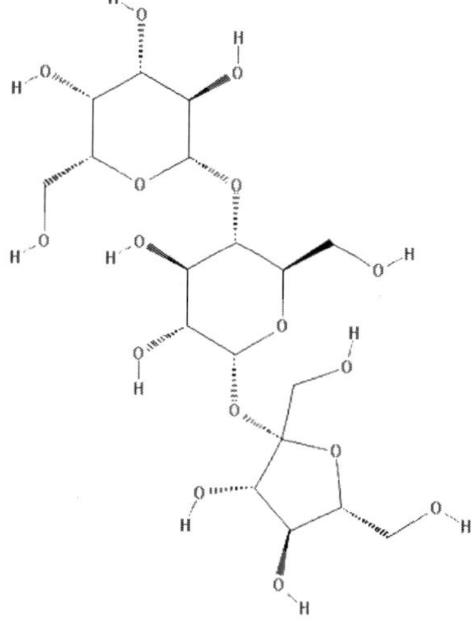	Prebiotic, calcium absorption	[59]
Tagatose	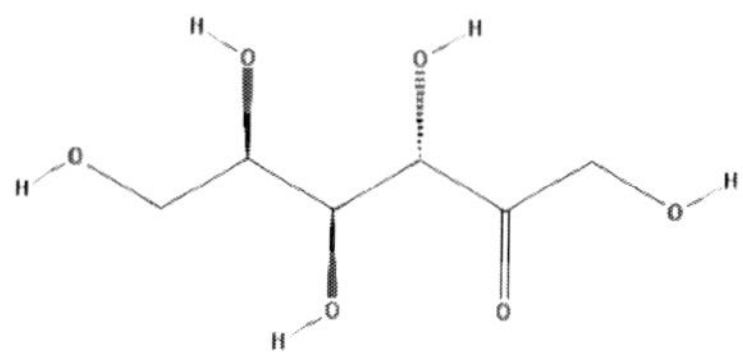	Low-calorie sweetener, toothpaste ingredient, ingredient in cosmetic and pharmaceutical industry	[63]
Lactitol	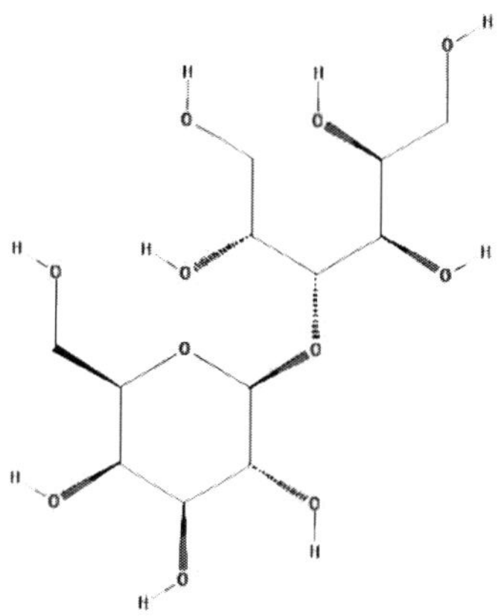	Laxative, low-calorie sweetener	[59]

Industrial production of lactic acid can be accomplished via chemical synthesis or by lactic acid fermentation [62, 65]. Currently, the latter is the primary production technique, accounting for 90% of lactic acid production [66]. Chemical synthesis requires harsh reaction conditions and a need for efficient cleaning procedures to reduce heavy metal catalysts [67]. Furthermore, chemical synthesis yields a racemic mixture of both L-(+)- and D-(-)-lactic acid. With appropriate microorganism selection, fermentation can produce either pure L- or D-lactic acid, a feature which is required for producing biodegradable plastics, where only L-(+)-lactic acid serves as precursor. Additional features of fermentation processes include the possibility of using milder reaction conditions and not using catalysts. In addition, inexpensive raw materials could be used as starting materials [67]. Despite the advantages of fermentation compared to chemical synthesis, which include low energy costs, some drawbacks include low productivity, low purity in the final product, and the formation of large quantities of gypsum as a byproduct [66, 68].

Raw materials containing either mono- or disaccharides and polymers such as starchy or lignocellulosic materials can be used for the fermentative production of lactic acid. A hydrolytic pretreatment of the polymeric material is required to yield fermentable sugars for the bacteria [62]. Whey permeate is therefore a promising raw material due to its high lactose content [67]. Depending on the microorganism chosen, whey permeate can be used as-is or after the hydrolysis of lactose by β-galactosidases. As all microorganisms have specific nutrient requirements which must be considered for bioprocess optimization, nutrients may need to be added [62, 69]. Various growth supplements have been investigated for lactic acid production from whey permeate due to its low nitrogen content [60, 70]. Adding readily available nitrogen sources like casein hydrolysate or ammonium citrate to whey permeate increased the lactic acid yield from 1.2g to 24.3g [70]. Although a purity of more than 98% was achieved, the addition of casein hydrolysate significantly increases processing costs. While the addition of nitrogen sources effectively sped up fermentation rate, adding complex matrices to an already complex food matrix, such as whey permeate make the isolation of small metabolites a very difficult task [71].

For both chemical and fermentative production, the lactic acid produced has to be further separated and purified. The isolation of small metabolites from whey permeate is challenged by the presence of naturally occurring contaminants such as minerals and peptides. Techniques such as nanofiltration, chromatography, and ion exchange are necessary to improve the purity of the recovered compounds. These steps have a significant impact on the overall process costs but are necessary to reach high yields and purities. Possible additional methods include reactive distillation followed by hydrolysis, crystallization or membrane technologies with bipolar membrane electrodialysis [62, 65, 72].

An innovative process has been validated using mathematical modeling where the fermentation broth was treated by ultrafiltration, followed by ion exchange, reverse osmosis and vacuum evaporation. The final product obtained by this procedure is 50% w/w lactic acid, which is comparable to commercial products available, but production costs are lower [73]. Research on lactic acid production from whey permeate on a large-scale is needed, and downstream processing must become cost-effective to become commercially viable.

2.2.3.2. Lactulose

Lactulose is a synthetic disaccharide composed of one galactose and one fructose molecule. It is used primarily as a sweetener in different products and as a functional food ingredient. It has been demonstrated to be a laxative, a prebiotic, hepatic encephalopathy and a prebiotic food additive [59, 74, 75]. Due to the increasing use of lactulose by the food and pharmaceutical industries, the interest in developing feasible industrial processes for the production of lactulose has prompted.

Lactulose can be produced either via semi-synthetic isomerization of lactose under alkaline conditions or by enzymatic isomerization and transgalactosylation reaction of lactose. Currently, lactulose is mainly produced via chemical isomerization. This process involves the presence of catalysts such as sodium hydroxide and boric acid. These catalysts help maximize the level of isomerization, minimize byproduct formation, and are environmentally safe and inexpensive. Large amounts of catalysts are needed to obtain high lactulose yields, which results in costly separation and purification steps [76]. Current research focuses on the development of catalysts that are environmental-friendly and lead to a minimum amount of byproduct formation [75].

Whey permeate is the main source for the industrial production of lactulose by chemical isomerization. The process involves isomerization of the glucose residue in the lactose molecule into fructose at alkaline pH values, which forms lactulose. This step is followed by a rapid degradation which leads to the formation of undesired byproducts including colored compounds which are difficult to remove from the lactulose solution [77]. Catalysts can be removed by chromatography or nanofiltration. Furthermore, isomerization leads to various breakdown products, such as galactose, epi-lactose, tagatose and fructose which must be separated to increase the lactulose yield and purity [78]. Depending on the catalyst used, yields vary from < 1% to 87%, with boric acid yielding 75% and calcium carbonate yielding only 20% [79].

The production via enzymatic isomerization is promising, as no chemical catalysts are needed. Lactose is hydrolyzed into glucose and galactose by the use of β-galactosidases. The galactosyl group, in the presence of fructose, is transferred to fructose, and lactulose is formed via rapid transgalactosylation [75]. Major challenges are the improvement of lactulose yield, purity, safety and process cost reduction. The yield of lactulose produced via enzymatic reaction depends on the origin of beta-galactosidase used, and are reported to be nearly 30% for enzyme isolated from *Aspergillus oryzae* [80].

2.2.3.3. Tagatose

Tagatose is a monosaccharide whose main application is as a sweetener in products for diabetics. It received GRAS status in 2001. Due to its sweetness and sucrose-like taste, it can be used as supplement in a wide variety of foods. Tagatose can also be used as an additive in detergents, cosmetics, and pharmaceutical formulations [77]. D-Tagatose can be produced via oxidation of D-galactitol. However, due to the limited availability of the substrate, large-scale productions are not feasible. The monosaccharide can be obtained both by chemical isomerization and by enzymatic synthesis. The major disadvantage of chemical isomerization is the lack of specificity, which leads to the loss of sweetness and functionality due to side product formation. However, it is so far the most economical method [81].

Hydrolyzed whey permeate is a good substrate for tagatose production, as it is rich in the substrates glucose and galactose. For the chemical reaction, basic catalysts such as calcium hydroxide or aluminates are necessary. [77]. Enzymatic synthesis has been explored to avoid reduced purity associated with chemical isomerization procedures. Current research focuses on the isomerization of D-galactose with the enzyme L-arabinose isomerase, although those procedures have their disadvantages. Wanarska et al. [82] constructed both a recombinant β-D-galactosidase and a recombinant L-arabinose isomerase from a yeast strain and psychrotolerant bacterium, respectively. They focused on the developing of a single-step method for the production of D-tagatose from whey permeate where hydrolysis of lactose, glucose utilization and galactose isomerization into D-tagatose take place simultaneously. This process led to a 90% yield of lactose hydrolysis and overall 30% yield of D-galactose bioconversion. Although the tagatose yield is low, the process is promising as it is more economically viable and less complicated than multistep techniques.

2.2.3.4. Galactooligosaccharides

Galactooligosaccharides (GOS) are prebiotic carbohydrates that have been shown to have bifidogenic properties and are often added to infant formula as a supplement to match the composition of human milk. As such, they garner a 10-fold cost in the marketplace and so offer an attractive value addition for the dairy industry. These compounds are usually synthesized via trans-galactosylation of lactose. GOS are typically simple structures and comprise mainly tri- and tetra-saccharides made up of glucose and galactose residues. Typically the trans-galactosylation reaction is incomplete, and so glucose, galactose, and lactose remain at the end [59]. Trans-galactosylation is catalyzed by β–galactosidases as these enzymes possess simultaneous hydrolytic and transgalactolytic activities, and so yield a heterogeneous mixture of carbohydrates having different chain lengths [80]. Cho et al. [83] compared the GOS formation in both pure lactose solution and whey permeate. GOS concentrations of 41% and 34% of total carbohydrates from lactose and whey permeate were achieved, respectively. Hereby, the degree of polymerization (DP) in cheese whey permeate was primarily three, with some tetrasaccharides and pentasaccharides as well. The results indicate that whey permeate is a suitable inexpensive raw material for the production of simple GOS.

2.2.3.5. Other Lactose Derivatives

In addition to the aforementioned lactose derivatives, there are several minimally-investigated compounds worth mentioning. These include lactosucrose, lactobionic acid, and lactitol. Current research has focused on the production of those derivatives with pure lactose feedstock with practically no research utilizing whey permeate as feedstock. It is necessary to optimize both upstream and downstream processes if processes are to be adapted for whey permeate as an inexpensive raw material for the production of these derivatives.

2.2.4. Naturally Occurring Oligosaccharides

Human milk oligosaccharides (HMO) have been shown to exhibit beneficial properties with respect to the development of the infant gut microbiota. They have a non-nutritive function as they are indigestible by the host and serve as growth factors for gut bacteria, i.e., *Bifidobacterium infantis* [84]. Breastfed infants show less diarrhea and fewer intestinal infections, benefits which have been attributed to the presence of HMO. As such, there is a heightened interest in finding sources of oligosaccharides similar to those present in human milk that could be used to improve the current infant formula oligosaccharide (OS) profile.

Recent research has demonstrated that dairy streams such as whey permeate contain complex bovine milk oligosaccharides (BMO) similar to HMOs (Figure 4).

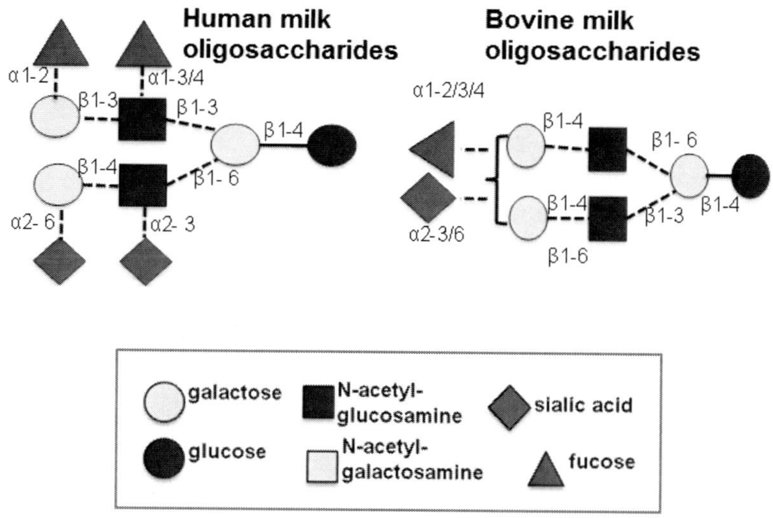

Figure 4. Structural diversity of human milk oligosaccharides (HMO) and bovine milk oligosaccharides (BMO). Adapted from (86).

It is expected that, due to similarities in their structures, BMO might have similar prebiotic functions as the ones observed in HMO (Barile, 2009). Thus far, it has to be shown that BMO support bacterial growth, prevent pathogen growth, exhibit antiviral properties, and have brain development and immunomodulatory effects [85]. Barile et al. [11] first identified 15 BMO in whey permeate, of which eight are neutral and seven are acidic. Later it was shown that whey permeate obtained by pilot-scale fractionation contains 24 oligosaccharide compositions [12]. Due to its high market value and recognized health benefits, an increasing interest in the direct isolation of sialic acids from whey permeate is arising [59].

The concentration of OS in both colostrum and mature bovine milk (1.6 g/L and 0.06 g/L, respectively) is 20 fold lower than in human colostrum and mature milk (20-23 g/L and 5-16 g/L, respectively). However, the vast quantities of whey and whey permeate produced by the cheese industry make whey permeate a promising source of bioactive OS [87].

The biggest challenge associated with the development of large-scale or industrial processes to recover OS from mammalian milks is to achieve a high degree of purity; more specifically to remove simple sugars that lack prebiotic activities such as lactose, glucose, and galactose, while simultaneously maximizing product recovery. Downstream isolation of OS can be simplified by enzymatic hydrolysis of lactose into glucose and galactose, with subsequent nanofiltration and extensive diafiltration [88–91]. Alternatively, removal of monosaccharides can be accomplished with graphitized carbon chromatography or size-exclusion gel filtration with greater degrees of purity [85]. However, for an economically viable application of large-scale chromatography to recover OS from dairy streams, resin costs must be minimized, which may represent up to 70% of the processing costs [92].

Bovine milk oligosaccharides have immediately tangible value in the market place. As of publishing, 1mg of 6' siallylactose costs $226.50 and 1mg of 3'siallylactose costs $173.00 [93], and therefore could provide a value stream for the cheese processing industry.

The development of more environmentally friendly processes that are also economically viable is a key step towards the production of sufficient quantities of OS for clinical trials and better elucidation of the functionality of those compounds *in vivo*. In addition to producing a new generation of prebiotics, environmental issues associated with the disposal of oligosaccharide-containing waste streams and poor economic viability will be mitigated.

2.2.5. Naturally Occurring Peptides

In current research, peptides have been identified and characterized in bovine milk and bovine colostrum [12, 94]. Some peptides are able to perform specific biological functions (bioactive peptides), which makes them useful in the growing market of functional foods and as nutraceuticals [3]. Such purported bioactive properties include anti-cancer effects against certain tumors, cell growth activity, wound healing effects, ACE-inhibitory effects, ileum stimulation and contraction, anti-thrombotic and antimicrobial effects [3]. Membrane technology seems to be a promising industrial scale recovery technique for peptides, as they can permeate ultrafiltration membranes thus being separated from macromolecules. Nanofiltration is then applied to concentrate peptides based on their molecular weight and electrostatic properties [95].

2.2.6. Whey Permeate As Substrate for Bacteriocin Production

Both nisin and pediocin are of great interest as food preservatives due to their antimicrobial activities and further characteristics like heat and pH stability. The bacteriocin nisin is produced by lactic acid bacteria (LAB), whereas pediocin is produced by *Pediococcus* strains [96]. Currently, bacteriocins can be produced with high yields. However, the growth media used are expensive and often not food-grade. As LAB grows good on whey permeate, this makes whey permeate a suitable medium for food-grade bacteriocin production [96, 97]. Research has been conducted using whey permeate as a food-grade medium for the production of bacteriocins [98–100], with yields as high as $5x10^4$ AU/mL in continuous culture [97]. Another advantage of using whey permeate as substrate for bacteriocin production is the presence of minerals and vitamins, which are important co-factors for *Lactobacillus lactis*. However, additional nutrients such as casein hydrolysates must be added to increase the yield due to the low nitrogen content of whey permeate (97). In general, bacteriocin production can be conducted on food-grade material, but the yields are lower and their purification is expensive. Lab-scale experiments demonstrated that with optimal process conditions and proper nutrient addition, respectable product yields can be achieved. However, pilot-scale validation is necessary for commercialization of bacteriocin production using whey permeate.

2.2.7. Minerals: A Need of Developing Feasible Desalination Processes

An important limiting factor in whey permeate applications is the high mineral content, especially for WPP. Jaros et al. [44] examined the replacement of sucrose in stirred yogurt both with WPP and nanofiltered whey permeate. Sensory studies showed that 25% of sucrose can be substituted with WPP and 30% with nanofiltered WPP. Further increase in sucrose substitution was accompanied by decreased consumer acceptance due to the high mineral content, being associated with a more pronounced salty taste. Partial removal of mineral salts by nanofiltration enabled a slight increase in the addition of WPP (25 vs. 30%).

In many applications, the minerals have a negative impact on either the process itself or also on the application and functionality of the final product. For example, during the evaporation to concentrate whey, calcium salts can precipitate on heat exchanger surfaces and furthermore contaminate lactose crystals during crystallization [56]. Desalination technologies could be utilized to increase the purity of lactose solutions obtained from whey permeate, or to improve the water reclamation from delactosed permeate, all across the food processing industry.

Electrodialysis has been investigated as a method to increase the purity of lactose in whey permeate (up to 98.5% on a solids basis) since the 1970s (101). Although a 90% reduction in conductivity has been obtained in approximately one hour, high energy costs for electrodialysis render the process uneconomical [102]. Further investigations concerning optimal feedstock concentration are needed for the process to become feasible at an industrial scale [103].

Emerging technologies offer promising resource-efficient alternatives to energy-intensive reverse osmosis for desalination of whey permeate. In particular, the development of microbial desalination cells (MDCs) holds potential to replace or reduce the need for reverse osmosis [104, 105]. In their most basic format, MDCs consist of a three chamber system wherein a central desalination chamber is flanked by anode and cathode chambers that operate similarly to a microbial fuel cell. Anion- and cation-exchange membranes separate the desalination chamber from the anode and cathode chambers, respectively.

In the anode chamber, exoelectrogenic bacteria consume organic matter, producing electrons and protons as products. Electrons are delivered to the cathode chamber on the opposite side of the desalination chamber via an external circuit, creating electricity that can be used to do work. Moreover, the electric field generated between the anode and cathode chambers imposes electrostatic forces on any ions within the solution located in the desalination chamber.

These forces drive anions and cations out of the desalination chamber, across the ion exchange membranes, and into anode and cathode chamber, respectively, resulting in desalination of the central chamber without any externally supplied pressure or energy. In addition to whey permeate, cheese processing generates other materials that could promote the use of MDCs. Specifically, the organic nutrient-rich wastewater from processing facilities might serve as the anode chamber substrate, ultimately providing the energy for desalination. The prospective three-fold benefit of renewable energy generation, wastewater biochemical oxygen demand reduction, and whey permeate desalination warrants further research to study the application of MDCs to whey permeate management.

Microbial desalination cell studies have primarily used synthetic wastewater solutions as substrate in the anode chamber and simple salt solutions in the desalination chamber. For salt solutions with 20 to 35 g/L of sodium chloride, near complete desalination was achieved over a period of several days when ample refreshing of anode substrate was used [104]. Although these studies have demonstrated the efficacy of using microorganism-derived electrical current to drive desalination, additional work is needed to examine MDC compatibility with streams relevant to cheese processing and whey permeate desalination. In particular, while domestic wastewater has been used successfully in MDCs [89], no work to date has studied cheese processing wastewater as anode substrate for MDCs. The organic matter content, microbial load, salinity, pH, and bacterial inhibitor content of cheese processing wastewater will affect MDC power output and desalination potential.

Furthermore, the composition of whey permeate will also influence desalination. Transfer of material other than salts through the ion exchange membranes separating the desalination chamber from the cathode and anode chambers must be considered. This is especially important because whey permeate desalination may be coupled with downstream purification of high-value compounds, such as bioactive oligosaccharides. Retention of these compounds during desalination in an MDC must be investigated. Similarly, potential contamination of whey permeate in the desalination chamber due to diffusion of material from the flanking anode and cathode chambers must be understood, particularly when wastewater is used in the anode chamber.

Aside from the technical considerations specific to whey permeate desalination, additional research is needed to make MDCs suitable for industrial use in general. While MDCs have been run continuously at scales up to one liter (106), much larger scales will be required to process commercially relevant volumes of wastewater and whey permeate. Concurrent with scale-up work, improvements in desalination kinetics are necessary to be comparable to reverse osmosis, as desalination in MDCs currently requires retention times of several days.

Advancements in MDC materials and designs will be needed to address these challenges while making MDCs cost-effective at larger scales [105]. The accelerating pace of research for MDCs and bioelectrochemical systems in general, along with growing incentives for energy efficiency and sustainable wastewater management, will likely drive the translation of these technologies to the dairy processing industry.

3. Whey Permeate As a Fuel and Energy Source

By virtue of its high carbohydrate content (approximately 75% of solids), whey permeate has been evaluated as a renewable energy source. Investigations have primarily focused on fermentations utilizing various phyla of yeast to produce bioethanol. However, other fuel sources including renewable diesel, biodiesel, hydrogen gas, and methane have been proposed and investigated. The gold standard for a wastewater stream such as whey permeate involves the mitigation of environmental impact, while adding value to the industry to make the process more cost-effective in all aspects. As evidenced by the cheese industry, each step in the process creates yet another waste stream. While cheese manufacturing produces whey and whey protein recovery produces whey permeate, the focus in the future must be on utilizing the total content of all by-product streams generated, with clean water as the ultimate effluent. Many technologies that propose to harvest the chemical energy from whey permeate as fuel advance this goal.

3.1. Biodiesel and Renewable Diesel

Currently, biodiesel can be produced from vegetable oil or animal-derived fat. In 2007, nearly half of all biodiesel came from canola oil as a feedstock, while the next two most used were soybean (22%) and palm (11%) oils, with approximately 19% being other plant-based oils and animal-derived fats [107]. Additionally, oil accumulated in microorganisms, known as microbial or single cell oil, can be used as a feedstock for biodiesel production. Practically any triglyceride source can undergo esterification in the presence of methanol or ethanol to form an alkyl ester known as biodiesel [108].

While all industrial biodiesel in 2014 is derived from animal or plant lipids, microbial oil might become a promising alternative in the near future. Up to 85% of the operating costs in biodiesel production stem from the feedstock cost, making microbial oil a promising alternative [109]. Although the biodiesel industry relies heavily upon subsidies to remain profitable, with biodiesel garnering a $0.29 per energy equivalent liter in the United States [110], alternative fuel sources are attractive for the current energy landscape. An alternative to producing typical biodiesel composed of methyl esters is to produce a "drop-in" biofuel commonly known as renewable diesel. "Drop in" biofuels are preferred for the long term as they require no change to the existing liquid fueling infrastructure, making market and consumer adoption of these kinds of fuels easier due to renewable diesel meeting the same specification as petroleum derived diesel fuel [108].

There are clear benefits to using oleaginous microorganisms as a feedstock for biodiesel or renewable diesel production. For example, many microorganisms have a fast growth rate and short life cycles, there is a small amount of human labor involved in cultivating the biomass, and culturing the microorganisms does not use valuable arable land [111].

Whey permeate has been recognized as a suitable growth medium for lipogenesis due to its vitamin and mineral content. Additionally, its high carbon to nitrogen (C:N) ratio leads to high lipid accumulation, making whey permeate a suitable substrate for biodiesel production [112]. Oleaginous microorganisms will typically begin to accumulate oil upon depletion of the nitrogen source or when put under some other environmental stress [113]. The three types of organisms studied for lipid accumulation from conversion of whey permeate compounds have been yeast, fungi, and algae. A commercial plant producing polyunsaturated fatty acid-enriched microbial oil using *Mucor circinelloides* was operated for 6 years in the late 1980s, but shut down for economic reasons. With proper process optimization and feedstock choice, microbial oil production at industrial scale could become a reality [113]. Over the past two decades, there have been several investigations into whey permeate as a feedstock for microbial oil production.

Recent investigation on the use of mixotrophic growth of algae to produce a lipid-rich biomass for use as a substrate for biodiesel production has been reported [114]. By using local lake water combined with whey permeate as a source of dissolved organic carbon to culture the microalgae, the investigators were able to obtain a biomass yield of up to 3.5 g/L, and a total lipid production of 28.3 mg/L/d. If all lactose derived from whey worldwide were converted into biodiesel by this process it would have represented over 8% of the total global biodiesel production in 2011 [115].

Additionally, fungi have been evaluated for their ability to produce single cell oil for human consumption, animal feed, or to be used as a precursor for biodiesel production. The organism, *Mortierella isabellina* has been shown to accumulate microbial oil from various sugar sources [116]. Whey permeate shows very promising results when used as a cultivation medium for this microorganism. Yields up to 17 g oil/L can be obtained when this fungus is cultivated in a medium containing approximately 16% lactose previously hydrolyzed into glucose and galactose [117].

Table 2. Summary of oil accumulation by various microorganisms on whey permeate (Operated in batch fermentation mode unless otherwise noted)

Type	Strain	Reference	Lactose (g/L)	pH	g oil/L feedstock	g oil/100 g lactose	Biomass (g/L) dry weight	Oil in dry biomass (%)	Size of reactor
Algae	*Chlorella prototheroides*	[120]	182	5.5	3.8	2.09	9.1±0.2	42±6.6	5L
	C.prototheroides	[120][a]	182	5.5	3.5	1.92	17.2±1.3	20.5±0.3	5L
	C. prototheroides	[120]	182	5.5	3.6	1.98	7.3±1.3	49.9±3.3	5L
	Scenedesmus obliquus	[114][b]	23	6.4	0.49	2.13	4.9±0.2	10.5±2.5	n/a
Yeast	*Apotrichum curvatum*	[119][c]	n/a	4.8	7.78	>20	21.6	36	1L
	Apotrichum curvatum	[22][d]	36.4	6.29	1.12	3.08	18.6	6	1L
	Cryptococcus albidus	[22][e]	36.4	6.29	0.35	0.96	13.8	2.5	1L
	Lipomyces starkeyii	[22][d]	36.4	6.29	2.77	7.61	9.9	28	1L
	Rhodosporidium toruloides	[22][d]	36.4	6.29	0.07	0.19	0.9	8	1L
	Candida curvata D	[112][f]	45	5.4	5.9	13.1	13.1	45	350 mL
	Candida curvata R	[27]	50	5.4-5.8	7.3	14.6	n/a	n/a	14 L
	Candida curvata D	[27]	50	5.4-5.8	9.3	18.6	n/a	n/a	14 L
	Trichosporon cutaneum 24	[27]	50	5.4-5.8	6.3	12.5	n/a	n/a	14 L
	Trichosporon cutaneum 40	[27]	50	5.4-5.8	7.8	15.6	n/a	n/a	14 L
Fungi	*Mortierella isabellina*	[117][g]	160	6	17.13	10.71	26.55	64	250mL

[a]Fed-batch mode, [b]Mixotrophic culture with 60% lake water, [c]Optimal conditions shown in table, C:N ratio 40:1, [d]C:N ratio 60:1, [e]C:N ratio 30:1, [f]Continuous fermentation (dilution rate 0.05, 0.3 g/L-h lipid production), [g]Lactose hydrolyzed.

Several yeasts have been examined for their ability to accumulate oil on whey permeate. These so-called oleaginous yeasts can accumulate up to 37% lipids on a dry basis of the cells, of which 50.8% were triacylglycerols [22]. Since the 1970s, yeasts have been validated for their oil accumulation on whey permeate [27, 118]. As early as 1985, various cell culture modes including continuous and cell recycle have been used to grow yeasts on whey permeate [112]. The optimal conditions were obtained in a chemostat-like setup with a nearly 1 g/L/h of oil productivity being achieved (119). Table 2 summarizes research articles from 1978 to 2014 regarding the accumulation of oil on whey permeate by yeasts, algae, and fungi.

Biomass harvesting and oil extraction from cells/biomass represent a major challenge in the microbial oil production downstream process. To date, unfeasible separation costs have hindered most of the microbial processes [121]. As an example, small cellular morphology such as non-flocculating yeast and algae present challenges during the harvesting and drying steps, usually requiring the use of energy-intensive separations techniques such as centrifugation and evaporation. However, some microorganisms (i.e filamentous fungi) have the ability to form pellets or agglomerates which facilitates the harvesting of the cells, which can be accomplished by simpler screening procedures [122]. Additionally, interest in natural or engineered secretion of lipids into the growth media has grown considerably due to the simplification of downstream processing and subsequent minimization of costs associated with solvent extraction [123]. Alternatively, extraction techniques such as hydrothermal liquefaction (HTL) are being investigated for microbial oils cultivated in dilute aqueous streams and have shown great promise [124].

3.2. Biohydrogen

Hydrogen is often regarded as the cleanest renewable fuel because its exhaust products are simply water and heat. Currently, hydrogen for fuel cells or combustion is produced from several main sources which include electrolysis of water, hydrocarbon reforming from natural gas and oil processing, as well as hydrogen formation from biomass. The most common method for hydrogen production for both fuel cell and chemical uses is steam/fossil fuel reformation [125]. Although fossil fuel derived hydrogen gas is more energy-efficient than electrolytically produced gas, investigations over the past several decades have suggested that the pollutants released from the combustion of petroleum-based fuels can have an impact on human health and have negative implications for the environment [126]. Therefore, energy derived from zero-emission processes, such as wind, solar, and hydroelectric, could be used to power the electrolysis of water to create energy-dense hydrogen fuel cells.

Alternatively, there has been interest in producing hydrogen gas from bioconversion of organic waste streams. Due to the high costs of producing hydrogen for fuel or chemicals via steam reformation of hydrocarbons and electrolysis of water, organic waste has been investigated as a low-cost substrate for producing hydrogen gas. Food processing waste streams are typically rich in carbohydrates, either cellulosic or starch-based, or, in the case of whey permeate, rich in simple sugars. The major microbiological processes associated with biohydrogen production from organic matter include dark fermentation during anaerobic digestion, photo fermentation, as well as a sequential dark fermentation/photo-fermentative production [127].

Several anaerobic microorganisms have been investigated in recent years for their ability to grow on either lactose or whey permeate. For example, *Clostridium stercorarium subsp. thermolacticum* (ATCC 43739) has been shown to have a production of 3 mol H_2/mol lactose of a theoretical yield of 8 mol/mol lactose during dark fermentation [128].

Additionally, mixed anaerobic cultures have been used to produce biohydrogen from whey permeate. Yang et al. [129] have utilized cultures directly from a commercial sludge anaerobic digester and have discovered how various processing systems (batch vs. continuous) and pH have a profound effect on the yield of biohydrogen gas production. A hydrogen yield of 2 mM/g of COD was achieved at a loading rate of 10 g/L COD per day in a continuous culture. Additionally, analysis of the 16S rRNA gene sequences in bacterial genomic DNA extracted during peak H_2 production showed that there were approximately 50% *Lactobacillus* species and 5% *Clostridia* species. Additionally, microbial electrolysis cells (MECs) have been validated for high efficiency production of hydrogen gas from organic matter, but the concept has never been adapted for use of whey permeate [130].

3.3. Biogas (Methane)

Methane in the form of natural gas is a very common fuel source. While methane in natural gas is a fossil fuel, it can also be formed from biomass via thermal gasification as well as anaerobic digestion. Anaerobic digestion can be effectively implemented with dilute solutions of organic material with high water content, and the process can also be scaled economically for any purpose. However, the nutrient source for the digesting bacteria must offer complete nutrition for optimal conversion. Even under optimal conditions, residual organic matter may remain after anaerobic digestion [131].

While this resource has been investigated over the past several decades [132], there has been very little recent research concerning the production of methane/biogas from whey permeate. Rather, recent research has focused more on two-stage digestion processes utilizing sequential hydrogenic and methanogenic steps or the co-digestion of whey permeate with cow manure due to the nutrient content supplied by the manure [133, 134]. Anaerobic digestion is widely used in the treatment of municipal wastewater, and has been examined for its ability to ferment whey permeate under methanogenic as well as acidogenic conditions. Under methanogenic conditions, a methane yield of 63 mMol/g COD was achieved on whey permeate [135].

Although the biogas recovered can be used with minimal processing, such as desulfurization in industrial boilers and heaters, it has a heating value of about half of what is found in natural gas [136]. Although no studies have directly investigated the heating value of recovered biogas from anaerobic digestion of whey permeate, it is likely slightly greater than the heating value of landfill biogas due to the higher methane content of the whey permeate anaerobic digestion biogas (60-70% in whey permeate biogas compared to 45-60% in landfill biogas) [135, 136]. In order to increase the heating value, the methane in the biogas must be isolated. The cost of this isolation, or "scrubbing" process, must be taken into account when considering the feasibility of this technology. Various scrubbing systems exist, which include solid surface adsorption, physical and chemical absorption, membrane separation, and cryogenic separation. There are advantages and disadvantages to each, and they can be applied for various processing operations [137].

3.4. Bioethanol

Ethanol has been used as a fuel source in automobiles since the late 19th century, and has increased in popularity as "gasahol" to increase the octane rating of the fuel and extend its volume. Its use has increased by a large margin in the last decade, due in large part to the heavy restrictions on using methyl tertiary butyl ether in gasoline as an oxygenate [138] Adding ethanol has been found to improve engine performance in a typical automobile four stroke internal combustion engine, while significantly reducing the exhaust emissions of carbon monoxide and hydrocarbons [139].

Corn grain is the source for 90% of the ethanol produced in the United States while whey permeate lactose is a largely untapped source of carbohydrates that could be fermented to produce ethanol [140]. Due to its availability and accessibility compared to cellulosic biomass, many have investigated the potential to use whey permeate as a primary feedstock for ethanol production. If all of the whey lactose globally were fermented and distilled into ethanol at a conversion efficiency of 85%, the worldwide volume of ethanol would be 4.6 billion liters, which represents approximately 16% of the world's production of bioethanol in 2011 [115, 141].

Lactose has been investigated for its ability to be fermented into ethanol as early as 1887, but was not suggested for use as a whey treatment until 1941 [142, 143]. Much of the work regarding ethanol fermentation of whey and whey permeate were conducted in the late 20th century [144–154]. In recent years, the idea has been validated and optimized using various microorganisms, conditions, reactor types, and cell states. Table 3 summarizes recent investigations into different organisms and fermentation conditions on whey permeate for ethanol production.

There is a multitude of yeasts that have been used to ferment whey permeate into ethanol. However, a major consideration before choosing a yeast strain is its ability to metabolize lactose. In yeasts, the genes associated with the ability to metabolize lactose have been identified as LAC4, which codes for the enzyme β-D-1,4 galactosidase, and LAC12, which codes for a membrane-bound lactose permease. These two genes allow for a eukaryotic microorganism to metabolize lactose, which has been demonstrated in *Saccharomyces cerevisiae* as early as 1985 [155]. However, recent efforts have focused on those yeasts which can inherently metabolize and ferment lactose into ethanol. Specifically, those of the *Kluyveromyces* genus which have been isolated from fermented dairy products [156, 157]. This genus of yeasts can most often be found on the surface of ripened cheeses [158, 159].

As of late, *K. marxianus* has been examined in many biotechnological roles, including the production of ethanol and single cell protein (SCP) [160]. It can ferment at high temperatures, exhibits Crabtree-negative metabolic characteristics, and as stated before, readily and rapidly assimilates lactose [28, 141].

Importantly, different studies have determined various productivities of ethanol production. There are many reasons for these disparities. First, in some studies whey permeate is used as-is, without supplements or added nutrients. Due to the inherent lack of nitrogen sources from the removal of protein, many microorganisms may not grow well in whey permeate [161].

Table 3. Recent (2000-2014) investigations of ethanol production from whey permeate

Ref	Organism	Mode of reactor	T(°C)	ph	Lactose concentration	Conversion efficiency[a]
[163]	*Candida kefyr* ATCC 14245	Batch, 250 mL	30	4.8	98 g/L[b]	86.5
[163]	*Kluyveromyces marxianus* ATCC 8554	Batch, 250 mL	30	4.8	49 g/L[b]	94.9
[156]	*K. marxianus* CCT 4086 Immobilized in Ca alginate	Continuous fluidized bed, 0.1 hr^{-1} dilution rate 370 mL	30	n/a	90 g/L	89.2
[164]	*K. marxianus* UFV-3	Batch, hypoxic 250 mL	30	n/a	50-170 g/L	>95%
[165]	*K. marxianus* UFV-3	Batch, quiescent, anoxic 50 mL	33.3-38.5	4.7-5.7	50-108 g/L	>90%
[166]	*Saccharomyces cerevisiae* NCYC869-A3/T1, expresses LAC4 and LAC12, flocculating phenotype	Continuous airlift 0.23 hr^{-1} dilution rate 5.5 L	30	4	50 g/L	80%

[a] Conversion efficiency is the percent of theoretical yield achieved in the process. The theoretical conversion ratio is 0.538 kg ethanol per kg lactose consumed [167].

[b] 0.7% NH_4OH, 0.25% $NH_4H_2PO_4$ added.

These nitrogen supplements include casein hydrolysates, whey protein hydrolysates, yeast extract, malt combing nuts, as well as nitrogen containing salts [28, 69–71]. Additionally, the cells which metabolize whey permeate could be free or immobilized in some manner. Methods of immobilization include natural or induced flocculation, cell entrapment in a gel or matrix, and adsorption of cells onto a surface [162].

The type of reactor and its mode of reaction is also an important operating parameter in various studies. For example, the dynamics of a batch fermentation in a shake flask using supplemented, unconcentrated whey permeate cannot be compared with a continuous fermentation in an airlift reactor using concentrated whey permeate supplemented with yeast extract.

Conclusion and Future Directions

Whey permeate, as a very low-cost co-product of the dairy industry, causes many environmental and economic issues today. While there are many proposed outlets for adding value to this waste stream, currently the majority of whey permeate is concentrated and used to produce lactose. Wastewater management in all industries focuses on minimizing environmental impact while also considering the economic implications simultaneously [56].

Much of the research surrounding whey permeate utilization takes advantage of the fact that this stream has a high lactose content on a solids basis, and many industries require a low-cost carbohydrate feedstock to remain economically competitive.

While whey permeate is mostly composed of lactose, the other soluble components pose issues for the use of whey permeate as a substrate for microorganisms or enzymatic reactions. While the endogenous peptides may provide a valuable nitrogen source for microorganisms, they could complicate downstream isolation processes, as could the high quantity of mineral salts. The individual components of whey permeate are much more valuable and simple to manage when separated, but as such, there is no effective way to accomplish a total separation economically. While the water can be removed by reverse osmosis or evaporation, and the lactose can be mostly removed by crystallization, minerals and the remaining lactose typically end up in the delactosed permeate, which is an even lower-value product. With efficient optimization of upstream synthesis and reaction conditions for lactose derivatives such as lactic acid and galactooligosaccharides, downstream isolation remains the major barrier to commercial production of pure products. Therefore, clean and novel separation techniques must be applied to better fractionate whey permeate compounds and its derivatives. While membrane filtration is very effective, the technology tends to be only available to large-scale processors, as operating costs are high and a significant capital investment is required [168].

There are many potential fermentation products that could add value to whey permeate and find use in the food, chemical, energy, and animal feed industries. As stated above, these applications simply utilize lactose in whey permeate as a low-cost source of soluble sugar. However, as milk is a biological fluid, and whey permeate represents almost the entire suite of soluble compounds in milk, it provides a unique opportunity to harvest the full biological potential of the original milk. High-value naturally occurring oligosaccharides and peptides could be isolated at industrial scale to provide a consistent and reliable source of oligosaccharides for therapeutics or supplements, while economically revitalizing the cheesemaking industry.

ACKNOWLEDGMENTS

The authors acknowledge all of the researchers in the UC Davis Foods for Health Institute and the Milk Bioactives Program for their enthusiasm, imagination, and collective contribution to this subject matter. Work by the Milk Bioactives Program has been supported by the UC Davis Research Investments in the Sciences and Engineering Program; the UC Discovery Grant Program; the California Dairy Research Foundation; the Dairy Research Institute; the Bill & Melinda Gates Foundation; and the National Institutes of Health awards R01HD059127, R01HD065122, R01HD061923, R21AT006180, and R01AT007079. We also acknowledge support from the Peter J. Shields Endowed Chair in Dairy Food Science.

REFERENCES

[1] Tunick MH. Whey Protein Production and Utilization: A Brief History. Whey Processing, Functionality and Health Benefits. John Wiley & Sons; 2009.

[2] Jelen P. Industrial whey processing technology: An overview. *J Agric Food Chem.* 1979;27(4):658–61.

[3] Smithers GW. Whey and whey proteins—From "gutter-to-gold."*Int Dairy J.* 2008 Jul;18(7):695–704.

[4] Bos C, Gaudichon C, Tomé D. Nutritional and physiological criteria in the assessment of milk protein quality for humans. *J Am Coll Nutr.* 2000 Apr;19(2 Suppl):191S–205S.

[5] Ramchandran L, Vasiljevic T. *Whey processing.* Membr Process. 2013;193–207.

[6] Schmidt RH, Packard VS, Morris HA. Effect of Processing on Whey Protein Functionality. *J Dairy Sci.* 1984 Nov;67(11):2723–33.

[7] Wingerd WH. Lactalbumin as a Food Ingredient. *J Dairy Sci.* 1971 Aug;54(8):1234–6.

[8] Marcelo PA, Rizvi SSH. Physicochemical properties of liquid virgin whey protein isolate. *Int Dairy J.* 2008 Mar;18(3):236–46.

[9] Reference Manual for U.S. Whey and Lactose Products. U.S. Dairy Export Council; 2004 Jun.

[10] Palidwor GA, Andrade-Navarro MA. MLTrends: Graphing MEDLINE term usage over time. *J Biomed Discov Collab.* 2010 Jan 25;5:1–6.

[11] Barile D, Tao N, Lebrilla CB, Coisson J-D, Arlorio M, German JB. Permeate from cheese whey ultrafiltration is a source of milk oligosaccharides. *Int Dairy J.* 2009 Sep;19(9):524–30.

[12] Dallas DC, Weinborn V, de Moura Bell JMLN, Wang M, Parker EA, Guerrero A, et al. Comprehensive peptidomic and glycomic evaluation reveals that sweet whey permeate from colostrum is a source of milk protein-derived peptides and oligosaccharides. *Food Res Int.* 2014 Sep;63, Part B:203–9.

[13] Berry RA. The production, composition and utilisation of whey. *J Agric Sci.* 1923;13(02):192–239.

[14] Dixon EM. Whey Permeate, Delactosed Permeate, and Delactosed Whey as Ingredients to Lower Sodium Content of Cream Based Soups [Internet]. North Carolina State University; 2009 [cited 2014 Oct 16]. Available from: http://repository.lib.ncsu.edu/ir/handle/1840.16/2886

[15] Peterson AE, Walker WG, Watson KS. Effect of whey applications on chemical properties of soils and crops. *J Agric Food Chem.* 1979;27(4):654–8.

[16] Watson KS, Peterson AE, Powell RD. Benefits of spreading whey on agricultural land. *J Water Pollut Control Fed.* 1977;24–34.

[17] Ghaly AE, Mahmoud NS, Rushton DG, Arab F. Potential environmental and health impacts of high land application of cheese whey. *Am J Agric Biol Sci.* 2007;2(2):106.

[18] Westendorf ML. Food Waste as Animal Feed: An Introduction. *Food Waste as Animal Feed.* John Wiley & Sons; 2000. p. 3–16.

[19] Kim BG, Lee JW, Stein HH. Energy concentration and phosphorus digestibility in whey powder, whey permeate, and low-ash whey permeate fed to weanling pigs. *J Anim Sci.* 2012 Jan 1;90(1):289–95.

[20] Naranjo VD, Bidner TD, Southern LL. Comparison of dried whey permeate and a carbohydrate product in diets for nursery pigs. *J Anim Sci.* 2010 May 1;88(5):1868–79.
[21] Cotanch KW, Darrah JW. The Effect of Feeding Lactose in the Form of Whey Permeate on the Productivity of Lactating Dairy Cattle. Chazy, NY: W. H. Miner Agricultural Research Institute; 2006.
[22] Akhtar P, Gray JI, Asghar A. Synthesis of lipids by certain yeast strains grown on whey permeate. *J Food Lipids.* 1998;5(4):283–97.
[23] Fernández C, Carracedo B, Martínez EJ, Gómez X, Morán A. Application of a packed bed reactor for the production of hydrogen from cheese whey permeate: Effect of organic loading rate. *J Environ Sci Health Part A.* 2014 Jan 28;49(2):210–7.
[24] Hargrove RE, McDonough FE, Lacroix DE, Alford JA. Production and properties of deproteinized whey powders. *J Dairy Sci.* 1976;59(1):25–33.
[25] Affertsholt T. Market developments and industry challenges for lactose and lactose derivatives [Internet]. IDF Symposium: "Lactose and its Derivatives"; 2007 May 15; Moscow, Russia. Available from: http://lactose.ru/present/1Tage_Affertsholt-Allen.pdf
[26] Giec A, Skupin J. Single cell protein as food and feed. *Nahr.* 1988;32(3):219–29.
[27] Moon NJ, Hammond EG, Glatz BA. Conversion of cheese whey and whey permeate to oil and single-cell protein. *J Dairy Sci.* 1978;61(11):1537–47.
[28] Schultz N, Lifung Chang, Hauck A, Reuss M, Syldatk C. Microbial production of single-cell protein from deproteinized whey concentrates. *Appl Microbiol Biotechnol.* 2006 Jan;69(5):515–20.
[29] Dornburg V, Hermann BG, Patel MK. Scenario Projections for Future Market Potentials of Biobased Bulk Chemicals. *Environ Sci Technol.* 2008;42(7):2261–7.
[30] Jones DT, Woods DR. Acetone-butanol fermentation revisited. *Microbiol Rev.* 1986 Dec;50(4):484–524.
[31] Ni Y, Sun Z. Recent progress on industrial fermentative production of acetone–butanol–ethanol by < i>Clostridium acetobutylicum</i> in China. *Appl Microbiol Biotechnol.* 2009 Jun 1;83(3):415–23.
[32] Ennis BM, Maddox IS. Use of Clostridium acetobutylicum P262 for production of solvents from whey permeate. *Biotechnol Lett.* 1985;7(8):601–6.
[33] Qureshi N, Maddox IS. Reduction in butanol inhibition by perstraction: utilization of concentrated lactose/whey permeate by Clostridium acetobutylicum to enhance butanol fermentation economics. *Food Bioprod Process.* 2005;83(1):43–52.
[34] Dürre P. Fermentative Butanol Production: Bulk Chemical and Biofuel. *Ann N Y Acad Sci.* 2008 Mar 1;1125(1):353–62.
[35] Danese PN, Pratt LA, Kolter R. Exopolysaccharide production is required for development of Escherichia coli K-12 biofilm architecture. *J Bacteriol.* 2000;182(12):3593–6.
[36] Hess SJ, Roberts RF, Ziegler GR. Rheological Properties of Nonfat Yogurt Stabilized Using< i> Lactobacillus delbrueckii</i> ssp.< i> bulgaricus</i> Producing Exopolysaccharide or Using Commercial Stabilizer Systems. *J Dairy Sci.* 1997;80(2):252–63.
[37] Roukas T. Pullulan production from deproteinized whey by *Aureobasidium pullulans. J Ind Microbiol Biotechnol.* 1999;22(6):617–21.

[38] Savvides AL, Katsifas EA, Hatzinikolaou DG, Karagouni AD. Xanthan production by Xanthomonas campestris using whey permeate medium. *World J Microbiol Biotechnol.* 2012 Aug 1;28(8):2759–64.

[39] Pacwa-Plociniczak M, Plaza GA, Piotrowska-Seget Z, Cameotra SS. Environmental Applications of Biosurfactants: Recent Advances. *Int J Mol Sci.* 2011 Jan 18;12(1):633–54.

[40] Bogaert INAV, Saerens K, Muynck CD, Develter D, Soetaert W, Vandamme EJ. Microbial production and application of sophorolipids. *Appl Microbiol Biotechnol.* 2007 Aug 1;76(1):23–34.

[41] Daverey A, Pakshirajan K. Sophorolipids from Candida bombicola using mixed hydrophilic substrates: Production, purification and characterization. *Colloids Surf B Biointerfaces.* 2010 Aug;79(1):246–53.

[42] Beucler J, Drake M, Foegeding EA. Design of a beverage from whey permeate. *J Food Sci.* 2005;70(4):S277–S285.

[43] Bradley RL, Rexroat TM. Acceptance and Stability of Chocolate and Strawberry Fountain Syrups Made with Concentrated, Decolorized, Deionized Hydrolyzed Whey Permeate 1. *J Food Process Preserv.* 1988 Jun 1;12(2):163–70.

[44] Jaros D, Spieler C, Kleinschmidt T, Rohm H. Using whey permeate powders for partial sucrose substitution in flavored yogurt. *Milchwissenschaft.* 2008;63(2):174–8.

[45] Gänzle MG, Haase G, Jelen P. Lactose: Crystallization, hydrolysis and value-added derivatives. *Int Dairy J.* 2008 Jul;18(7):685–94.

[46] Paterson AHJ. Production and Uses of Lactose. In: McSweeney P, Fox PF, editors. *Advanced Dairy Chemistry* [Internet]. Springer New York; 2009 [cited 2014 Nov 21]. p. 105–20. Available from: http://link.springer.com/chapter/10.1007/978-0-387-84865-5_4

[47] Schaafsma G. Lactose and lactose derivatives as bioactive ingredients in human nutrition. *Int Dairy J.* 2008 May;18(5):458–65.

[48] Fox PF. Lactose: Chemistry and Properties. In: McSweeney P, Fox PF, editors. *Advanced Dairy Chemistry* [Internet]. New York, NY: Springer New York; 2009 [cited 2014 Aug 7]. Available from: http://link.springer.com/10.1007/978-0-387-84865-5

[49] McSweeney P, Fox PF, editors. *Advanced Dairy Chemistry* [Internet]. New York, NY: Springer New York; 2009 [cited 2014 Aug 7]. Available from: http://link.springer.com/10.1007/978-0-387-84865-5

[50] Patel SR, Murthy ZVP. Ultrasound assisted crystallization for the recovery of lactose in an anti-solvent acetone. *Cryst Res Technol.* 2009 Aug;44(8):889–96.

[51] Patel SR, Murthy ZVP. Waste valorization: Recovery of lactose from partially deproteinated whey by using acetone as anti-solvent. *Dairy Sci Technol* [Internet]. 2010 Aug 2 [cited 2014 Aug 7]; Available from: http://link.springer.com/10.1051/dst/2010036

[52] Zisu B, Sciberras M, Jayasena V, Weeks M, Palmer M, Dincer TD. Sonocrystallisation of lactose in concentrated whey. *Ultrason Sonochem.* 2014;21(6):2117–21.

[53] Wong SY, Hartel RW. Crystallization in Lactose Refining-A Review: Review on crystallization in lactose refining. *J Food Sci.* 2014 Mar;79(3):R257–R272.

[54] Pisponen A, Pajumägi S, Mootse H, Sats A, Poikalainen V, Karus A. Effect of cooling rates and low crystallization temperatures on morphology of lactose crystals obtained from Ricotta cheese whey. *Agron Res.* 2014;12(3):787–92.

[55] Atra R, Vatai G, Bekassy-Molnar E, Balint A. Investigation of ultra-and nanofiltration for utilization of whey protein and lactose. *J Food Eng.* 2005;67(3):325–32.

[56] Patel SR, Murthy ZVP. Lactose Recovery Processes from Whey: A Comparative Study Based on Sonocrystallization. *Sep Purif Rev.* 2012 Oct;41(4):251–66.

[57] Burrington K. Permeate for sodium reduction. *U.S. Dairy Export* Council; 2011.

[58] Liang B, Bund RK, Hartel RW. Effect of composition on moisture sorption of delactosed permeate. *Int Dairy J.* 2009 Oct;19(10):630–6.

[59] Lifran EV, Hourigan JA, Sleigh RW. Lactose derivatives: Turning waste into functional foods. *Aust J Dairy Technol.* 2009;64(1):89–93.

[60] Panesar P, Kennedy J, Gandhi D, Bunko K. Bioutilisation of whey for lactic acid production. *Food Chem.* 2007;105(1):1–14.

[61] Paseephol T, Small DM, Sherkat F. Lactulose production from milk concentration permeate using calcium carbonate-based catalysts. *Food Chem.* 2008 Nov;111(2): 283–90.

[62] Castillo Martinez FA, Balciunas EM, Salgado JM, Domínguez González JM, Converti A, Oliveira RP de S. Lactic acid properties, applications and production: *A review. Trends Food Sci Technol.* 2013 Mar;30(1):70–83.

[63] Oh D-K. Tagatose: properties, applications, and biotechnological processes. *Appl Microbiol Biotechnol.* 2007 Aug 1;76(1):1–8.

[64] Abdel-Rahman MA, Tashiro Y, Sonomoto K. Recent advances in lactic acid production by microbial fermentation processes. *Biotechnol Adv.* 2013 Nov;31(6):877–902.

[65] Ren J. Lactic Acid. Biodegradable Poly (Lactic Acid): *Synthesis, Modification, Processing and Applications* [Internet]. Springer; 2011 [cited 2014 Oct 27]. p. 4–14. Available from: http://link.springer.com/chapter/10.1007/978-3-642-17596-1_2

[66] Datta R, Henry M. Lactic acid: recent advances in products, processes and technologies — a review. *J Chem Technol Biotechnol.* 2006 Jul;81(7):1119–29.

[67] Gao C, Ma C, Xu P. Biotechnological routes based on lactic acid production from biomass. *Biotechnol Adv.* 2011 Nov;29(6):930–9.

[68] Mäki-Arvela P, Simakova IL, Salmi T, Murzin DY. Production of Lactic Acid/Lactates from Biomass and Their Catalytic Transformations to Commodities. *Chem Rev.* 2014 Feb 12;114(3):1909–71.

[69] Fitzpatrick JJ, O'Keeffe U. Influence of whey protein hydrolysate addition to whey permeate batch fermentations for producing lactic acid. *Process Biochem.* 2001;37(2):183–6.

[70] Prasad S, Srikanth K, Limaye AM, Sivaprakasam S. Homo-fermentative production of d-lactic acid by Lactobacillus sp. employing casein whey permeate as a raw feed-stock. *Biotechnol Lett.* 2014 Jun;36(6):1303–7.

[71] Pauli T, Fitzpatrick JJ. Malt combing nuts as a nutrient supplement to whey permeate for producing lactic by fermentation with< i> Lactobacillus casei</i>. Process Biochem. 2002;38(1):1–6.

[72] Wang K, Li W, Fan Y, Xing W. Integrated Membrane Process for the Purification of Lactic Acid from a Fermentation Broth Neutralized with Sodium Hydroxide. *Ind Eng Chem Res.* 2013 Feb 13;52(6):2412–7.

[73] González MI, Álvarez S, Riera F, Álvarez R. Economic evaluation of an integrated process for lactic acid production from ultrafiltered whey. *J Food Eng.* 2007 May;80(2):553–61.

[74] Foster KJ, Lin S. Current and emerging strategies for treating hepatic encephalopathy. *Crit Care Nurs Clin North Am.* 2010;22(3):341–50.

[75] Wang H, Yang R, Hua X, Zhao W, Zhang W. Enzymatic production of lactulose and 1-lactulose: current state and perspectives. *Appl Microbiol Biotechnol.* 2013 Jul;97(14):6167–80.

[76] Aider M, Halleux D de. Isomerization of lactose and lactulose production: review. *Trends Food Sci Technol.* 2007 Jul;18(7):356–64.

[77] Villamiel M, Montilla A, Olano A, Corzo N. Production and Bioactivity of Oligosaccharides Derived from Lactose. In: Moreno FJ, Sanz ría L, editors. *Food Oligosaccharides* [Internet]. John Wiley & Sons, Ltd; 2014 [cited 2014 Nov 20]. p. 135–67. Available from: http://onlinelibrary.wiley.com/doi/10.1002/9781118817360.ch9/summary

[78] Playne MJ, Crittenden RG. Galacto-oligosaccharides and Other Products Derived from Lactose. In: McSweeney P, Fox PF, editors. *Advanced Dairy Chemistry* [Internet]. Springer New York; 2009 [cited 2015 Jan 15]. p. 121–201. Available from: http://link.springer.com/chapter/10.1007/978-0-387-84865-5_5

[79] Panesar PS, Kumari S. Lactulose: Production, purification and potential applications. *Biotechnol Adv.* 2011 Nov;29(6):940–8.

[80] Matella N j., Dolan K d., Lee Y s. Comparison of Galactooligosaccharide Production in Free-Enzyme Ultrafiltration and in Immobilized-Enzyme Systems. *J Food Sci.* 2006;71(7):C363–C368.

[81] Manzo RM, Simonetta AC, Rubiolo AC, Mammarella EJ. Screening and selection of wild strains for L-arabinose isomerase production. *Braz J Chem Eng.* 2013;30(4): 711–20.

[82] Wanarska M, Kur J, others. A method for the production of D-tagatose using a recombinant Pichia pastoris strain secreting β-D-galactosidase from Arthrobacter chlorophenolicus and a recombinant L-arabinose isomerase from Arthrobacter sp. 22c. *Microb Cell Fact.* 2012;11:113.

[83] Cho Y-J, Shin H-J, Bucke C. Purification and biochemical properties of a galactooligosaccharide producing β-galactosidase from Bullera singularis. *Biotechnol Lett.* 2003 Dec 1;25(24):2107–11.

[84] Zivkovic AM, Barile D. Bovine Milk as a Source of Functional Oligosaccharides for Improving Human Health. *Adv Nutr Int Rev J.* 2011 May 1;2(3):284–9.

[85] Dallas DC, Meyrand M, Barile D. Production and Bioactivity of Bovine Milk Oligosaccharides. In: Moreno FJ, Sanz ría L, editors. Food Oligosaccharides [Internet]. John Wiley & Sons, Ltd; 2014 [cited 2014 Nov 20]. p. 21–34. Available from: http://onlinelibrary.wiley.com/doi/10.1002/9781118817360.ch2/summary

[86] Pacheco AR, Barile D, Underwood MA, Mills DA. The Impact of the Milk Glycobiome on the Neonate Gut Microbiota. *Annu Rev Anim Biosci.* 2015 Mar 15;3(1): 141111105737004.

[87] Crisà A. Milk Carbohydrates and Oligosaccharides. In: Park YW, Haenlein GFW, editors. *Milk and Dairy Products in Human Nutrition* [Internet]. John Wiley & Sons; 2013 [cited 2015 Jan 12]. p. 129–47. Available from: http://onlinelibrary.wiley.com/doi/10.1002/9781118534168.ch7/summary

[88] Douglas B. Sarney CH. A novel approach to the recovery of biologically active oligosaccharides from milk using a combination of enzymatic treatment and nanofiltration. *Biotechnol Bioeng*. 2000;69(4):461–7.

[89] Luo H, Xu P, Roane TM, Jenkins PE, Ren Z. Microbial desalination cells for improved performance in wastewater treatment, electricity production, and desalination. *Bioresour Technol*. 2012;105:60–6.

[90] Martinez-Ferez A, Rudloff S, Guadix A, Henkel CA, Pohlentz G, Boza JJ, et al. Goats' milk as a natural source of lactose-derived oligosaccharides: Isolation by membrane technology. *Int Dairy J*. 2006 Feb;16(2):173–81.

[91] Oliveira DL, Wilbey RA, Grandison AS, Duarte LC, Roseiro LB. Separation of oligosaccharides from caprine milk whey, prior to prebiotic evaluation. *Int Dairy J*. 2012 Jun;24(2):102–6.

[92] Ottens M, Chilamkurthi S. Advances in process chromatography and applications in the food, beverage and nutraceutical industries. In: Rizvi SSH, editor. Separation, Extraction and Concentration Processes in the *Food, Beverage and Nutraceutical Industries* [Internet]. Woodhead Publishing; 2013 [cited 2015 Jan 23]. p. 109–47. Available from: http://www.sciencedirect.com/science/article/pii/B9781845696450500049

[93] 3′-Sialyllactose from bovine milk or colostrum, ≥97% (HPLC) | Sigma-Aldrich [Internet]. [cited 2014 Nov 14]. Available from: http://www.sigmaaldrich.com/catalog/product/sigma/a8681?lang=en&region=US

[94] Dallas DC, Guerrero A, Parker EA, Garay LA, Bhandari A, Lebrilla CB, et al. Peptidomic Profile of Milk of Holstein Cows at Peak Lactation. *J Agric Food Chem*. 2014 Jan 8;62(1):58–65.

[95] Butylina S, Luque S, Nyström M. Fractionation of whey-derived peptides using a combination of ultrafiltration and nanofiltration. *J Membr Sci*. 2006 Sep 1;280(1–2):418–26.

[96] Garsa AK, Kumariya R, Sood SK, Kumar A, Kapila S. Bacteriocin Production and Different Strategies for Their Recovery and Purification. *Probiotics Antimicrob Proteins*. 2014 Mar;6(1):47–58.

[97] Liu X, Chung Y-K, Yang S-T, Yousef AE. Continuous nisin production in laboratory media and whey permeate by immobilized Lactococcus lactis. *Process Biochem*. 2005 Jan;40(1):13–24.

[98] Enan G, Amri AAA. Novel plantaricin UG1 production by Lactobacillus plantarum UG1 in enriched whey permeate in batch fermentation processes. *J Food Agric Environ*. 2006;4(2):85.

[99] Nicolas G, Auger I, Beaudoin M, Hallé F, Morency H, LaPointe G, et al. Improved methods for mutacin detection and production. *J Microbiol Methods*. 2004 Dec;59(3):351–61.

[100] Somkuti GA, Gilbreth SE. Influence of Organic Buffers on Bacteriocin Production by Streptococcus thermophilus ST110. *Curr Microbiol*. 2007 Aug;55(2):173–7.

[101] Short JL, Doughty RK. Demineralization of deproteinated wheys by electrodialysis. *N Z J Dairy Sci Technol*. 1977;12:156–9.

[102] Pérez A, Andrés L j., Álvarez R, Coca J, Hill C g. Electrodialysis of Whey Permeates and Retentates Obtained By Ultrafiltration. *J Food Process Eng*. 1994;17(2):177–90.

[103] Higgins JJ, Short JL. Demineralization by electrodialysis of permeates derived from ultrafiltration of wheys and skim milk. *N Z J Dairy Sci Technol.* 1980;15(3):277–88.

[104] Cao X, Huang X, Liang P, Xiao K, Zhou Y, Zhang X, et al. A new method for water desalination using microbial desalination cells. *Environ Sci Technol.* 2009;43(18):7148–52.

[105] Mehanna M, Saito T, Yan J, Hickner M, Cao X, Huang X, et al. Using microbial desalination cells to reduce water salinity prior to reverse osmosis. *Energy Environ Sci.* 2010;3(8):1114–20.

[106] Jacobson KS, Drew DM, He Z. Use of a liter-scale microbial desalination cell as a platform to study bioelectrochemical desalination with salt solution or artificial seawater. *Environ Sci Technol.* 2011;45(10):4652–7.

[107] Moser BR. Biodiesel Production, Properties, and Feedstocks. In: Tomes D, Lakshmanan P, Songstad D, editors. Biofuels [Internet]. Springer New York; 2011 [cited 2014 Sep 9]. p. 285–347. Available from: http://link.springer.com/chapter/10.1007/978-1-4419-7145-6_15

[108] Westfall PJ, Gardner TS. Industrial fermentation of renewable diesel fuels. *Curr Opin Biotechnol.* 2011 Jun;22(3):344–50.

[109] Paulson ND, Ginder RG, others. The growth and direction of the biodiesel industry in the United States [Internet]. Center for Agricultural and Rural Development, Iowa State University; 2007 [cited 2014 Sep 9]. Available from: http://core.kmi.open.ac.uk/download/pdf/11348306.pdf

[110] Hill J, Nelson E, Tilman D, Polasky S, Tiffany D. Environmental, economic, and energetic costs and benefits of biodiesel and ethanol biofuels. *Proc Natl Acad Sci.* 2006 Jul 25;103(30):11206–10.

[111] Li Q, Du W, Liu D. Perspectives of microbial oils for biodiesel production. *Appl Microbiol Biotechnol.* 2008 Oct 1;80(5):749–56.

[112] Floetenmeyer MD, Glatz BA, Hammond EG. Continuous culture fermentation of whey permeate to produce microbial oil. *J Dairy Sci.* 1985;68(3):633–7.

[113] Ratledge C. Fatty acid biosynthesis in microorganisms being used for Single Cell Oil production. *Biochimie.* 2004 Nov;86(11):807–15.

[114] Girard J-M, Roy M-L, Hafsa MB, Gagnon J, Faucheux N, Heitz M, et al. Mixotrophic cultivation of green microalgae< i> Scenedesmus obliquus</i> on cheese whey permeate for biodiesel production. Algal Res [Internet]. 2014 [cited 2014 Jun 17]; Available from: http://www.sciencedirect.com/science/article/pii/S2211926414000265

[115] International Energy Statistics - EIA [Internet]. [cited 2014 Sep 9]. Available from: http://www.eia.gov/cfapps/ipdbproject/IEDIndex3.cfm?tid=79&pid=79&aid=1

[116] Ruan Z, Zanotti M, Wang X, Ducey C, Liu Y. Evaluation of lipid accumulation from lignocellulosic sugars by Mortierella isabellina for biodiesel production. *Bioresour Technol.* 2012 Apr;110:198–205.

[117] Demir M, Turhan I, Kucukcetin A, Alpkent Z. Oil production by Mortierella isabellina from whey treated with lactase. *Bioresour Technol.* 2013 Jan;128:365–9.

[118] Meng X, Yang J, Xu X, Zhang L, Nie Q, Xian M. Biodiesel production from oleaginous microorganisms. *Renew Energy.* 2009 Jan;34(1):1–5.

[119] Ykema A, Verbree EC, Kater MM, Smit H. Optimization of lipid production in the oleaginous yeastApiotrichum curvatum in wheypermeate. *Appl Microbiol Biotechnol.* 1988;29(2-3):211–8.

[120] Espinosa-Gonzalez I, Parashar A, Bressler DC. Heterotrophic growth and lipid accumulation of Chlorella protothecoides in whey permeate, a dairy by-product stream, for biofuel production. *Bioresour Technol.* 2014 Mar;155:170–6.

[121] Xia C, Zhang J, Zhang W, Hu B. A new cultivation method for microbial oil production: cell pelletization and lipid accumulation by Mucor circinelloides. *Biotechnol Biofuels.* 2011;4(1):15.

[122] Mitra D, Rasmussen ML, Chand P, Chintareddy VR, Yao L, Grewell D, et al. Value-added oil and animal feed production from corn-ethanol stillage using the oleaginous fungus Mucor circinelloides. Bioresour Technol. 2012 Mar;107:368–75.

[123] Mercer P, Armenta RE. Developments in oil extraction from microalgae. *Eur J Lipid Sci Technol.* 2011;113(5):539–47.

[124] Bell PJ. The Fungi-to-Fuel Presentation [Internet]. International Fuel Ethanol Workshop & Expo; 2014 Jun 11; Indianapolis, IN. Available from: http://www.fuelethanolworkshop.com/files/docs/2014/Bell_Peter.pdf

[125] Holladay JD, Hu J, King DL, Wang Y. An overview of hydrogen production technologies. *Catal Today.* 2009 Jan;139(4):244–60.

[126] Kampa M, Castanas E. Human health effects of air pollution. *Environ Pollut.* 2008 Jan;151(2):362–7.

[127] Kapdan IK, Kargi F. Bio-hydrogen production from waste materials. *Enzyme Microb Technol.* 2006 Mar;38(5):569–82.

[128] Collet C. Hydrogen production by Clostridium thermolacticum during continuous fermentation of lactose. *Int J Hydrog Energy.* 2004 Nov;29(14):1479–85.

[129] Yang P, Zhang R, Mcgarvey J, Benemann J. Biohydrogen production from cheese processing wastewater by anaerobic fermentation using mixed microbial communities. *Int J Hydrog Energy.* 2007 Dec;32(18):4761–71.

[130] Logan BE, Call D, Cheng S, Hamelers HVM, Sleutels THJA, Jeremiasse AW, et al. Microbial Electrolysis Cells for High Yield Hydrogen Gas Production from Organic Matter. *Environ Sci Technol.* 2008 Dec;42(23):8630–40.

[131] Chynoweth DP, Owens JM, Legrand R. Renewable methane from anaerobic digestion of biomass. *Renew Energy.* 2001;22(1):1–8.

[132] Yang S-T, Guo M. Kinetics of methanogenesis from whey permeate in packed bed immobilized cells bioreactor. *Biotechnol Bioeng.* 1990;36(4):427–36.

[133] Hagen LH, Vivekanand V, Linjordet R, Pope PB, Eijsink VGH, Horn SJ. Microbial community structure and dynamics during co-digestion of whey permeate and cow manure in continuous stirred tank reactor systems. *Bioresour Technol.* 2014 Nov;171:350–9.

[134] Kisielewska M, Wysocka I, Rynkiewicz M r. Continuous biohydrogen and biomethane production from whey permeate in a two-stage fermentation process. *Environ Prog Sustain Energy.* 2014;33(4):1411–8.

[135] Lee C, Kim J, Hwang K, O'Flaherty V, Hwang S. Quantitative analysis of methanogenic community dynamics in three anaerobic batch digesters treating different wastewaters. *Water Res.* 2009 Jan;43(1):157–65.

[136] Bade Shrestha SO, Narayanan G. Landfill gas with hydrogen addition – A fuel for SI engines. *Fuel*. 2008 Dec;87(17–18):3616–26.

[137] Kapdi SS, Vijay VK, Rajesh SK, Prasad R. Biogas scrubbing, compression and storage: perspective and prospectus in Indian context. *Renew Energy*. 2005;30(8):1195–202.

[138] Solomon BD, Barnes JR, Halvorsen KE. Grain and cellulosic ethanol: History, economics, and energy policy. *Biomass Bioenergy*. 2007 Jun;31(6):416–25.

[139] Al-Hasan M. Effect of ethanol–unleaded gasoline blends on engine performance and exhaust emission. *Energy Convers Manag*. 2003;44(9):1547–61.

[140] De Oliveira MED, Vaughan BE, Rykiel EJ. Ethanol as fuel: energy, carbon dioxide balances, and ecological footprint. *BioScience*. 2005;55(7):593–602.

[141] Guimarães PMR, Teixeira JA, Domingues L. Fermentation of lactose to bio-ethanol by yeasts as part of integrated solutions for the valorisation of cheese whey. *Biotechnol Adv*. 2010 May;28(3):375–84.

[142] Browne HH. Ethyl alcohol from fermentation of lactose in whey. Ind *Eng Chem News Ed*. 1941;19:1272–6.

[143] Duclaux E. Fermentation alcoolique du sucre de lait. *Ann Instit Pasteur*. 1887;1:573.

[144] Dale MC, Eagger A, Okos MR. Osmotic inhibition of free and immobilized K. marxianus anaerobic growth and ethanol productivity in whey permeate concentrate. *Process Biochem*. 1994;29(7):535–44.

[145] Ferrari MD, Loperena L, Varela H. Ethanol production from concentrated whey permeate using a fed-batch culture ofKluyveromyces fragilis. *Biotechnol Lett*. 1994;16(2):205–10.

[146] Gawel J, Kosikowski FV. Improving alcohol fermentation in concentrated ultrafiltration permeates of cottage cheese whey. *J Food Sci*. 1978;43(6):1717–9.

[147] Hahn-Hägerdal B. Comparison between immobilized Kluyveromyces fragilis and Saccharomyces cerevisiae coimmobilized with β-galactosidase, with respect to continuous ethanol production from concentrated whey permeate. *Biotechnol Bioeng*. 1985;27(6):914–6.

[148] Mahmoud MM, Kosikowski FV. Alcohol and single cell protein production by Kluyveromyces in concentrated whey permeates with reduced ash. *J Dairy Sci*. 1982;65(11):2082–7.

[149] Marwaha SS, Kennedy JF. Ethanol production from whey permeate by immobilized yeast cells. *Enzyme Microb Technol*. 1984;6(1):18–22.

[150] Marwaha SS, Kennedy JF, others. Alcohol production from whey permeate by immobilized and free cells of Kluyveromyces marxianus NCYC 179. *Process Biochem*. 1984;19(2):79–80.

[151] Mehaia DMA, Cheryan DM. Ethanol from hydrolyzed whey permeate using Saccharomyces cerevisiae in a membrane recycle bioreactor. *Bioprocess Eng*. 1990 Mar 1;5(2):57–61.

[152] Moulin G, Guillaume M, Galzy P. Alcohol production by yeast in whey ultrafiltrate. *Biotechnol Bioeng*. 1980;22(6):1277–81.

[153] O'leary VS, Sutton C, Bencivengo M, Sullivan B, Holsinger VH. Influence of lactose hydrolysis and solids concentration on alcohol production by yeast in acid whey ultrafiltrate. *Biotechnol Bioeng*. 1977;19(11):1689–702.

[154] Vienne P, Von Stockar U. An investigation of ethanol inhibition and other limitations occurring during the fermentation of concentrated whey permeate byKluyveromyces fragilis. *Biotechnol Lett.* 1985;7(7):521–6.

[155] Sreekrishna K, Dickson RC. Construction of strains of Saccharomyces cerevisiae that grow on lactose. *Proc Natl Acad Sci.* 1985;82(23):7909–13.

[156] Gabardo S, Rech R, Rosa CA, Ayub MAZ. Dynamics of ethanol production from whey and whey permeate by immobilized strains of Kluyveromyces marxianus in batch and continuous bioreactors. *Renew Energy.* 2014 Sep;69:89–96.

[157] Viljoen BC. The interaction between yeasts and bacteria in dairy environments. *Int J Food Microbiol.* 2001 Sep 19;69(1–2):37–44.

[158] Bockelmann W, Hoppe-Seyler T. The surface flora of bacterial smear-ripened cheeses from cow's and goat's milk. *Int Dairy J.* 2001;11(4):307–14.

[159] Roostita R, Fleet GH. The occurrence and growth of yeasts in Camembert and blue-veined cheeses. *Int J Food Microbiol.* 1996;28(3):393–404.

[160] Fonseca GG, Heinzle E, Wittmann C, Gombert AK. The yeast Kluyveromyces marxianus and its biotechnological potential. *Appl Microbiol Biotechnol.* 2008 Jun 1;79(3):339–54.

[161] Haast JD, Britz TJ, Novello JC, Verwey EW. Anaerobic digestion of deproteinated cheese whey. *J Dairy Res.* 1985;52(03):457–67.

[162] Woodward J. Methods of immobilization of microbial cells. *J Microbiol Methods.* 1988 Jun;8(1–2):91–102.

[163] Koushki M, Jafari M, Azizi M. Comparison of ethanol production from cheese whey permeate by two yeast strains. *J Food Sci Technol.* 2012;49(5):614–9.

[164] Silveira WB, Passos FJV, Mantovani HC, Passos FML. Ethanol production from cheese whey permeate by Kluyveromyces marxianus UFV-3: A flux analysis of oxido-reductive metabolism as a function of lactose concentration and oxygen levels. *Enzyme Microb Technol.* 2005 May;36(7):930–6.

[165] Diniz RHS, Rodrigues MQRB, Fietto LG, Passos FML, Silveira WB. Optimizing and validating the production of ethanol from cheese whey permeate by Kluyveromyces marxianus UFV-3. *Biocatal Agric Biotechnol.* 2014 Apr;3(2):111–7.

[166] Domingues L, Lima N, Teixeira JA. Alcohol production from cheese whey permeate using genetically modified flocculent yeast cells. *Biotechnol Bioeng.* 2001;72(72):507–14.

[167] Mawson AJ. Bioconversions for whey utilization and waste abatement. *Bioresour Technol.* 1994;47(3):195–203.

[168] Bund RK, Pandit AB. Rapid lactose recovery from paneer whey using sonocrystallization: A process optimization. *Chem Eng Process Process Intensif.* 2007 Sep;46(9):846–50.

In: Agricultural Wastes
Editor: Camille N. Foster

ISBN: 978-1-63482-359-3

Chapter 3

THERMOCHEMICAL AND BIOCHEMICAL CONVERSION OF OLIVE STONES

Manuel Cuevas[1], Sebastián Sánchez[1] and Juan Francisco García[2]
[1]Department of Chemical, Environmental and Materials Engineering, University of Jaén, Jaén, Spain
[2]Instituto de la Grasa, Spanish National Research Council. Seville, Spain

ABSTRACT

The olive tree is extensively cultivated in countries of the Mediterranean basin; the area currently under cultivation covers roughly 5.5 10^6 ha in the EU and 11.0 10^6 ha in the world. By-products from olive culture and related industries, such as prunings, leaves, olive pomace, and olive stones, are interesting materials for the production of energy, food, fertilizers and other chemicals due to the large available feedstock and their chemical composition. Olive stones are by-products derived from the olive oil extraction industry and from manufacturing of pitted-table olives. Basically, there are two current ways for valorization of olive stones: thermochemical (energy source by combustion, gasification or pyrolysis) and biochemical (ethanol and xylitol production) conversion. Bioconversion of olive stones can also provide other high added-value products such as xylooligosaccharides or natural antioxidants (tyrosol and hydroxytyrosol). Finally, comparison of the different procedures and potential future applications will be discussed as well.

INTRODUCTION

The olive tree is a tree native to the Mediterranean whose fruit, olives, yields two important foods: olive oil, one of the healthiest vegetable fats [1], and table olives. Olive cultivation has spread in recent decades to areas far from the Mediterranean Sea such as the West coast of the United States, Argentina, Chile or Australia. This expansion has led to a rapidly increase in the area under olive cultivation, which currently cover roughly 10 million hectares [2]. Spain is the world's largest producer of olives. The average production in the period 2002-2012 was 5.8 Mt [3] Most of these fruits are intended for the extraction of olive oil, and only about 8% is used for olives production.

The management of olive orchard as well the industrial processing of the fruit of the olive tree generates lignocellulosic by-products whose uses are under research (Figure 1). Among them we can highlight the olive stones due to not only their production volume and geographic concentration but also their composition. The main generating facilities of this biomass are the olive mills and the pomace oil mills. In olive mills the oil olive is separated from the rest of the olive components by centrifugation (pomace oil). Some centrifuges are called three-phase because oil, water and solids are separately recovered. The two-phase centrifuges separate the oil from a wet paste (pomace oil). Fragmented olive stones are then separated from pomace oil by boning machines. In the pomace oil industry, olive stones are recovered as orujillo along with remnants of the fruit pulp.

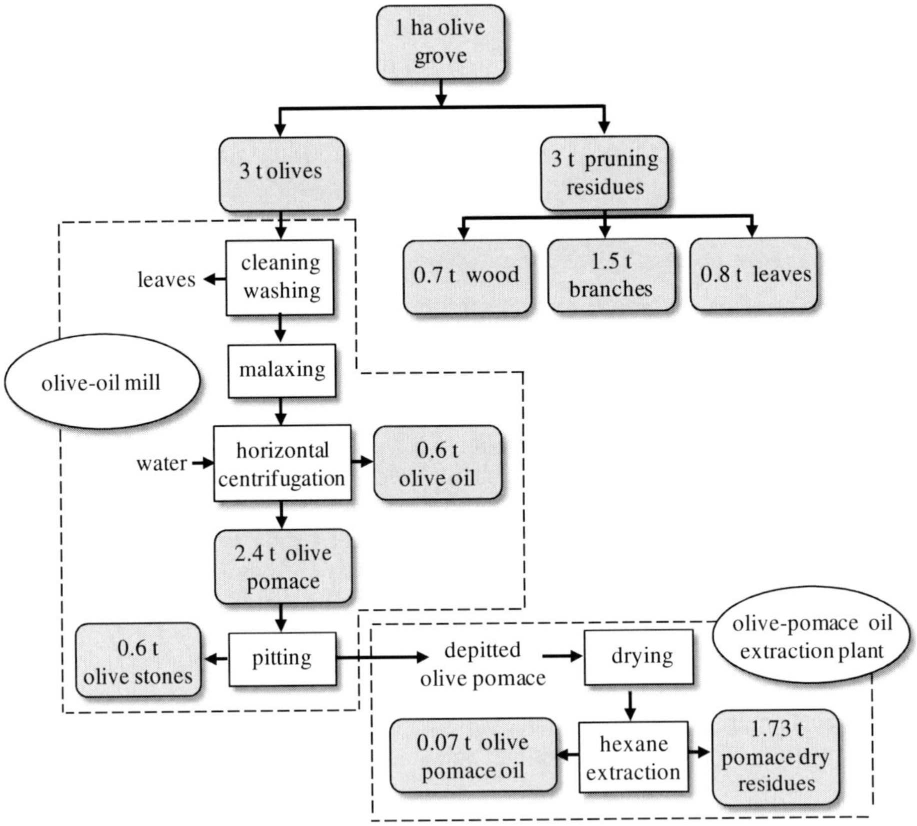

Figure 1. Biomass potential from 1 hectare of olive trees in Spain.

In both cases the crushed endocarps take different shapes and sizes, although particles never exceed 7 mm in length. Most of the solids have sizes greater than 1 mm, although the percentage of finer materials increases with increased weight of pulp and inorganic matter that soil endocarp. Saleh et al. (2014) reported that 97.3% of the weight of a sample of olive stones collected in an olive mill had more than 1.2 mm size [4] while Mata-Sánchez et al. (2015) found that 96.3% of the sample had a size greater than 1.4 mm [5]. Fernández-Bolaños et al. (1999) and Barreca and Fichera (2013) pointed out maximum olive stones dimensions ranging between 6.2 mm and 1.6 mm, respectively [6, 7]. The small particle size of this raw material is an advantage over other biomasses with higher dimensions that need to be subjected to costly milling stages for subsequent use.

The endocarp represents approximately 20% of the weight of the drupe. According to Rodriguez et al. (2008), 100 kg fresh olives lead to about 22 kg of olive stones containing 4 kg seeds and 18 kg endocarps [8]. From the data above, it can be calculated that the average annual production of olive endocarps in Spain is roughly 1.2 Mt. With regard to the composition of the biomass, from the point of view of the elemental analysis, the contents in carbon, oxygen and hydrogen can be highlighted (Table 1). Nitrogen percentage is generally very low. Finally, sulfur is barely detected, as it is characteristic of olive tree by-products. Among the trace elements present, chlorine and copper are the main, with concentrations ranging from 90 to 435 mg/kg and from 0.6 to 2.3 mg/kg, respectively [5]. Furthermore, Pattara et al. (2010) reported a chlorine content of 340 mg/kg [9]. Ash percentages are very low, typically below 2%. Various inorganic compounds are usually accumulated in ash, mainly CaO, Fe_2O_3, Al_2O_3, MgO, K_2O and SiO_2 [10, 11, 12].

Table 1. Elemental analysis and ash content of olive stones

Composition (%, w/w)						
C	H	O	N	S	Ash	References
51.2	6.0	41.9	0.15	0.02	0.78	[5]
50.1	5.9	42.0	0.6	0.02	1.33	[13]
46.6	6.3	45.2	1.8	0.10	1.40	[14]
48.6	5.7	44.1	1.6	0.05	1.90	[15]

In relation to the molecular composition of olives' endocarps, in scientific literature can be found different values for the contents of cellulose, hemicellulose and lignin (Table 2). The wide dispersion of results in Table 2 could be explained taking into account the different varieties of olive fruit, the presence of traces of pulp and pericarp, and the analytical techniques used. Concerning analytical procedures, the existence of non-structural material in the samples to be tested (free sugars, proteins, oils ...) generally leads to overestimate the lignin percentage, so initially olive stones should be subjected to extraction steps both with water as organic solvents [16]. Thus the content of extractives in the raw material could be also calculated, which normally is below 13% (5.0%, [17]; 5.5%, [11]; 6.0%, [4]; 12.5% [18]).

Table 2. Fiber composition of olive stones

Composition (%, w/w)			
Cellulose	Hemicellulose	Lignin	References
29.9	28.1	27.7	[4]
33.5	24.5	23.1	[7]
27.1[a]	32.2[a]	40.4	[19]
36.4	26.8	26.0	[20]

[a] Free-extractives material.

Coimbra et al. (1995) published one of the first detailed analysis of the structural components of olive endocarp [21]. These researchers concluded that the studied biomass contained 62% carbohydrates, 31% of which were α-cellulose. The hemicellulose fraction was rich in D-xylose (> 80% molar) and poor in D-glucose (< 1% molar). The main components of the total fibber fraction were D-xylose (48%, mol), D-glucose (41%, mol), uronic acids (8%, mol) and L-arabinose (2%, mol). The above results are quite close to those published by Montane et al. (2002), who reported percentages of α-cellulose and holocellulose of 28.0 and 65.3%, respectively [18]. Furthermore, Nabarlatz et al. (2007) pointed out that the percentages of L-arabinose, D-glucose, D-xylose, and acetyl groups in olive stones were 24.0, 23.3, 1.4 and 3.1%, respectively [22]. Finally, Cuevas et al. (2013) also reported a high percentage of D-xylose (23.4%) in olive stones [23].

Non-energy Use of Olive Stones

Non-energy uses for olive stones have been reported in literature. Some of the most important are described below.

Adsorbent

The possibility of using olive stones as adsorbent, either directly [24] or transforming them into activated carbon [25, 26, 27, 28], has been investigated. Non-activated endocarps, finely divided, allow to separate heavy metals in wastewater. For example, in literature it can be found recoveries higher than 80% for Cr^{3+} and Cr^{6+} [29], approximately 90% for cadmium ions [30] and above 70% for Fe^{3+} [31]. In addition, olive stones have been reported as adsorbent for drin pesticides [32].

Source of Natural Antioxidants

The two-phase pomace is interesting to produce some natural compounds with high antioxidant power, such as hydroxytyrosol or 3,4-dihydroxyphenylglycol [33]. The treatment of olive endocarps with the so-called technique 'steam explosion' (operating at 200-235°C maintained for 2 minutes) allows to retrieve large amounts of hydroxytyrosol in the liquid hydrolysate [6], which has led to a patent currently under exploitation (Ref OEPM Spanish Patent. P200100346). Meanwhile, Cuevas et al. (2007) verified that when increasing the operational temperature from 190 to 225°C in the hot water treatment of olive stones, the amount of hydroxytyrosol, vanillin and syringic acid was increased by 48, 151 and 369%, respectively [34].

Furfural Production

The major fraction of xylans from olive endocarp (about 20% w/w) can be used for furfural production by acid hydrolysis. According to Montané et al. (2002) 95 kg furfural could be obtained from one ton of dry olive stones working with 0.25 M sulfuric acid at high temperatures (240°C) and short reaction times (150 s) [18].

Xilooligosaccharide Production

Partial hydrolysis of hemicellulose from olive stones allows to obtain xylooligosaccharides, which are carbohydrates with low degree of polymerization (DP between 2 and 7) generated by the binding of molecules of D-xylose by β-1,4-glycosidic bonds. These compounds have a high economic potential due to their interesting applications, among which may be mentioned the production of monosaccharides [35], the production of biodegradable thermoplastics [36] and hydrogels [37], and synthesis of antileukemic drugs [38]. However, the most interesting alternative for xylooligosaccharide manufacture is in the formulation of prebiotics to enable selective growth of bifidobacteria in the human colon [39]. To this end it is necessary to develop strategies to ensure not only a correct separation of products, which can be achieved by solvent extraction, adsorption, gel filtration [40], ultrafiltration [41] or nanofiltration [42], but also achieve optimum profile molecular weights using, for example, enzyme membrane reactors [43]. Cuevas et al., (2013) studied the effect of the hydrothermal treatment on olive stones (temperature ranging from 195°C to 225°C) for the production of oligosaccharides [23]. The highest yield (16.9 kg product / 100 kg of raw material, equivalent to 60% of the original hemicellulose) was obtained at 210°C. These oligosaccharides were mainly composed of D-xylose (89% w/w). Furthermore, Nabarlatz et al. (2007) investigated the composition of oligosaccharides obtained hydrothermally (179°C-23 min) from the same raw material, and reported that its contents in D-xylose and acetyl groups were 86.5 and 13.5%, respectively [22].

Manufacture of Plastic Materials

A suitable fractionation of olive stones would allow to recovery most of the lignin. Thus in literature it is described a procedure consisting in a steam explosion treatment (227°C, 4 min) followed by extraction of the resulting solid with NaOH (2%, w/w, at room temperature) and subsequent precipitation with sulfuric acid (pH 2–3), which resulted in an alkali-extracted lignin.

The low carbohydrate content and the properties of this extracted lignin make it an excellent prepolymer for the manufacture of plastics or resins [6]. Olive endocarps could also be subjected to liquefaction in the presence of phenol to generate monomers useful in the production of phenol-formaldehyde resins [44], or be subjected to oxypropylation to produce biopolyols [45]. Moreover, olive stones have been assayed as plastic filled [46, 47, 48, 49].

ENERGY USE OF OLIVE STONES

Biomass is a source of energy used by mankind since immemorial time which has a potential to play a large role in the future in comparison to fossil fuels. Currently, biomass represents roughly 14% of the world's energy consumption. It has been estimated that our planet stored approximately $4 \cdot 10^{18}$ kJ green biomass [50], and its production rate is $1.46 \cdot 10^{11}$ t/year [51]. Biomass is a renewable energy source which has a low emissions footprint as it produces no net increases in atmospheric CO_2 levels. What is more, most of biomass can re-grow quickly and therefore feedstocks are again available after a relatively short period of time.

Among the various biomass sources, lignocellulosic materials (those which synthesize high amounts of lignin) have a great potential for the production of energy and chemicals. Two other advantages of these systems are their low price in origin, especially when they have residual character, and that they are not related to the food market. This last point is noteworthy because it seems logical to decouple the generation of energy using products with high nutritional value (e.g., cereals and edible oils) which should be used only to guarantee human consumption [52]. One of the main problems of biomass as an energy source is its geographical spread which decreases its feasibility when increasing the distance of collection and transport centers to the processing plant. Figure 2 depicts a decision diagram for the main routes for energy production from biomass wastes.

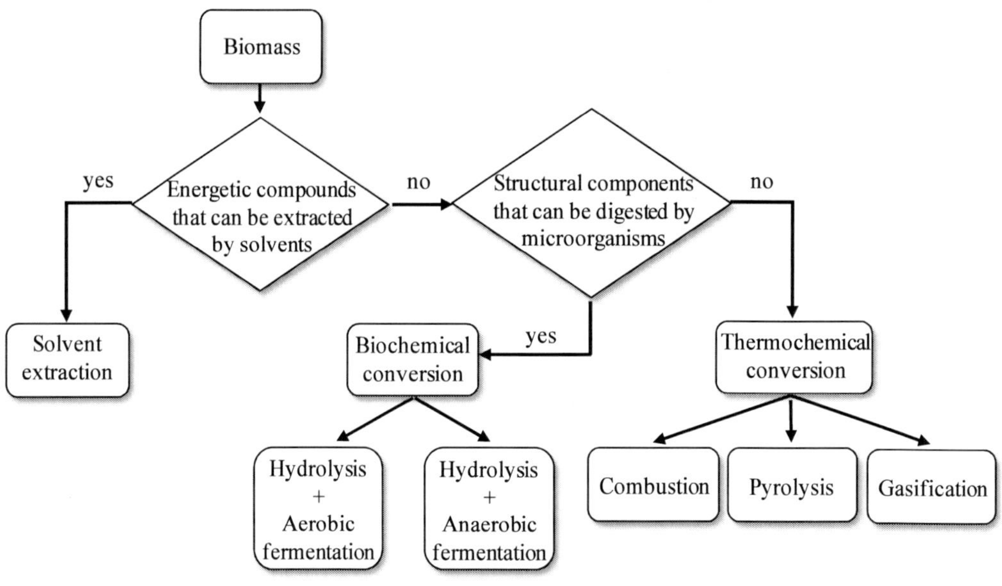

Figure 2. Pathways for converting biomass wastes into energy.

EXTRACTION

This procedure is solely suitable for plants containing oils or other high-energy compounds that can be extracted by solvents. In a previous section, the small losses of olive endocarp during the extraction at low temperature with water or organic solvents were discussed. This biomass contains a low percentage of fat (about 1%, [53]) compared to the entire olive stones (5%, [6]) or to the mesocarp (28%, [54]). As a result, olive stones can hardly be used as substrate for fat extraction. The fat content will depend on the greater or lesser presence of pulp, since the fat percentage is between 0.1 and 1.0% in completely clean olive stones while the fat percentage is roughly 6% in the pulp [5]. With respect to phenolic compounds derived from lignin fraction, Fernandez-Bolaños et al. (1999) pointed out that their calorific values ranged between 5600 and 6000 kcal/kg [6]. These polyaromatic compounds, composed mainly of guaiacyl-syringyl lignin (typical of hardwoods), can be easily separated from the aqueous phase by acid precipitation, and could be used for the formulation of fuels [55], provided that chemical modifications occur in these compounds that reduce oxygen content (for example by hydroprocessing).

THERMOCHEMICAL CONVERSION

The thermochemical conversion of biomass is based on processes of combustion, gasification and pyrolysis, which involve temperatures higher than 400°C in oxidizing environments. These operating conditions cause severe changes in the molecular structures of the raw material and the generation of various gaseous products; the more oxidizing the reaction medium is and higher the temperature is, the simpler the gaseous compounds.

Combustion

The combustion involves the oxidation of carbon to carbon dioxide and hydrogen to water. The combustion of biomass takes place under very oxidizing atmosphere. The reaction products are primarily gases, leaving a solid residue whose composition depends on the mineral material content of the fuel. The main technical characteristics that must be taken into account for the combustion of a biomass are the calorific value, ignition temperature, combustion rate, content and characteristics of ash, and moisture content. In regard to olive stones:

- Calorific value.
  Olive stones have a relatively high calorific value when compared with other lignocellulosic materials. Some data of higher heating value (HHV), in descending order, are provided by Jenkin et al. (1998)–5200 kcal/kg [56], Mata-Sánchez et al. (2013)–4895 kca/kg [57], Yanes Duran (1985)–4700 kcal/kg [58], Doymaz et al. (2004)–4679 kcal/kg [59] González et al. (2004)–4600 kcal/kg [60] and García et al. (2012)–4279 kcal/kg [14]. That is, it can be considered a HHV value close to 4800 kcal/kg. Mata-Sánchez et al. (2013) calculated an average lower calorific value of 4581 kcal/kg from 30 different olive stones [57]. The calorific value is affected by the presence of pulp, whose HHV is 5307 kcal/kg [59].
- Ignition temperature and combustion rate
  Olive stones have a relatively low ignition point (about 215°C) while the maximum combustion rate (0.341 1/min) is achieved at 284°C [12].

- Ash content and melting temperature

  Olive stones not only contain little mineral matter (Table 1) but also have a melting temperature ash exceeding 1400ºC [5], which helps to reduce costs associated with cleaning the burners.

- Moisture percentage

  This parameter is very important, since an excess of water in fuel causes a decrease in combustion efficiency [61] and problems during transport and storage of the material. Gómez de la Cruz et al. (2014) reported initial and equilibrium moisture of 23% and 8%, respectively, in olive stones provided by olive mills [62]. These authors also analyzed the drying kinetics of olive stones at four temperatures (100, 150, 200 and 250ºC) and three thicknesses of the layer of solids (10, 20 and 30 mm).

At present, most olive stones end up being burned to generate heat energy because besides presenting high calorific values and low percentages of sulfur, ash and chlorides, the bulk density is relatively high (between 573 and 709 kg/m^3 [5, 13, 63]) without applying particle size reduction. According to Manyà et al. (2007) solids milled to 0.3–0.5 mm size would have a bulk density of 749 kg/m^3 [17].

Furthermore, the fuel supply to the burners is simpler than when softer and more fibrous wastes are used, thus reducing clogging metering devices. The average price of fragmented endocarps derived from olive mills in Spain was between 150 and 180 euros per ton in the past two years.

Pyrolysis

Pyrolysis of biomass is a technique that involves heating the material in a non-oxidizing environment at temperatures above 500°C. Thus, water and volatile matter removal results in a char with higher percentage of carbon and lower percentages of oxygen and hydrogen than the raw material. The volatile content in olive endocarps ranges from 72% to 83% [11, 14, 57, 60].

Blanco López et al. (2002) investigated the pyrolysis of olive stones at a maximum temperature of 600°C. According to these authors the process generated 55% (w/w) liquid fraction (of which, half corresponded approximately to aqueous phase), 35% (w/w) coke and 10% (w/w) gas. Regarding the gas stream, the concentrations of CO and CO_2 were maximized when the pyrolysis was performed at temperatures between 300 and 400°C, whereas CH_4 and H_2 were increased with increasing temperature up to concentrations close to 10 and 17%, respectively. The highest calorific values for liquid, solid and gaseous fraction were 1750, 2920 and 450 kcal/kg, respectively [11]. Manyà et al. (2007) reported on the pyrolysis of olive endocarps at 650ºC. Product yields depended on the heating rate and particle size, and were 25-35% (char), 25-40% (gas fraction) and 35-48% (liquid fraction), respectively. There was a percentage of CO between 16 and 25% in the gas stream [17]. Mendu et al. (2011) also conducted the pyrolysis of olive stones and their lignin fraction at 650°C. For the former, the two main chemical species detected were acetic acid and 2-methoxyphenol, while 2-methoxy-4-methylphenol and 2-methoxy-4-(2-propenyl)-phenol were primarily obtained in the pyrolysis of lignin [64]. Finally, Marcilla et al. (2013) investigated the effect of particle size on the non-isothermal pyrolysis of olive stones [65].

In recent years, the possibility of applying pyrolysis at low temperatures generally below 350°C (biomass torrefaction) to biomass is under research, as it is performed with lower energy consumption. Torrefaction can be carried out in the presence (wet torrefaction) or absence (dry torrefaction) of water. The resulting solids are very interesting for subsequent thermochemical use (combustion or gasification) for reasons such as the increase of heating value and stability against microbial attack, the improvement in the grindability of the solids and the reduction of equilibrium moisture because of the higher hydrophobic character of the material [66]. In a recent publication the effect of torrefaction on whole olive stones (endocarp and seed) at two temperatures (260 and 280ºC) maintained for 18 to 30 minutes [67] is described. Under the best conditions, the calorific value of the material was increased by 17%, and the carbon content increased from 51.3% to 61.6%.

Gasification

Gasification of biomass is achieved when biomass reacts with a much lower amount of oxygen (between 10 and 50%) than necessary to burn all the material completely. The process can be carried out with or without steam, and in the presence or absence of catalysts. The ultimate goal is to transform most of the raw material in a gas stream with high calorific containing chemical species such as CO, H_2 or CH_4. The synthesis gas (or syngas) also contains N_2, CO_2, H_2O, tars and solid particles [68]. The separation of the last two fractions is necessary for the subsequent use of the gas, and can be performed by using equipment such as cyclones or scrubbers. If the ultimate goal is the production of electricity, fuel may be burned in a gas turbine.

Gasification of olive stones has been investigated using both downdraft gasifier [63, 69, 70] and bubbing-fluidized-bed gasifier [15, 71]. These authors reported synthesis gas with lower heating values (LHV) of up to 6.54 MJ/kg and hot-synthesis-gas efficiencies of 88.5%. Skoulou et al. (2008) calculated the percentage of gas (56%), char (21%) and tar (23%) obtained in the process [63]. Table 3 shows composition and calorific value of the generated synthesis gas by gasification of olive endocarps.

Table 3. Composition of syngas obtained from olive stones gasification

Reactor type	T (ºC)	Gas composition (v/v %)				Gas LHV	Ref.
		CO	CO_2	H_2	HC[a]		
FB	850	4.8	19.5	7.8	3.6	2.89 MJ/Nm^3	[63]
FBG	824	7.5	19.7	7.6	2.9	3.00 MJ/Nm^3	[72]
FB-D	1007	21.6	11.6	16.8	3.1[b]	>4.5 MJ/kg	[73]
FB-D	1050-1100	22.8	5.1	16.9	3.4[b]	5.45 MJ/kg	[13]
BFB	750	14.3	19.4	24.0	5.6	6.54 MJ/Nm^3	[15]

[a]HC. hydrocarbons, [b]CH_4, FB: fixed bed; FBG: fluidized bed gasifier; FB-D: downdraft; BFB, bubbing fluidized bed.

According to Vera et al. (2014), the amortization period of a gasification plant of olive stones located in Andalusia (southern region of Spain with great presence of olive orchard) would be very short, and would range between 5 and 6 years [13]. However, the main actual hindrances of biomass gasification are the difficulty of automating the process [74] and the high percentage of tars in some syngas [71] and of unprocessed carbon in the solid waste. The first drawback could be improved by incorporating control systems based on fuzzy logic. Regarding the generation of tar in the syngas, its percentage could be reduced by making a suitable selection of gasifier, increasing the reactor temperature and the air equivalence ratio, [63] and selecting a suitable size particle [75], which can be achieved pretreating biomass (for example, by torrefaction) to reduce the volatile content [66], or by incorporating compounds catalyzing the destruction of tars, either within the gasifier or in a second reactor [76].

Biochemical Pathway

The biochemical conversion of lignocellulosic material involves its transformation into various products as a result of direct or indirect action of microorganisms. The main biochemical conversion products generated from olive stones are ethanol and xylitol. The interest of ethanol is in its use as renewable motor fuel [77], while xylitol is a low-calorie sweetener (its sweetener power is similar to that of sucrose and its caloric content is equivalent to 2.4 cal/g) that can be consumed by diabetics [78]. The conversion of biomass to ethanol or xylitol generally includes the following four steps: pre-treatment, hydrolysis of polysaccharides and oligosaccharides into monomer sugars, fermentation of sugars and, finally, separation of bioproducts.

Olive Stones Pretreatment

Lignocellulosic biomass is a complex mixture of carbohydrate polymers from plant cell walls known as cellulose and hemicellulose, plus lignin, ash and a smaller amount of other compounds generally known as extractives. Pretreatment is intended to change the structure of the biomass to improve subsequent hydrolysis of polysaccharides, mainly of cellulose [79]. This can be achieved by physical, chemical or physico-chemical transformations of the material. Physical pretreatment is, for example, the particle size reduction, which brings about no changes in the composition of the biomass but increases its specific surface, thus enhancing the subsequent hydrolysis.

Hydrothermal Pretreatments: Liquid Hot Water (LHW) and Steam Explosion (SE)

They are physico-chemical pretreatments based on the use of water at high temperature (150-250°C). Steam explosion ends with a sudden decompression (explosion) of the system. High temperature steaming makes cellulose and lignin degrade. A quick drop of pressure damages the fibber structure. These changes make the biomass easier to hydrolyze and ferment. By contrast, liquid hot water pretreatment involves heating the biomass into pressurized water. The process is similar to that of an autoclave (batch process). In both techniques, the acetyl groups contained in the hemicellulose are released, which provide an acidic medium to carry out the hydrolysis. Furthermore, the addition of diluted acid as catalyst allows the total solubilization and hydrolysis of hemicellulose into monomers. What is more, the use of an acid catalyst can reduce the processing temperature. Due to its nature, steam explosion is more complex and expensive than LHW.

Figure 3 shows the evolution of the pH of a hydrolysate obtained after subjecting olive stones to LHW treatment (50 g raw material and 300 mL of water). The volume of the liquid phase, after filtration, was 1 L. The meaning of the factor of severity (Log R_0) has been explained in various publications [80]. An appreciable decrease in pH which occurs with increasing Log R_0 is observed. Thus, for extreme conditions (Log R_0 = 0, 21°C maintained for 2 min; and Log R_0 = 6.34, 250ºC maintained for 0 minutes) the pH of the hydrolysate was 5.6 and 3.0, respectively [34].

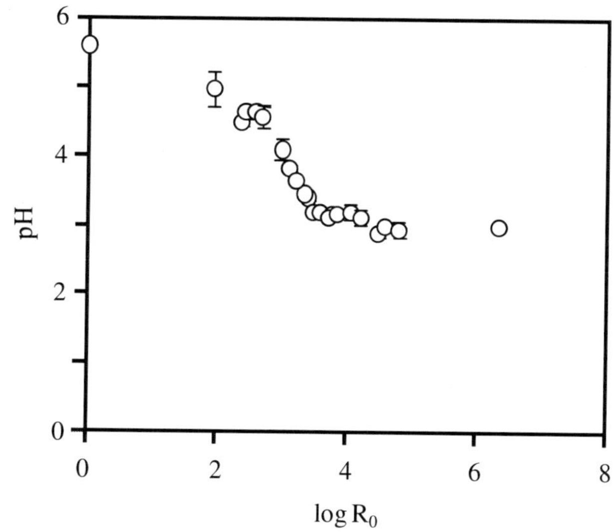

Figure 3. Effect of the severity parameter (Log R_0) on the pH of hydrolysate after LHW pretreatment of olive stones [34].

LHW pretreatments are very interesting for the following reasons:

- Low price of reagents.
- Almost nil problems in hydrolysate neutralization.
- Compared to other pretreatments, corrosion problems in the equipment are reduced.
- Complete hydrolysis of the hemicellulose fraction along with low degradation of monosaccharides is achieved [81].

Full differences between the use of liquid hot water and steam explosion are still unknown, although it seems that LHW can solubilize hemicellulose with greater selectivity towards oligomers, while the SE yields more monosaccharides [82]. This fact is observed in Table 4, which shows yields of D-xylose and L-arabinose in hydrothermal prehydrolysate obtained from olive stones using similar severity factor for the pretreatment.

Crystalline cellulose is barely attacked at low temperatures during LHW pretreatment. In some studies conducted on pure cellulose, it has been reported that this material is only hydrolyzed above 230°C [85]. Besides, from this temperature to 350°C significant pyrolysis of hemicelluloses takes place [86], and the decomposition rate of cellulose is lower than those of the degradation of D-glucose and cellobiose [87], which implies a low selectivity towards the more desirable compounds in the synthesis of ethanol. Fernández-Bolaños et al. (2001) pointed out that D-glucose production from hydrolysates obtained by steam explosion of olive stones solely occurs at high severities (Log $R_0 = 4.34$), [20]. Under these conditions, the degree of polymerization of the cellulose was 600, [6].

Lignin presents complex behavior during hydrothermal treatment since, depending on working conditions, can undergo processes of depolymerization or repolymerization. Lignin losses should be taken into account if the ultimate goal is the production of ethanol due to the inhibitory effect of the phenol-based compounds in the fermentation [88]. Laser et al. (2002) pointed out that hydrolysates generated by steam explosion contain more fermentation inhibitors than those obtained by LHW pretreatment [89].

Table 4. Effect of hydrothermal pretreatments on hemicelluloses hydrolysis of olive stones

Pretreatment	T (ºC)	t (min)	Hemicellulose solubilization (%)	D-Xylose yield (%)	L-Arabinose yield (%)	References
LHW	210	0	78	2.3	0.7	[83]
SE	215	2	63	13.6	0.6	[20]
SE	210	4	75	13.4	0.9	[84]

LHW: liquid hot water; SE: steam explosion.

Dilute Sulfuric Acid Pretreatment (DSA)

Hydrolysis with dilute sulfuric acid (0.5–3.0%) at high temperatures (150–210ºC) for short reaction times (few minutes) is a well-established process for conversion of hemicelluloses into their corresponding monomeric sugars [79]. This pretreatment can convert xylan to xylose with a maximum yield of 70–90%. The main drawback of dilute acid hydrolysis is the significant formation of sugar degradation compounds (furfural, 5-hydroxy-methyl-furfural, etc.) when increasing the severity of the pretreatment (e.g., when the target is the hydrolysis of cellulose). Therefore, when the target is to recover pentosans for subsequent fermentation, optimization of the conditions of acid hydrolysis (acid concentration, reaction temperature and process time) is required to maximize the solubilization of hemicellulose and the production of hemicellulosic sugars, thus minimizing cellulose hydrolysis and formation of degradation products of monosaccharides.

Comprehensive description of pretreatment with dilute sulfuric acid applied to olive olives can be found in literature [4]. As aforementioned, in some cases it consists in steam explosion pretreatments in which the lignocellulosic material has been previously impregnated with acid sulfuric [20, 84].

Saleh et al. (2014) applied response surface methodology to find the optimum temperature and reaction time to maximize recovery of D-xylose in the acid hydrolysis of olive stones [4]. The maximum yield of D-xylose (89.7% of D-xylose available in the raw material) was achieved working with 0.025 M sulfuric acid at 195ºC for 5 minutes. Under these conditions hemicellulose was virtually hydrolyzed. By contrast, the hydrolysate contained noticeable quantities of microbial inhibitors (e.g., 2.7 g acetic acid/100 g raw material). However, these inhibitors could be significantly reduced by vacuum distillation.

The pretreated solid was mainly composed of cellulose (39.1%) and lignin (41.7%) and was subjected to enzymatic hydrolysis to assess the generation of D-glucose.

Regarding SE pretreatment, impregnation with sulfuric acid has been carried out for 1 h under vacuum and with two concentrations of sulfuric acid: 0.1% (w/w) [84] and 0.5% (w/v) [20]. Ballesteros et al. (2001) were able to recover 14.2 g of D-xylose from 100g of raw material after pretreating olives stones at 210°C for 4 minutes, which was not a significant improvement over performing the SS pretreatment without acid impregnation. Under the above conditions, the pretreated solid contained 42.3% cellulose, 7.7% hemicellulose and 44.0% lignin. Furthermore, Fernández-Bolaños et al. (2001) reported that the acid impregnation did achieve increasing generation of monosaccharides instead oligomers at lower pretreatment severities [20]. These authors pointed out that pretreated solid after SE at 215°C for 2 min was composed of 42.7% cellulose, 6.8% hemicellulose and 35.7% lignin, values quite close to those reported by Ballesteros et al. (2001) [84]. Finally, it is worth noting that steam explosion pretreatment with or without acid impregnation did not modify either the crystallinity index or the degree of polymerization of cellulose [6].

Enzymatic Hydrolysis of Pretreated Cellulose

The first assays of enzymatic hydrolysis of cellulose were performed by Seillière (1906) and Reese et al. (1950), [90, 91]. In the seventies of the last century cellulases began to be applied to lignocellulosic materials due to the availability of commercial preparations of these biocatalysts. Early research demonstrated that enzymatic hydrolysis had notable advantages over the acid or alkaline hydrolysis, such as milder conditions (temperatures between 45 and 50°C and pH around 5.0) which reduce energy costs and corrosion problems [92]. However, the use of enzymes has shown that these macromolecules follow complex and unfamiliar patterns, especially if used in heterogeneous systems.

Although there have been significant advances in the production of cellulolytic enzymes in recent decades, they remain expensive. Therefore, the feasibility of enzymatic hydrolysis for bioethanol production from lignocellulosic materials depends on the D-glucose yield achieved, which is difficult with biomass with high percentage of lignin.

Enzymatic hydrolysis of cellulose takes place solely if there are several types of enzymes with endoglucanase, exoglucanase and β-glucosidase activity, acting synergistically [93]. This phenomenon depends on the proportion of each individual species in the enzyme complex [94]. Most of the current researches are based on commercial cellulase preparations supplemented with β-glucosidase to reduce the possibility of inhibition by cellobiose.

The relationship between the amounts of catalyst and substrate also plays a vital role in the reaction. Gregg and Saddler (1996) reported that while the protein loadings that are normally used are about 10 FPU/g cellulose, it is suggested using lower loadings to reduce the final costs of D-glucose [95]. Sun and Cheng (2002) established the concentrations range, depending on the type of substrate, between 7 and 33 FPU/g substrate [77].

Certain compounds existing in the hydrolysate can also act as enzyme inhibitors, so the tendency so far has been to avoid working directly with the hydrolysate generated in the pretreatments.

Lignin reduces the capacity of the hydrolytic enzymes in two ways: by acting as a physical barrier that isolates the substrate and, secondly, by binding the catalyst to immobilize it. Lignin has a high capacity to form bonds with proteins [96]. Sewalt et al. (1997) observed significant increases in cellulase inhibition by adding lignin to the reaction medium [97]. However, it seems that not all lignins have the same inactivating capacity. Table 5 shows data about the effect of the presence of lignin on the enzymatic hydrolysis of cellulose. Cellulase loading providing 60 FPU/g cellulose activity was used in all cases.

Table 5. Influence of lignin percentage on enzymatic hydrolysis of cellulose

Material	Lignin (%)	Enzymatic hydrolysis yield (%)	References
'Kraft' pulp	3.4	100	[98]
'Kraft' pulp	4.4	90	[99]
Delignified mechanical pulp	8.2	100	[99]
Mechanical pulp	27.3	20	[99]
Mechanical pulp	27.9	20	[98]

Table 6. Impact of pretreatment methods on enzymatic hydrolysis of olive stones

Pretreatment			Enzymatic hydrolysis		
Type	T(°C) – t(min)	Lignin in pretreated solids (%)	Cellulose loading (FPU/g)	Yield (%)	References
DSA	195 – 5	41.7	50	43.2	[4]
			100	56.0	
SE	210 – 4	45.7	15	40.5	[84]
			30	55.3	
SE-AI	210 – 4	44	15	52.7	[84]
			30	61.0	
SE	215 – 2	48.7	70	55–60[a]	[6, 20]
				30–35[b]	
SE-AI	215 – 2	35.7	70	85–88[a]	[6, 20]
				60–65[b]	

DSA: dilute sulfuric acid; SE: steam explosion; SE-AI: steam explosion with acid impregnation.
[a] Wet solids.
[b] dry solids.

The enzymatic hydrolysis of olive stones has been studied by different authors [4, 20, 84, 100], which have illustrated how the hydrolysis of the cellulose is limited by the high content of lignin in the biomass, by the strong crystallinity of the substrate, and by the low porosity of the solid. Thus, for example, Fernández-Bolaños et al. (2001) could only hydrolyze 15% cellulose of olive stones that were previously subjected to particle size reduction [20]. To increase performance, these authors used steam explosion pretreatment with or without acid impregnation of acid. Data of cellulose hydrolysis after DSA and SE pretreatments are shown in Table 6. Performance increases when the acid pretreatment is used, when the material is hydrolyzed without being previously dried, and when enzyme loading increases. 90% conversion of the pretreated cellulose was not achieved under any condition. Only when the pretreated solids were efficiently delignified, for combining alkali and chlorite bleaching, the pretreated cellulose was almost completely hydrolyzed [20].

On the other hand, Cuevas et al. (2009) analyzed the enzymatic hydrolysis of cellulose contained in the slurry (pretreated solid and liquid prehydrolysate) resulting after LHW pretreatment [101]. The increase of the pretreatment severity provoked an augmentation in cellulose conversion. Despite this increase, only 23.3% hydrolysis was achieved when working at 225ºC–0 min. This low value could be attributed to the fact that these authors (Cuevas et al., 2009) performed enzymatic hydrolysis of pretreated liquid and solid together. The maximum sugar concentration was 29.9 g/L. Furthermore, Cuevas et al. (2013) assessed different enzyme mixtures and biocatalyst loadings on the enzymatic hydrolysis of oligosaccharides obtained from hydrothermal fractionation of olives stones [23].

Sugar Fermentation

Once a hydrolysate with a high content of sugars and a low level of microbial inhibitors is obtained, this medium may be inoculated with microorganisms to produce ethanol, xylitol or other metabolites. Most of the industrial ethanol is nowadays produced by fermentation of carbohydrates. This process is influenced by many variables. First, the microorganism used. The most commonly microorganisms used for industrial production of ethanol are *Saccharomyces* yeasts as they quickly ferment hexoses into the bioproduct and they have high ethanol tolerance. However, these yeasts do not metabolize pentoses, thus limiting their application to hydrolysates rich in D-xylose. On the other hand, yeasts capable of assimilating pentoses are known, such as *Pachysolen tannophilus*, *Candida tropicalis*, *Candida shehatae*, *Candida parapsilosis*, *Pichia stipitis*, *Candida guilliermondii*, *Kluyveromyces marxianus*, *Kluyveromyces fragilis* and *Debaryomyces hansenii*. Nevertheless, it should be noted that some of them provide mainly ethanol, others generate xylitol while some yeasts synthesize products or mixtures thereof. *P. stipitis* and *C. shehatae* have been reported as the most suitable yeasts for the transformation of D-xylose to ethanol. Roberto et al. (1991) analyzed the performance of four microorganisms (*P. stipitis*, *P. tannophilus*, *C. utilis*, and *C. tropicalis*) on a lignocellulosic hydrolysate with 20 kg/m^3 of D-xylose obtained by subjecting sugar cane bagasse to SE pretreatment (190°C-5 min) with acid impregnation [102]. Fermentation with *P. tannophilus* NRRL 2460 achieved ethanol concentrations lower than 3 kg/m^3, whereas with the yeast *P. stipitis* CBS 5773 the ethanol concentration reached 9 kg/m^3. Furthermore, Toivola et al. (1984) investigated the capacity of 200 types of yeast to ferment D-xylose to ethanol, and found that *C. shehatae*, *P. stipitis* and *P. tannophilus* were among the six species capable of generating ethanol with yields above 5% [103]. Moreover, while the first two yeasts (*C. shehatae* and *P. stipitis*) generated ethanol concentrations between 5.9 and 6.6 g/L (starting from 20 g/L D-xylose) the third solely produced 2.1 g/L ethanol.

Initial inoculum concentration also affects the course of fermentations. This variable should be set according to the substrate concentration and, especially, to microbial inhibitors [104]. With respect to the substrate concentration, it is desired to be as high as possible (as far as it does not affect the microorganism) to further reduce bioproduct separation costs. Concentrations of D-xylose higher than 100 kg/m^3 are not desirable for fermentation with *P. stipitis* [105], although Roberto et al. (1991) obtained 30 kg/m^3 ethanol from 145 kg/m^3 D-xylose [102].

Numerous compounds have been described that inhibit yeast fermentation. Their influence on a given microorganism depends on the specie to be inoculated, the amount of biomass introduced, and concentrations of inhibitors and other components of the culture medium. Synergetic effects among species are important in explaining inhibitions. HLW and SE pretreatment can produce toxic compounds as depicted in Figure 4.

Numerous articles highlight the inhibitor effect of acetic acid. The toxic character of the acid is more pronounced the lower the pH of the culture [107]. It has been reported that furfural and 5-hydroxy-methyl-furfural are not critical microbial inhibitors because of their relatively low concentrations in hydrolysates. What is more, increasing the amount of inoculum is an effective way of reducing the inhibiting effect of these compounds. With respect to lignin phenolic derivatives, the most toxic are the less substituted acids since their lipophilic character favors its transport into the cell thereby causing the subsequent release of protons.

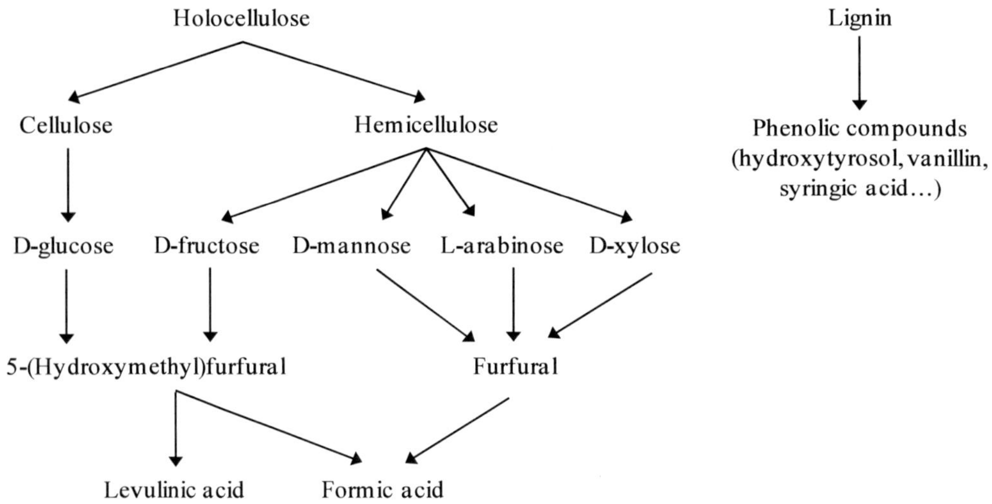

Figure 4. Products and by-products obtained during the acid hydrolysis of lignocellulosic biomass [4, 106].

The biochemical use of biomass to ethanol can be performed using sequential hydrolysis and fermentation (SHF) or simultaneous saccharification and fermentation (SSF). The latter could bring about higher yields and lower ethanol production costs (by reducing investment) also avoiding the enzyme inhibition by D-glucose [108, 109]. Another advantage of SSF schemes is the resistance to microbial contamination. However, thermotolerant yeasts (*Kluyveromyces marxianus*, *Kluyveromyces fragilis*, *Candida acidothermophilum*, etc.) are required for SSF. These yeasts must be capable to work at sufficiently high temperatures (even above 40ºC), close to the optimum temperatures of enzyme functions.

Scarce literature deals with the fermentation of sugars from olive endocarps to produce ethanol or xylitol. Ballesteros et al. (2001) focused on the utilization of cellulosic fraction of this biomass [84]. After a SE pretreatment step, the resulting solid was subjected to simultaneous saccharification and fermentation (SSF) using a thermally acclimated yeast (*K. marxianus* CECT 10895) capable of fermenting D-glucose. The highest yield of ethanol (58.8%) was achieved by pretreating (steam explosion, 210°C-4 min) acid-impregnated olive stones. The above conditions resulted in a culture medium with a concentration of 12.9 g/L ethanol. The presence of lignin in the pretreated solid could explain the low ethanol yield. To solve this problem, Cuevas et al. (2015) tested Organosolv pretreatments using mixtures of ethanol and water with or without the presence of acid catalyst (H_2SO_4) [110]. The pretreated solid was composed of 83.3% cellulose and 17% lignin. The SSF of this solid with the thermotolerant yeast *S. cerevisiae* IR2-9a completely hydrolyzed the cellulose fraction, and the final ethanol concentration was greater than 45 g/L.

Regarding sequential saccharification and fermentation (SHF) schemes, Cuevas et al. (2009) applied them for ethanol and xylitol production after LHW pretreatment of olive stones [101]. The variation in the severity of pretreatment (Log R_0 between 3.23 and 4.39) had no significant effects on overall yields of ethanol, but affected those of xylitol. The highest yield was 0.25 g ethanol/g sugar.

The SHF scheme was also assayed by Saleh et al. (2014) with olive stones pretreated with dilute sulfuric acid (195°C-5 min-0.025 M), [111]. After pretreatment, the prehydrolyzed liquid (rich in D-xylose) was separated from the solid, and this was subjected to enzymatic hydrolysis. Subsequently both hydrolysates were fermented separately with *P. tannophilus* ATCC 32 691. As a whole, 9.2 g xylitol and 10.3 g ethanol were produced from 100 g of olive stones.

Conclusion

Olive stones are an abundant lignocellulose material in countries of the Mediterranean basin, especially in Spain. As an agro-product, its geographical concentration is much higher than that of other materials derived from olive oil industry, such as prunings and leaves, thus reducing transport costs. Furthermore, the endocarps fragmented during the industrial processing of olives have maximum lengths of 7 mm. These small dimensions avoid the need to resort to subsequent milling steps. From the compositional point of view, olive stones are characterized by their significant content in hemicellulose, cellulose and lignin, along with low percentages of extractives and mineral matter. The high percentage of lignin explains the high hardness of olive stones.

Combustion (thermochemical process) is probably the method that leads to the best results using simple, available, inexpensive technology, due to the high calorific value and reduced ash generation. However, if the goal is electricity production, yields of gasification of olive stones could be improved if the syngas is cleaned and then burnt into a turbine. Although the gasification of olive stones is under research, optimum reactor designs, proper control of equipment and a full utilization of the biomass are needed. Regarding pyrolysis, the best alternative may be the use of torrefaction equipment, which would provide a solid with better characteristics for subsequent use in combustion or gasification assuming moderate energy expenditure.

Regarding the biochemical pathway, it has been illustrated the potential of fractionating olive stones to convert their polysaccharides to ethanol or xylitol. Because of the crystallinity of cellulose and the high D-xylose content of hemicellulose, pretreatments may be applied to achieve enhanced enzymatic hydrolysis of cellulose (for subsequent ethanol production) along with D-xylose-rich hydrolysates (which would provide xylitol by fermentation). If D-xylose is intended to be fermented to ethanol, future research using non-traditional fermentative yeasts such as *Candida shehatae* and *Pichia stipitis* is required. Finally, lignin fraction resulting from the fractionation could be used for power generation, manufacture of synthetic polymers or separation of natural antioxidants.

Acknowledgments

This work was financially supported by the Projects 01272/2005 and AGR/6509 (Andalusia Regional Government, Spain).

References

[1] Oliveras-López, MJ; Berná, G; Jurado-Ruiz, E; López-García de la Serrana, H; Martín, F. Consumption of extra-virgin olive oil rich in phenolic compounds has beneficial antioxidant effects in healthy human adults. *J. Funct. Foods*. 2014 10, 475-484.

[2] FAOSTAT (2014). http://faostat3.fao.org/ (last accessed 10 December, 2014).

[3] Ministry of Agricultural, Food and Environment, Spain (2014). http://www.magrama.gob.es (last accessed 10 December, 2014).

[4] Saleh, M; Cuevas, M; García, JF; Sánchez, S. Valorization of olive stones for xylitol and etanol production from dilute acid pretreatment via enzymatic hydrolysis and fermentation by *Pachysolen tannophilus*. *Biochem. Eng. J.* 2014 90, 286-293.

[5] Mata-Sánchez, J; Pérez-Jiménez, JA; Díaz-Villanueva, MJ; Serrano, A; Núñez-Sánchez, N; López-Giménez, FJ. Development of olive quality system base on biofuel energetic parameters study. *Renew. Energ*. 2014 66, 251-256.

[6] Fernández-Bolaños, J; Felizón, B; Heredia, A; Guillén, R; Jiménez, A. Characterization of the lignin obtained by alkaline delignification and of the cellulose residue from steam-exploded olive stones. *Bioresource Technol.* 1999 68, 121-132.

[7] Barreca, F; Fichera, CR. Use of olive stone as an additive in cement lime mortar to improve thermal insulation. *Energ. Buildings* 2013 62, 507-513.

[8] Rodríguez, G; Lama, A; Rodríguez, R; Jiménez, A; Guillén, R; Fernández-Bolaños, J. Olive stone an attractive source of bioactive and valuable compounds. *Bioresource Technol.* 2008 99, 5261-5269.

[9] Pattara, C; Cappelletti, GM; Cichelli, A. Recovery and use of olive stones: Commodity, environmental and economic assessment. *Renew. Sustain. Energy Rev*. 2010 14, 1484-1489.

[10] Blázquez García, G; Calero de Hoces, Mónica; Martínez García, Carmen; Cotes Palomino, María Teresa; Ronda Gálvez, Alicia; Martín-Lara, María Ángeles. Characterization and modeling of pyrolysis of the two-phase olive mill solid waste. *Fuel Process. Technol.* 2014 126, 104-111.

[11] Blanco López, MC; Blanco, CG; Martínez-Alonso, A; Tascón, JMD. Composition of gases released during olive stones pyrolysis. *J. Anal. Appl. Pyrol.* 2002 65, 313-322.

[12] Skodras, G; Grammelis, P; Basinas, P; Kakaras, E; Sakellaropoulos, G. Pyrolysis and combustión characteristics of biomass and waste-derived feedstock. *Ind. Eng. Chem. Res.* 2006 45, 3791-3799.

[13] Vera, D; Jurado, F; Margaritis, NK; Grammelis, P. Experimental and economic study of a gasification plant fuelled with olive industry wastes. *Energy Sustain. Dev.* 2014 23, 247-257.

[14] García, R; Pizarro, C; Lavín, AG; Bueno, JL. Characterization of Spanish biomass wastes for energy use. *Bioresource Technol.* 2012 103, 249-258.

[15] Skoulou, V; Koufodimos, G; Samaras, Z; Zabaniotou, A. Low temperature gasification of olive kernels in a 5-kW fluidized bed reactor for H_2-rich producer gas. *Int. J. Hydrogen Energy* 2008 33, 6515-6524.

[16] Sluiter, A; Ruiz, R; Scarlata, C; Sluiter, J; Templeton, D. 2008. Determination of extractives in biomass. In Laboratory Analytical Procedure NREL/TP-510-42619 National Renewable Energy Laboratory Colorado (USA) 1617 Cole Boulevard, Golden, Colorado 80401-3393.

[17] Manyà, J; Ruiz, J; Arauzo, J. Some peculiarities of conventional pyrolysis of several agricultural residues in a packed bed reactor. *Ind. Eng. Chem. Res.* 2007 46, 9061-9070.

[18] Montané, D; Salvadó, J; Torras, C; Farriol, X. High-temperature dilute-acid hydrolysis of olive stones for furfural production. *Biomass Bioenerg.* 2002 22, 295-304.

[19] Martín-Lara, MA; Hernáinz, F; Calero, M; Blázquez, G; Tenorio, G. Surface chemistry evaluation of some solid wastes from olive-oil industry used for lead removal from aqueous solutions. *Biochem. Eng. J.* 2009 44, 151-159.

[20] Fernández-Bolaños, J; Felizón, B; Heredia, A; Rodríguez, R; Guillén, R; Jiménez, A. Steam-explosion of olive stones: Hemicellulose solubilization and enhancement of enzymatic hydrolysis of cellulose. *Bioresource Technol.* 2001 79, 53-61.

[21] Coimbra, MA; Waldron, KW; Selvendran, RR. Isolation and characterization of cell wall polimers from the heavily lignified tissues of olive (*Olea europaea*) seed hull. *Carbohyd. Polym.* 1995 27, 285-294.

[22] Nabarlatz, D; Ebringerová, A; Montané, D. Autohydrolysis of agricultural by-products for the production of xylo-oligosaccharides. *Carbohyd. Polym.* 2007 69, 20-28.

[23] Cuevas, M; García, JF; Cruz, N; Sánchez, S. Generación de D-xilosa por tratamiento hidrotérmico de endocarpios de aceitunas e hidrólisis enzimática de los oligosacáridos. *Afinidad* 2013 70(562), 99-106.

[24] Calero, M; Hernáinz, F; Blázquez, G; Tenorio, G; Martín-Lara, MA. Study of Cr (III) biosorption in a fixed-bed column. *J. Hazard. Mater.* 2009 171, 886-893.

[25] El-Sheikh, AH; Newman, AP; Al-Daffaee, HK; Phull, S; Cresswell, N. Characterization of activated carbon prepared from a single cultivar of Jordanian Olive stones by chemical and physicochemical techniques. *J. Anal. Appl. Pyrol.* 2004 71, 151-164.

[26] Martínez, ML; Torres, MM; Guzmán, CA; Maestri, DM. Preparation and characteristics of activated carbon from olive stones and walnut shells. *Ind. Crops Prod.* 2005 23, 23-28.

[27] Budinova, T; Petrov, N; Razvigorova, M; Parra, J; Galiatsatou, P. Removal of arsenic(III) from aqueous solution by activated carbons prepared from solvent extracted olive pulp and olive stones. *Ind. Eng. Chem. Res.* 2006 45, 1896-1901.

[28] Alslaibi, TM; Abustan, I; Ahmad, MA; Fou, AA. Application of response surface methodology (RSM) for optimization of Cu2+, Cd2+, Ni2+, Pb2+, Fe2+, and Zn2+ removal from aqueous solution using microwaved olive stone activated carbon. *J. Chem. Technol. Biotechnol.* 2013 88, 2141-2151.

[29] Blázquez, G; Hernáinz, F; Calero, M; Martín-Lara, MA; Tenorio, G. The effect of pH on the biosorption of Cr (III) and Cr (VI) with olive stone. *Chem. Eng. J.* 2009 148 473-479.

[30] Blázquez, G; Hernáinz, F; Calero, M; Ruiz-Núñez, LF. Removal of cadmium ions with olive stones: the effect of somes parameters. *Process Biochem.* 2005 40, 2649-2654.

[31] Martínez, L; Driss, SB; Hodaifa, G; Faur, C; Rodríguez, S; Giménez, JA; Ochando, J. Adsorption of iron on crude olive stones. *Ind. Crop. Prod.* 2010 32 467-471.

[32] El Bakouri, H; Usero, J; Morillo, J; Ouassini, A. Adsorptive features of acid-treated olive stones for drin pesticides: Equilibrium, kinetic and thermodynamic modeling studies. *Bioresource Technol.* 2009 100, 4147-4155.

[33] Rodríguez, G; Lama, A; Trujillo, M; Espartero, JL; Fernández-Bolaños, J. Isolation of a poweful antioxidant from *Olea europaea* fruti-mill waste: 3,4-Dihydroxyphenylglycol. *LWT – Food Sci. Technol.* 2009 42(2), 483-490.

[34] Cuevas, M. Hidrólisis enzimática del hueso de aceituna. Fermentación de hidrolizados con levaduras. PhD Thesis. University of Jaén (Spain), 2007.

[35] Kabel, MA; Carvalheiro, F; Garrote, G; Avgerinos, E; Koukios, E; Parajó, JC; Gírio, FM; Schols, HA; Voragen, AGJ. Hydrothermally treated xylan rich by-products yield different classes of xylo-oligosaccharides. *Carbohyd. Polym.* 2002 50, 47-56.

[36] Glasser, WG; Ravindran, G; Jain, RK; Samaranayake, G; Tood, J. Comparative enzyme biodegradability of xylan, cellulose, and starch derivatives. *Biotechnol. Progr.* 1995 11, 552-557.

[37] Palm, M; Zacchi G. Extraction of hemicellulosic oligosaccharides from spruce using microwave oven or steam treatment. *Biomacromolecules* 2003 4, 617-623.

[38] Ando, H; Ohba, H; Sasaki, T; Takamine, K; Kamino, Y; Morikaki, S; Bakalova, R; Uemura, Y; Hatate Y. Hot-compressed-water decomposed products from bamboo manifest a selective cytotoxicity against acute lymphoblastic leukemia cells. *Toxicol. Vitro* 2004 18, 765-771.

[39] Izumi, Y; Ikemizu, S; Shizuka, F. Intestinal environment improving agents containing acidic xylooligosaccharides. Patent, Japanese Publication Number: 2004182609 (2004).

[40] Palm, M; Zacchi, G. Separation of hemicellulosic oligomers from steam-treated spruce wood using gel filtration. *Sep. Purif. Technol.* 2004 36, 191-201.

[41] Nabarlatz, D; Torras, C; García-Valls, R; Montané, D. Purification of xylo-oligosaccharides from almond shells by ultrafiltration. *Sep. Purif. Technol.* 2007 53, 235-243.

[42] Vegas, R; Moure, A; Domínguez, H; Parajó, JC; Álvarez, JR; Luque, S. Purification of oligosaccharides from rice husk autohydrolysis liquors by ultra- and nano-filtration. *Desalination* 2006 199, 541-543.
[43] Yang, F; Fang, Z; Xu, Y; Yao, C; Yu, S; Zhu, Q. Separation of xylooligosaccharides from enzymatic hydrolytes using membrane reactor. *J. Central South University Technol.* 2003 10, 122-125.
[44] Tejeda-Ricardez, J; Vaca-García, C; Borredon, ME. Design of a batch solvolytic liquefaction reactor for the vaporization of residues from the agricultural foodstuff. *Chem. Eng. Res. Des.* 2003 81, 1066-1070.
[45] Matos, M; Barreiro, MF; Gandini, A. Olive stone as a renewable source of biopolyols. *Ind. Crop. Prod.* 2010 32, 7-12.
[46] Cristofaro, D. A process for the realization of plates and panels consisting of exhausted olive husk of crushed olives stones and polypropylene, and derived product. Patent, International Publication Number: WO 9738834 (1997).
[47] Siracusa, G; La Rosa, AD; Siracusa, V; Trovato, M. Eco-Compatible use of olive husk as filler in thermoplastic composites. *J. Polym. Environ.* 2001 9, 157-161.
[48] Perinovic, S; Andricic, B; Erceg, M. Thermal properties of poly(l-lactide)/olive stone flour composites. *Thermochim. Acta* 2010 510, 97-102.
[49] Koutsomitopoulou, AF; Bénézet, JC; Bergeret, A; Papanicolaou, GC. Preparation and characterization of olive pit powder as a filler to PLA-matrix bio-composites. *Powder Technol.* 2014 255, 10-16.
[50] Fischer, G; Schrattenholzer, L. Global bioenergy potentials through 2050. *Biomass Bioenerg.* 2001 20, 151-159.
[51] Demirbaş, A. Partly chemical analysis of liquid fraction of flash pyrolysis products from biomass in the presence of sodium carbonate. *Energ. Conv. Manage.* 2002 43, 1801-1809.
[52] Cornelissen, S; Koper, M; Deng, YY. The role of bioenergy in a fully sustainable global energy system. *Biomass Bioenerg.*2012 41, 21-33.
[53] Heredia, A; Guillen, R; Fernández-Bolaños, J; Rivas, M. Olives stone as a source of fermentable sugars. *Biomass* 1987 14, 143-148.
[54] Kiritsakis, A.K. El aceite de oliva. A. Madrid Vicente, Madrid (Spain), 1992.
[55] Rana, MS; Sámano, V; Ancheyta, J; Díaz, JAI. A review of recent advances on process technologies for upgrading of heavy oils and residua. *Fuel* 2007 86, 1216-1231.
[56] Jenkin; BM; Baxter, LL; Miles Jr, TR; Mile, TR. Combustion properties of biomass. *Fuel Process. Technol.* 1998 54, 17-46.
[57] Mata-Sánchez, J; Pérez-Jiménez, JA; Díaz-Villanueva, MJ; Serrano, A; Núñez-Sánchez, N; López-Giménez, FJ. Statistical evaluation of quality parameters of olive stone to predict its heating value. *Fuel* 2013 113, 750-756.
[58] Yanes Duran, C. Propiedades termoquímicas del orujo de aceitunas. Poder calorífico. *Grasas Aceites.* 1985 36, 45-47.
[59] Doymaz, I; Gorel, O; Akgun, NA. Drying characteristics of the solid by-product of olive oil extraction. *Biosyst. Eng.* 2004 88(2), 213-219.
[60] González, JF; González-García, CM; Ramiro, A; González, J; Sabio, E; Gañán, J; Rodríguez, MA. Combustion optimisation of biomass residue pellets for domestic heating with a mural boiler. *Biomass Bioenerg.* 2004 27, 145-154.

[61] Gebreegziabher, T; Oyedun, AO; Hui, CW. Optimum biomass drying for combustion - A modeling approach. *Energy* 2013 53, 67-73.

[62] Gómez-de la Cruz, FJ; Casanova-Peláez, PJ; Palomar-Carnicero, JM; Cruz-Peragón, F. Drying kinetics of olive stone: A valuable source of biomass obtained in the olive oil extraction. *Energy* 2014 75, 146-152.

[63] Skoulou, V; Zabaniotou, A; Stavropoulos, G; Sakelaropoulos, G. Syngas production from olive tree cuttings and olive kernels in a downdraft fixed-bed gasifier. *Int. J. Hydrogen Energy* 2008 33, 1185-1194.

[64] Mendu, V; Harman-Ware, AE; Crocker, M; Jae, J; Stork, J; Morton III, S; Placido, A; Huber, G; DeBolt, S. Identification and thermochemical analysis of high-lignin feedstocks for biofuel and biochemical production. Biotechnology for Biofuels 2011 4(43) 1-13.

[65] Marcilla, A; García, AN; Pastor, MV; León, M; Sánchez, AJ; Gómez, DM. Thermal decomposition of the different particles size fractions of almond shells and olive stones. Thermal behaviour changes due to the milling processes. *Thermochim. Acta* 2013 564, 24-33.

[66] Batidzirai, B; Mignot, APR; Schakel, WB; Junginger, HM; Faaij, APC. Biomass torrefaction technology: Techno-economic status and future prospects. *Energy* 2013 62, 196-214.

[67] Doassans-Carrère, N; Muller, S; Mitzkat, M. REVE – a new industrial technology for biomass torrefaction: pilot studies. *Fuel Process. Technol.* 2014 126, 155-162.

[68] Basu, P. Biomass gasification and pyrolysis. Practical design and theory. Academic Press, Burlington (USA), 2010.

[69] Vera, D; Jurado, F; Carpio, J. Study of a downdraft gasifier and externally fired gas turbine for olive industry wastes. *Fuel Process. Technol.* 2011 92, 1970-1979.

[70] Skoulou, V; Swiderski, A; Yang, W; Zabaniotou, A. Process characteristics and products of olive kernel high temperature steam gasification (HTSG). *Bioresource Technol.* 2009 100, 2444-2451.

[71] Gómez-Barea, A; Arjona, R; Ollero, P. Pilot-plant gasification of olive stone: A technical assesment. *Energ. Fuel* 2005 19, 598-605.

[72] Song, X; Guo, Z. Technologies for direct production of flexible H2/CO synthesis gas. *Energy Conv. Manage.* 2006 47, 560-569.

[73] Vera, D; de Mena, B; Jurado, F; Schories, G. Study of a downdraft gasifier and gas engine fueled with olive oil industry wastes. *Appl. Therm. Eng.* 2013 51, 119-129.

[74] Striūgas, N; Zakarauskas, K; Džiugys, A; Navakas, R; Paulauskas, R. An evaluation of performance of automatically operated multi-fuel downdraft gasifier for energy production. *Appl. Therm. Eng.* 2014 73(1), 1151-1159.

[75] Yin, R; Liu, R; Wu, J; Wu X; Sun, C; Wu, C. Influence of particle size on performance of a pilot-scale fixed-bed gasification system. *Bioresource Technol.* 2012 119, 15-21.

[76] González, JF; Román, S; Engo, G; Encinar, JM; Martínez, G. Reduction of tars by dolomite cracking during two-stage gasification of olive cake. *Biomass Bioenerg.* 2011 35, 4324–4330.

[77] Sun, Y; Cheng, J. Hydrolysis of lignocellulosic materials for ethanol production: A review. *Bioresource Technol.* 2002 83(1), 1-11.

[78] Sánchez, S; Bravo, V; García, JF; Cruz, N; Cuevas, M. Fermentation of D-glucose and D-xylose mixtures by *Candida tropicalis* NBRC 0618 for xylitol production. *World J. Microbiol. Biotechnol.* 2008 24, 709-716.

[79] Kumar, P; Barrett, DM; Delwiche, MJ; Stroeve, P. Methods for pretreatment of lignocellulosic biomass for efficient hydrolysis and biofuel production. *Ind. Eng. Chem. Res.* 2009 48, 3713-3729.

[80] Overend, RP; Chornet, E. Fractionation of lignocellulosics by steam-aqueous pre-treatment. *Phil. Trans. R. Soc. Lond. Ser. A.* 1987 321, 523-536.

[81] Mateo, S; Puentes, J G; Sánchez, S; Moya, A J. Oligosaccharides and monomeric carbohydrates production from olive tree pruning biomass. *Carbohyd. Polym.* 2013 93, 416-423.

[82] Allen, SG; Schulman, D; Lichwa, J; Antal, MJ; Jennings, E; Elander, R. A comparison of aqueous and dilute-acid single-temperature pretreatment of yelow poplar sawdust. *Ind. Eng. Chem. Res.* 2001 40, 2352-2361.

[83] Cuevas, M; Sánchez, S; Bravo, V; Cruz, N; García, JF. Autohydrolysis and diluted-acid treatment of olive stone: study of the composition of the hydrolyzates. 17th European Biomass Conference and Exhibition, Hamburg (Germany), 2009.

[84] Ballesteros, I; Oliva, JM; Saez, F; Ballesteros, M. Ethanol production from lignocellulosic byproducts of olive oil extraction. *Appl. Biochem. Biotechnol.* 2001 91:93, 237-252.

[85] Sasaki, M; Adschiri, T; Arai, K. Fractionation of sugarcane bagasse by hydrothermal treatment. *Bioresource Technol.* 2003 86, 301-304.

[86] Martínez, JM; Reguant, J; Montero, MA; Montané, D; Salvadó, J; Farriol, X. Hydrolytic pretreatment of softwood and almond shells. Degree of polymerization and enzymatic digestibility of the cellulose fraction. *Ind. Eng. Chem. Res.* 36 1997, 688-696.

[87] Sasaki, M; Kabyemela, B; Malaluan, R; Hirose, S; Takeda, N; Adschiri, T; Arai, K. Cellulose hydrolysis in subcritical and supercritical water. *J. Supercrit. Fluid.* 1998 13, 261-268.

[88] Palmqvist, E; Hahn-Hägerdal, B; Galbe, M; Zacchi, G. The effect of water-soluble inhibitors from steam-pretreated willow on enzymatic hydrolysis and ethanol fermentation. *Enzyme Microb. Technol.* 19, 1996 470-476.

[89] Laser, M; Schulman, D; Allen, SG; Lichwa, J; Antal, MJ; Lynd, LR. A comparison of liquid hot water and steam pretreatments of sugar cane bagasse for bioconversion to ethanol. *Bioresource Technol.* 2002 81, 33-44.

[90] Seillière, G. Sur un cas d´hydrolyse diastasique de la cellulose du coton, après dissolution dans la liqueur de Schweitzer. *C.R. Soc. Biol.* 1906 61, 205-206.

[91] Reese, E; Siu, R; Levinson, H. The biological degradation of soluble cellulose derivatives and its relationship to the mechanism of cellulose hydrolysis. *J. Appl. Bacteriol.* 1950 59, 485-497.

[92] Duff, SJB; Murray, WD. Ethanol production from pulp mill cellulosic wastes. *Bioresource Technol.* 1996 55, 1 33.

[93] Chanzy, H; Henrissat, B. Undirectional degradation of Valonia cellulose microcrystals subjected to cellulase action. *FEBS J.* 1985 184, 285-288.

[94] Nidetzky, B; Steiner, W; Claeyssens, M. Cellulose hydrolysis by the cellulases from *Trichoderma reesei*: Adsorptions of two cellobiohydrolases, two endocellulases and their core proteins on filter paper and their relation to hydrolysis. *Biochem. J.* 1994 303, 817-823.

[95] Gregg, DJ; Saddler, JN. Factor affecting cellulose hydrolysis and the potential of enzyme recycle to enhance the efficiency of an integrated wood to ethanol process. *Biotechnol. Bioeng.* 1996 51, 375-383.

[96] Berlin, A; Gilkes, N; Kurabi, A; Bura, R; Tu, MB; Kilburn, D. Weak lignin-binding enzymes – A novel approach to improve activity of cellulases for hydrolysis of lignocellulosics. *Appl. Biochem. Biotechnol.* 2005 121, 163-170.

[97] Sewalt, VJH; Glasser, WG; Beauchemin, KA. Lignin impact on fiber degradation. 3. Reversal of inhibition of enzymatic hydrolysis by chemical modification of lignin and by additives. *J. Agr. Food Chem.* 1997 45, 1823-1828.

[98] Boussaid, AL; Saddler, JN. Adsorption and activity profiles of cellulases during the hydrolysis of two Douglas-fir pulps. *Enzyme Microb. Technol.* 1999 24, 138-143.

[99] Mooney, CA; Mansfield, SD; Touhy, MG; Saddler, JN. The effect of initial pore volume and lignin content on the enzymatic hydrolysis of softwoods. *Bioresource Technol.* 1998 64, 113-119.

[100] Abdi, N; Hamdache, F; Belhocine, D; Grib, H; Lounici, H; Piron, DL; Mameri, N. Enzymatic saccharification of solid residue of olive mill in a batch reactor. *Biochem. Eng. J.* 2000 6, 177-183.

[101] Cuevas, M; Sánchez, S; Bravo, V; Cruz, N; García, JF. Fermentation of enzymatic hydrolysates from olive stones by *Pachysolen tannophilus*. *J. Chem. Technol. Biotechnol.* 2009 84, 461-467.

[102] Roberto, IC; Lacis, LS; Barbosa, MFS; de Mancilla IM. Utilization of sugar cane bagasse hemicellulosic hydrolisate by *Pichia stipitis* for the production of ethanol. *Process Biochem.* 1991 26, 15-21.

[103] Toivola, A; Yarrow, D; van den Bosch, E; van Dijken, JP; Scheffers, WA. Alcoholic fermentation of D-xylose by yeasts. *Appl. Environ. Microbiol.* 1984 47(6), 1221-1223.

[104] Sassner, P; Galbe, M; Zacchi, G. Bioethanol production based on simultaneous saccharification and fermentation of steam-pretreated Salix at high dry-matter content. *Enzyme Microb. Technol.* 2006 39, 756-762.

[105] Slininger, PJ; Bothast, RJ; Okos, MR; Ladisch, MR. Comparative evaluation of ethanol production by xylose-fermenting yeasts presented high xylose concentrations. *Biotechnol. Lett.* 1985 7, 431-436.

[106] Palmqvist, E; Grage, H; Meinander, NQ; Hahn-Hagerdal, B. Main and interaction effects of acetic acid, furfural and p-hidroxibenzoic acid on growth and ethanol productivity of yeasts. Biotechnol. Bioeng. 1999 63(1), 46-55.

[107] Helle, S; Cameron, D; Lam, J; White, B; Duff, S. Effect of inhibitory compounds founds in biomass hydrolysates on growth and xylose fermentation by a genetically engineered strain of *S. cerevisiae*. *Enzyme Microb. Technol.* 2003 33, 786-792.

[108] Alfani, F; Gallifouco, A; Saporosi, A; Spera, A; Cantarelle, M. Comparison of SHF and SSF processes for the bioconversion of steam exploded wheat straw. *J. Ind. Microbiol. Biotechnol.* 2000 25, 184-192.

[109] Wingren, A; Galbe, M; Zacchi, G. Techno-economic evaluation of producing ethanol from softwood comparison of SSF and SHF and identification of bottlenecks. *Biotechnol. Progr.* 2003 19, 1109-1117.

[110] Cuevas, M; Sánchez, S; García, JF; Baeza, J; Parra, C; Freer, J. Enhanced ethanol production by simultaneous saccharification and fermentation of pretreated olive stones. *Renew. Energ.* 2015 74, 839-847.

[111] Saleh, M. Tratamientos a presión e hidrólisis del hueso de aceituna. Fermentación de hidrolizados con *Pachysolen tannophilus*. PhD Thesis. University of Jaén (Spain), 2014.

In: Agricultural Wastes
Editor: Camille N. Foster

ISBN: 978-1-63482-359-3

Chapter 4

POTENTIAL USE OF NUT AGRICULTURAL BY-PRODUCTS IN POLYMER MATERIALS: A REVIEW

Arantzazu Valdés*, Ana Beltrán and María Carmen Garrigós
Analytical Chemistry, Nutrition & Food Sciences Department,
University of Alicante, Alicante, Spain

ABSTRACT

Billion metric tons of agricultural residues are generated every year from industry worldwide that may be considered one of the most abundant, cheap and renewable resources on earth. However, they are normally incinerated or dumped causing environmental problems such as air pollution, soil erosion and decreasing soil biological activity. The reuse of these residues not only prevents environmental concerns, but also can provide farmers the opportunity of a second income from plantation. The incorporation of agricultural residues into polymer matrices is currently a trending topic in research due to the relatively high strength, stiffness and low density of natural fibres present in these residues. Nut by-products, such as almonds (brown hulls, shells and seeds coating) or walnuts (shells), among others, have been used as reinforcement in polymeric materials due to their desirable properties: low density and cost, availability, recyclability, environmental friendliness, total degradation without emission of toxic compounds in composting conditions, and good mechanical properties. On the other hand, nut residues (peanuts, almonds, hazelnuts, chestnuts, walnuts, pecan nuts or pistachios) are rich in bioactive compounds which can be extracted and further used as potential natural additives in food packaging materials with antioxidant and/or antimicrobial activity. In this chapter, different strategies for reusing nut by-products in polymer materials obtaining high value-added materials either as reinforcement or as a source of active compounds are reviewed. Finally, the utilization of gums is currently in the spotlight of the chemical industry.

* Corresponding author:Tel: +34 965903400. Ext 1187. Fax: +34 965903527. E-mail: arancha.valdes@ua.es.

1. INTRODUCTION

Nuts are recommended as an important food constituent of a healthy diet in human populations throughout the world. Extensive research has been carried out on nuts and health outcomes during the last two decades. In this sense, the publication of a report from the pioneering Adventist Health Study has shown the association of nut consumption with a lower risk of coronary heart disease [1].

According to the Phenol-Explorer database [2], nuts in the *botanical sense* are produced by some families of the order *Fagales*: the *Juglandaceae* (walnut and pecan nut), the *Fagaceae* (chestnut), and the *Betulaceae* (hazelnut). In the *culinary sense*, the term nut is applied to a wide variety of dried seeds and fruits and to any large, oily kernel found within a shell [3]. These nuts belong to the *Fabaceae* (peanut), the *Rosaceae* (almond) and the *Anacardiaceae* (pistachio) families. From a *bromatology point of view*, nuts are defined as food with an edible portion having in its composition less than 50 % weight of water with the exception of Brazil nuts and coconut, which exceed this amount [3].

The most common popular edible tree nuts are almonds (*Prunus amygdalus*), hazelnuts (*Corylus avellana*), walnuts (*Juglans regia*), and pistachios (*Pistacia vera*). Other common edible nuts are pine nuts (*Pinus pinea*), cashews (*Anacardium occidentale*), pecans (*Carya illinoiensis*), macadamias (*Macadamia integrifolia*), and Brazil nuts (*Bertholletia excelsa*). This group also includes peanuts (*Arachis hypogea*), which botanically are groundnuts or legumes but are widely identified as part of the nuts food group, having a nutrient profile similar to the tree nuts.

Finally, although chestnuts (*Castanea sativa*) are tree nuts as well, they are different from all other common nuts by their starchier character and different nutrient profile, with higher amounts of water and carbohydrates and lower protein and fat contents [1].

The recent recognition of nuts as "heart-healthy" food by the U.S. Food and Drug Administration (FDA) and their role in the Mediterranean food patterns have resulted in their substantial intake into regular diets (up to 35-40 % of the total energy intake of dietary fat)[4]. In fact, based on data recorded by the Food and Agriculture Organization of the United Nations (FAO), the worldwide nut production has increased significantly in the last years from 8.8 Mtons in 2002 up to 14.4 Mtons in 2012 [5].

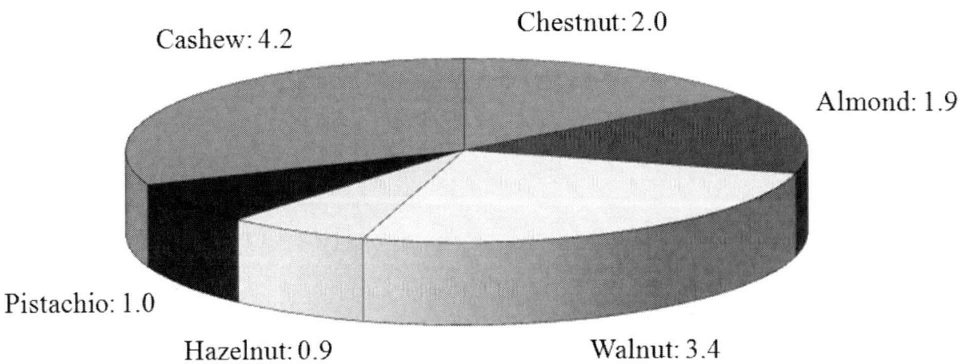

Figure 1. Worldwide nut production (Mtons) registered by FAO in 2012[5].

The major worldwide production was registered for cashew nut followed by walnut, chestnut, almond, and finally, pistachio and hazelnut (Figure 1).

Nuts are seeds showing a high content in proteins and lipids and they are widely added to some important food formulations such as ice creams, chocolates, confectioneries, cookies, cereal bars, and cakes. Nuts can also be eaten raw, roasted, or salted as snacks [6]. Industrial processing of nuts starts with the removal of the external coating among other parts as leaves. Up to now, the application of these agricultural residues has not received much attention, causing potential disposal problems. This bioresidues are considered to have high amount of fibre contents as well as other interesting compounds, such as flavonoids and phenolic acids with high antioxidant activity [7]. The increasing in energy demand coupled with the need to reduce greenhouse gas emissions and the threat of exhaustion of oil reserves have made to consider an eventual recourse to the use of biomass nut wastes as renewable energy source; and different efforts have been done in order to valorise nut agricultural by-products.

At the same time, the public concern about the environment, climate change and global warming while limited fossil fuel resources are available, have been important drivers for governments, companies and scientists to find alternatives to crude oil in plastics production. The durability of traditional plastics, which makes them ideal for applications such as packaging, can also lead to waste-disposal problems, since these materials are not biodegradable. Apart from the disposal, the fabrication of industrial products should also take into account raw materials from renewable resources in order to preserve fossil resources [8].

In this context, bio-composites obtained from biopolymers and reinforced with natural fibres may offer important contributions by reducing the dependence on fossil fuels and the related environmental impacts. In particular, the use and valorisation of agricultural wastes is currently a trending topic in research and a raising number of results in this area are being reported [9].

In this chapter, different strategies for reusing nut by-products in polymers obtaining high value-added materials are reviewed. Firstly, the nutritional value of nuts and the state of the art of nut agricultural by-products will be introduced.

Then, the use of nut by-products as reinforcement in polymeric materials will be discussed. Also, nut by-products as a potential source of bioactive compounds with antioxidant and/or antimicrobial activity in food packaging materials will be presented. Finally, the use of gums, currently in the spotlight of the chemical industry, will be mentioned.

2. Nutritional Value of Nuts

The edible part (seed) of nuts is surrounded by a shell, accounting for a proportion between 25 and 70 % of the total weight. The shape, size and flavour of the edible portion are extremely variable among different nuts and different varieties of the same nut.

Although they present a high sensory appeal and numerous health benefits, their consumption should be moderate due to their high caloric value, about 500-700 kcal per 100 grams.

However, tree nuts are highly nutritious and provide macronutrients such as fats, proteins, carbohydrates and fibres (Table 1).

Table 1. Macronutrients of 13 different nuts (average per 100 g portions) [10]

Nut	Energy (kcal)	Water (g)	Proteins (g)	Fat (g)	Ash (g)	CH (g)	Fibre (g)	Sugars (g)
Black walnut	618	4.6	24.1	59.0	2.5	9.9	6.8	1.1
English walnut	654	4.1	15.2	65.2	1.8	13.7	6.7	2.6
Pine nut	673	2.3	13.7	68.4	2.6	13.1	3.7	3.6
Pistachio	562	3.9	20.3	45.4	2.9	27.5	10.3	7.7
Pecan	691	3.5	9.2	71.9	1.5	13.7	9.6	4.0
Peanut	570	6.4	26.2	49.6	2.0	15.8	9.5	na
Macadamia	718	1.4	7.9	75.8	1.1	13.8	8.6	4.6
Hazelnut	628	5.3	15.0	60.8	2.3	16.7	9.7	4.3
Cashew	553	5.2	18.2	43.9	2.5	30.2	3.3	5.9
Brazil nut	656	3.5	14.3	66.4	3.5	12.3	7.5	2.3
Almond	575	4.7	21.2	49.4	3.0	21.7	12.2	3.9
Chestnut	213	48.7	2.4	2.26	1.1	45.5	8.1	na
Sunflower seed	584	4.7	20.8	51.4	3.0	20.0	8.6	2.6

na = not available in the USDA Nutrient Database, 2011.
CH = Carbohydrates.

Despite their similarities in water (less than 6.4 % total weight) and protein (13-26 % total weight) contents, nuts show a wide range of chemical compositions. The majority of nuts have 9-30 % of carbohydrates, less than 3.5 % of ash and 1.1-4.0 % of sugars.

Nuts are also a good source of dietary fibre, which ranges from 7 to 12 g per 100 g dry weight, and in standard servings providing 5–10 % of the daily fibre requirements.

For this reason, the regular consumption of nuts reduces pathological conditions, such as hypercholesterolemia, prevents atheroma formation and the development of diseases such as colon cancer due to their high fibre content that helps the gastrointestinal tract function [3, 11]. Nevertheless, pistachio and cashew nuts differ from the rest on its high content in carbohydrates (27.5 % for pistachio and 30.2 % for cashew) and sugars (7.7 and 5.9 %, respectively). Cashew also presents low fibre content (3.3 %) as well as pine nut (3.7 %). As it was previously mentioned, chestnuts are different from all the other common nuts showing a different nutrient profile with lower energy dense (213 Kcal) due to its lower fat content (2.26 g), with higher water and carbohydrates amounts (48.7 and 45.5 g, respectively).

As it can be observed in Table 1, nuts are energy dense food with a high-fat content (45–75 % of the total weight) [3]. This energy contribution is due, in part, to their high content in MUFAs (monounsaturated fatty acids) and PUFAs (polyunsaturated fatty acids) with low SFAs (saturated fatty acids) content for some nuts (Table 2). It is interesting to note that important differences can be observed among nuts regarding their fat profile. In this sense, walnut has about 47 % of PUFAs while macadamia nut shows the highest MUFAs content with 59 %. Regarding SFAs, Brazil nut has the higher content (15 %) compared to almond with only 4 %. In general, palmitic (C 16:0), stearic (C 18:0) and arachidic fatty acids (C 20:0) are the major SFAs present in nuts, with oleic (C 18:1), eicosenoic (C 20:1) and palmitoleic (C 16:1) acids being the dominant MUFAs. Finally, the most abundant PUFA present in nuts is linoleic acid (C 18:2) followed by α-linolenic acid (C 18:3).

Generally, C 18:1, 18:2, 16:0, and 18:0 are the dominant fatty acids in nuts, in descending order. In addition, English and Black walnuts are good sources of 18:3 (9.080 and 2.006 g per 100 g of portion, respectively).

Table 2. Fatty acid composition of 13 different nuts expressed as g of fatty acid per 100 g of nut [10]

Nuts	SFAs			MUFAs			PUFAs	
	C 16:0	C 18:0	C 20:0	C 16:1	C 18:1	C 20:1	C 18:2	C 18:3
Black walnut	1.923	1.445	0	0.063	14.533	0.408	33.072	2.006
English walnut	4.404	1.659	0.063	0	8.799	0.134	38.093	9.080
Pine nut	3.212	1.390	0.229	0.017	17.947	0.801	33.150	0.164
Pistachio	4.994	0.476	0.043	0.473	23.174	0.174	13.485	0.259
Pecan	4.366	1.745	0.069	0	40.594	0.207	20.628	0.986
Peanut	5.674	1.288	0	0.045	21.757	0.523	17.192	0.010
Macadamia	6.036	2.329	1.940	12.981	43.755	1.890	1.296	0.206
Hazelnut	3.097	1.265	0.102	0.116	45.405	0.131	7.833	0.087
Cashew	3.916	3.223	0.266	0.136	23.523	0.138	7.782	0.062
Brazil nut	9.085	5.794	0.160	0.226	24.223	0.052	20.543	0.035
Almond	3.044	0.013	0.658	0.243	30.611	0.010	12.061	0.006
Chestnut	0.384	0.021	0	0.021	0.749	0.010	0.798	0.095
Sunflower seed	2.210	1.690	0.115	0.020	18.380	0.085	23.050	0.060

Among nuts, higher oleic fatty acid content can be found as follows: hazelnut > macadamia > pecan > almond > Brazil nut > cashew > pistachio > peanut. However, macadamia has markedly higher palmitoleic fatty acid content (13 %) than the rest.

The major linoleic fatty acid content is present in English walnut composition (38.093 g per 100 g) followed by Pine nut (33.150 g), Black walnut (33.072 g) and sunflower seed (23.050 g). As it can be observed in Table 1, chestnuts contain the least amount of fat (2.26 g of fat per 100 g of portion). For this reason, only 0.749 g of C 18:1 per 100 g of portion and 0.798 g for C 18:2 are present in their composition.

Nuts are also rich sources of other bioactive macronutrients with potential to improve metabolic and cardiovascular outcomes1, being an excellent source of *proteins* (approximately 25 %) and often of a high content in arginine. As this aminoacid is the precursor of the endogenous vasodilator (nitric oxide), nut intake might help to improve vascular reactivity with beneficial cardiovascular effects. The contribution of vitamins B1 (thiamine), B6 (pyridoxine), B3 (niacin) and B9 (folic acid) is also noteworthy and highly recommended to prevent anemia and tiredness. Furthermore, the presence of *vitamin E* and antioxidants, such as *phytosterols* (stigmasterol, campesterol and sitosterol) and *polyphenols* (catechins, resveratrol, etc.), in nuts reduces the toxic effects of free radicals delaying aging processes [12]. Regarding polyphenols content [2], the highest amount (determined by Folin assay) is found in chestnut (2.8 g per 100 g); while walnut (1.6 g per 100 g), pecan nut (1.3 g per 100 g), pistachio (867 mg per 100 g), peanut (420 mg per 100 g), hazelnut (291 mg per 100 g) and almond (192 mg per 100 g) show lower values.

Finally, compared to other common foods, nuts have an optimal nutritional density in healthy minerals, such *as calcium, magnesium and potassium.* It has been proved that a high intake of these micronutrients, together with the low *sodium* content of nuts, is associated with the protection against bone demineralization, arterial hypertension, insulin resistance, and overall cardiovascular risks [13].

As a consequence of the high nutritional value of nuts, in 2003 the FDA approved a qualified health claim stating that the consumption of 1.5 ounces (42.5 g) per day of most nuts might reduce the risk of coronary heart disease [14, 15].

3. Nut Agricultural by-Products

Billions of metric tons of biomass are generated every year from agricultural processes, including liquid, solid and gaseous residues; that may be considered one of the most abundant, cheap and renewable resources on Earth [16]. Many of these residues are just incinerated or dumped without any kind of control causing several environmental problems, such as air pollution or soil erosion.

In this context, a more efficient utilization of agricultural wastes to yield a number of added-value resources is highly attractive to ensure sustainable and cleaner production processes that are economically viable, environmentally sound and socially beneficial.

As a consequence, the incorporation of agricultural wastes into different polymer matrices is a current trending topic in materials research with a raising number of results. The obtained biocomposites show relatively high strength, stiffness and low density by the incorporation of natural fibres [17]. In this context, Mande clearly defined two main categories of agricultural residues [18]:

Crop residues or primary biomass residues, generated in the farm, which are normally non-edible plant parts that are left in the field or orchard after the main crop has been harvested. These residues mainly include straw, stover, stubble, stalks, sticks, leaves, haulms, roots, branches, twigs, brushes, trimmings and pruning; and they are produced from different sources including seeds, fruits, nuts, vegetables and energy crops.

Agro-industrial residues or secondary biomass residues, generated during the post-harvest processing, are those by-products or sub-products obtained from the post-harvesting processes or the transformation of the crop into valuable products. These products include residues from wood and food processing industries in the form of husks, hulls, peels, dust, straws, bagasse, sawdust, corncobs, pomace, etc. Regarding nut wastes, a huge quantity of residues with high environmental impact and greenhouse effects are generated and disposed every year (Table 3).

Table 3. Yield and estimated potential production of primary and secondary residues from some nut crops [16]

Crop	Yield (MMT/year)	Main residue	Total residue production (MMT/year)[a]
Almonds	2.51	Brown hulls, shells, seeds coating	0.88 0.10
Chestnuts	1.96	Shells (outer, inner)	0.39
Hazelnuts	0.89	Shells, kernel	0.44
Walnuts	2.54	Shells	1.70

[a] Residue production = (Yield x % of residue after crop processing) per 100.

MMT: Millions metric tons.

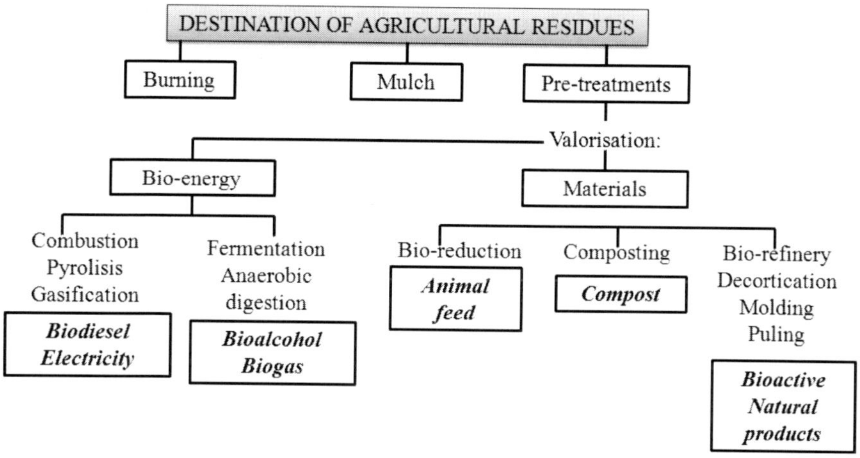

Figure 2. Flow diagram of final destination of agricultural residues [16].

The main current applications of these residues are in the production of bio-energy due to the characteristics (non-edible, high energy potential, etc.) of lignocellulosic materials, generation of biomass-based energy fuels, industrial materials such as animal feed and compost and a great variety of industrial products, such as wood-based panels, paper, cardboard and bio-fertilizers; among others [16]. Figure 2 shows a scheme of the main post-processing destination of agricultural residues.

The use of these residues for energy purposes contributes to avoiding disposal problems. As an example, a kinetic model based on the pyrolysis of Brazil nut shells between 350 and 850 ºC was reported showing an efficient conversion of the shells into useful products[19]. Cardozo et al. compared the combustion behaviour of selected agricultural residues: sugar cane bagasse, sunflower husks and Brazil nut hells with commercially available wood pellets to evaluate the impact of the fuel properties on emission levels, oxygen levels, temperatures in the combustion chamber, and efficiencies based on the conversion of fuel compounds into gaseous products [20]. As a result, the type of agricultural residue had an impact on the power input, oxygen levels and combustion chamber temperature when compared with wood pellets. The power input of Brazil nut shells and sunflower husk pellets due to their high ash content were reduced and therefore high oxygen levels were measured. Also, higher NO, CO and SO_2 emissions for Brazil nut shells when comparing with wood pellets were reported mainly due to their higher contents of nitrogen, sulphur and ash; and therefore, improvements in the combustion process are required. As a result, new applications of these residues as a valuable stock material to be used in a wide range of fields have been carried out in the industry including biomaterials, compost, fertilizers, food and feed or added value compounds.

Gasification technology has been reported as a suitable technique for biomass residues conversion and remains an economical alternative for the valorisation of cashew nut shells in small scale industries [21]. In order to evaluate the cashew net shells valorisation by gasification several experiments were satisfactorily conducted on a char obtained from the pyrolysis of this biomass and using a fixed bed reactor. As gasifying agents, carbon dioxide, steam and the mixture of carbon dioxide and steam were used at different temperatures for the gasification of fine particles.

In direct combustion applications these biomasses show some drawbacks: low combustion efficiency in spite of rather high calorific value; release of unhealthy smoke and soot deposits in closer environment. Pyrolysis might be an alternative valorisation pathway of such extractive rich feedstock, to convert it into either char (slow pyrolysis) or pyrolysis oils (fast pyrolysis). In this context, Melzer et al. studied the usability of cashew nut shells in energetic terms [22]. In contrast to lignocellulosic biomass, these residues are rich in extractives. The feedstocks were characterised in a first step upon their physical and chemical properties before they were pyrolysed in a thermogravimetric system and a tubular reactor under rapid pyrolysis conditions. A detailed study of obtained pyrolysis oils showed that the extractives of cashew nut shells are not entirely cracked while vegetable oils decompose almost entirely. Furthermore, pyrolysis oils obtained from this study would contain valuable products which could be extracted from cashew nut shells, in particular chemicals appreciated in green chemistry such as phenols and amines.

3.1. Revalorisation of Nut by-Products in Polymer Materials as Reinforcement

Around 32 million tonnes of natural fibres are extracted each year from a wide range of plants and animals, according to FAO [23]. Natural fibres basically consist of cellulose, lignin, and hemicelluloses. Pectins, pigments, and extractives can be found in smaller quantities.

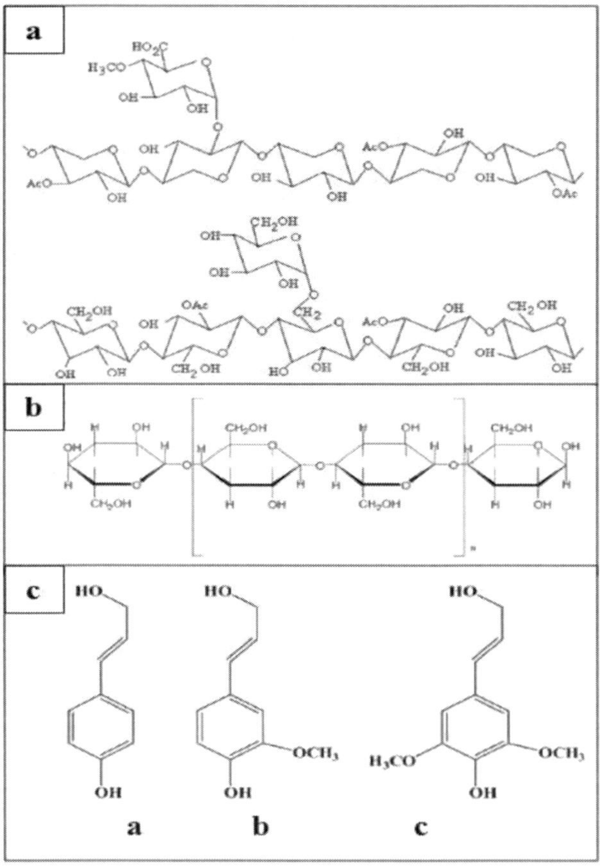

Figure 3. Hemicellulose (a), cellulose, (b) and lignin (c) structures characteristics of agricultural residues (with permission) [25].

Table 4. Potential use of different nut agricultural by-products in plastic materials as reinforcement

Nut	Nut by-product	Reinforced matrix	Reference
Macadamia	Shell	Polyester resin	Dong and Davies (2012) [27]
Walnut	Shell	Urea-formaldehyde resin	Pirayesh, Khazaeian and Tabarsa (2012) [28]
		Polyethylene	Salasinska and Ryszkouska, (2012) [29]
Walnut/almond	Shell	Urea-formaldehyde resin	Pirayesh and Khazaeian (2013) [30]
Almond	Shell	Polypropylene	Essabir et al. (2013) [31]
	Skin or tegument	Poly(ε-caprolactone)	Valdés, Ramos, Beltrán and Garrigós (2014) [33]
Peanut	Shell	Polyethylene	Caraschi, Leão and Chamma (2009) [34]

The chemical composition and cell structural of natural fibres is quite complicated. Each fibre is essentially a composite in which rigid cellulose microfibrils are embedded in a softer matrix mainly composed of lignin and hemicelluloses [24]. Hemicelluloses are build-up of a combination of 5- and 6-ring carbon polysaccharides (excluding pectin) having branched structure, with much lower molecular weight than cellulose (Figure 3a). Cellulose is defined as the non-branched macromolecule containing chains of variable length of 1–4 linked β-D-anhydroglucopyranose units. Cellulose is primarily composed of C, H, and O_2, having a general formula of $C_6H_{10}O_5$ (Figure 3b). Finally, lignin is the highly branched polymer which serves as the matrix material along with hemicellulose to embed cellulose fibres. Lignin structure is very complex consisting of phenyl propane units organized in a three-dimensional structure (Figure 3c). Different units in lignin are linked together through various types of carbon–carbon and ether bonds as opposed to linear or branched chains as in carbohydrates [25].Polymers reinforced with natural fibres are used to fabricate composites with specific applications; using fully sustainable, biodegradable, environmentally friendly and renewable fibres, particularly those derived from plants. Regarding the use of agricultural residues in polymer materials, the use of bamboo, rice husk, abaca, sisal or oil palm, among others, has been strengthened to exploit these natural materials as non-wood renewable fibres to reinforce different polyesters and polyolefins [26]. Table 4 summarizes the potential use of different nut agricultural by-products in plastic materials as reinforcement.

Regarding nuts, previous studies have reported the use of nut agricultural residues for reinforcement, such as macadamia [27] and walnut shells [28], improving physical and mechanical properties of several polymer matrices. According to Dong & Davies [27], the shells and other waste comprise almost 70 % by weight of the *macadamia* nuts. They can be burned as a wood substitute in coffee roasting, ground to produce organic waste for gardening, used for mulch in the nut tree orchards, or used for chicken litter that, after use, returns to the orchard for its use as fertilizer. Macadamia nutshell is hard and brittle with approximately the same fracture toughness as common ceramics and glass, and when compared on the basis of specific strength or modulus, it outperforms these materials due to its low density. In this study, the flexural properties of macadamia nutshell particle reinforced polyester composites (SR250 orthophthalic polyester resin) were studied with four weight fractions of macadamia nutshell particles (10, 20, 30 and 40 %). The process-induced voids were studied and it was shown that voids played an important role in flexural properties. In this sense, flexural modulus increased with the weigh fraction of macadamia nutshell particles, while decreased with increasing void content.

Since *walnut* shell comprises 67 % of the total weight of the fruit, around 1.8 million tons of walnut shell is left behind each year [16, 28]. In order to revalorize this by-product, Pirayesh, Khazaeian and Tabarsa [28] studied the suitability of walnut shell in the production of three-layer particleboard as a supplement, and to alleviate the shortage of raw material in forest industry as well as diminishing environmental problems regarding their burning. For this purpose, particleboards containing different walnut shell particle ratios (0, 10, 20, 30, 40 and 100 %) were obtained using urea–formaldehyde (UF) resin. The results showed that it is possible to produce particleboards using a mixture of walnut shell and wood particles while using urea–formaldehyde as binder. The amount of walnut shell particles at most should be 20 % in the mixture to meet the standards required for modulus of rupture. Also, a series of composites with different shares of finely ground walnut shells (10, 25 and 40% mass) were produced by Salasinska and Ryszkowska [29]. Polyethylene obtained from selective packaging waste collection was used as matrix, and the preparation process was in two stages-mixed and injection. As a result, the addition of the walnut residue increased the hardness and considerably improved the stiffness of the resulting composite.

The use of *almond* shells were tested combined with other nut residues such as walnut at different walnut/almond shell particle ratios (0, 10, 20, 30 and 100 %), incorporated in a resin [30] or alone as reinforcement in a thermoplastic matrix as polypropylene up to 30 % of particle content with and without different compatibilizers [31]. The results indicated that both residues can be considered as alternative raw materials or fillers in the manufacture of biocomposites. In this sense, results showed a clear improvement in mechanical and rheological properties reducing the formaldehyde emissions as well as highly improving water resistance of the panels.

Despite the outer shell, referred to as the hull, almond has an inner shell, known as the skin. Almond skins (AS) are industrially removed from the nut by hot water blanching and constitute 4–8 % of the total shelled almond weight. Almond-processing industries are interested in the valorisation of these by-products, which are considered to have one of the highest fibre contents of all the edible nuts; among other interesting compounds, such as flavonoids and phenolic acids with high antioxidant activity [32].

For this reason, the incorporation of low-cost AS residues into a biodegradable polymer, such as poly(ε-caprolactone) (PCL), is an attractive alternative to transform agricultural residues into useful industrial resources, with a positive benefit on environment, energy and economy. In fact, novel biodegradable composites based on PCL and almond skin residues were produced in a previous work [33]. A remarkable improvement in mechanical properties with the addition of almond skin particles was obtained indicating the potential use of this residue as reinforcement agent in PCL composites. Furthermore, the presence of AS by-products accelerated the degradation of the PCL matrix in the composite films, being this effect more pronounced with the increase in AS contents. This effect was explained in terms of the reduction in crystallinity of the polymer matrix and the high hydrophilicity of the natural fibres, promoting the water uptake and, consequently, the microbial attack and hydrolysis of the PCL matrix. The best performance regarding the studied properties was found for composite films with 10 wt% AS loading. Mechanical properties were also improved with good adhesion between AS and the PCL matrix, as observed by scanning electron microscopy. No significant differences were observed regarding thermal degradation and barrier properties compared to neat PCL. In conclusion, this formulation could be an interesting environmentally-friendly material to be used for food packaging applications showing a biodegradable nature and increasing the added-value potential of almond agricultural wastes. In this sense, it is clear that some reduction in transparency of the polymer matrix will be obtained with AS incorporation, but the obtained formulations could be suitable for the development of sustainable food trays and similar containers where transparency is not an issue.

Peanut shell (30, 40 and 60 wt%) was used by Caraschi, Leão and Chamma [33] to evaluate the utilization of solid residues in the preparation of panels for civil construction. Low density panels were obtained, suggesting their possible use as ceilings, partitions, decorative flooring and other applications requiring the same physical and mechanical properties.

The architectural element developed in this study meet the concept of eco-efficient product to be produced from waste and enabled its use in urban or rural environments.

In summary, several advantages of incorporating natural fibres into biopolymer matrices are interesting to note, such as low density and cost, availability, recyclability, environmental friendliness, total degradation in soil without emission of toxic compounds in composting conditions, and good mechanical properties; permitting the design of new biocomposites with adequate properties for particular applications.

3.2. Revalorisation of Nut by-Products in Polymer Active Packaging as Bioactive Natural Product Sources

It has been found that agricultural residues are rich in natural compounds making them attractive alternatives to synthetic additives in food packaging materials.

Table 5. High added-value products obtained from processing-based residues of different nuts [16]

Crop	Main residue	Bioactive compounds	Activity or application
Peanut	Skins, Seed coats	Polyphenols, proanthocyanidins and phenolic acids	Antioxidant, anticancer, blood vessels protector and antimicrobial
Almond	Hulls	Triterpenes and daucosterol	Anticancer
Hazelnut	Skins, Hard shells, Leafy covers	Phenolic acids (gallic, caffeic, p-coumaric and ferulic)	Antioxidant
Chestnut	Shells (outer, inner)	Tannins, polyphenols and tocopherols	Antioxidant
Walnut	Shells	Holocellulose, α-cellulose and lignin	Panel manufacture
Pecan nut	Shells (endocarp)	Poly and monomeric phenols	Antifungal
Pistachio	Hulls	Phenolic compounds	Antioxidant

Natural additives can be mainly obtained and classified into three categories [16]: (1) recovery of natural constituents by using conventional and advanced extraction processes, (2) optimization of lead compounds by means of chemical and/or enzymatic reactions to obtain analogues, and (3) development of bioprocess for the production of bioactive compounds. Table 5 summarizes the potential of nut residues as raw materials for the recovery of high added-value compounds. Main nut residues obtained from processing are the outer hull and inner shells, skin or endocarp and leafy covers; with polyphenols and phenolic acids showing antioxidant, antifungal and antimicrobial activity.

3.2.1. Antimicrobial and Antifungal Action of Nut By-Products

The use of antimicrobials in food has become increasingly necessary as the global economy boosts the production and transportation of food worldwide; however, to ensure the supply of high-quality food, the use of preservatives is essential. The potential application of natural antimicrobial compounds by the food industry is huge, and studies on the incorporation of antimicrobials in food and to maximize their bio-functionality have been conducted worldwide.

Regarding the use of nut residues as antimicrobial agents, Prado et al. [35] studied the use of peanut peels (Arachis hypogaea) waste as an antimicrobial agent against microorganisms commonly associated with food toxic-infections such as *Staphylococcus aureus*, *Listeria monocytogenes*, *Salmonella Enteritidis* and *Escherichia coli.*; showing antimicrobial activity against S. aureus and L. monocytogenes, which are important bacterial pathogens in humans. As a result, this constitutes a viable possibility for the use of such wastes by the food industry, being a natural alternative to synthetic preservatives, avoiding waste disposal into the environment and bringing benefits to both industry and consumers. On the other hand, phytopathogenic organisms cause a wide spectrum of diseases in plants including fungi, nematodes, bacteria, and viruses. Several fungi have been found to induce post-harvest spoilage of fruits and vegetables, which is associated with a decrease in nutritive elements. In this context, the antifungal activity of polyphenolic extracts from pecan nut shell (*Carya Illinoensis*) against plant pathogenic fungi was reported by Osorio et al. [36]. This study demonstrated the high antifungal capacity of poly and monomeric phenolic extracts, particularly of those prepared from agro-industrial residues (pomegranate husk and pecan nut shell) and creosote bush leaves. Low concentration extracts were capable to inhibit a great variety of fungal species. In particular, pecan nut shell extracts inhibited 80 % of the tested F. oxysporum strains.

3.2.2. Antioxidant action of Nut By-Products

Lipid oxidation is, after microbial growth, the main cause of food spoilage. In particular, food with high lipid content, especially those with a high grade of unsaturation, are susceptible to deterioration following this path. This is the case of nuts, vegetables, fish oils and meat or fishery products that have not been subjected to preservation treatments or to technologies to reduce microbial growth. The development of off-flavours, such as toxic aldehydes, typical of rancidity, renders products unacceptable for human consumption by their loss in nutritional quality because of PUFAs degradation.

Antioxidant active packaging technology is based on the incorporation of antioxidant agents to the packaging material to improve the stability of oxidation-sensitive food products reducing or even eliminating some of the main food spoilage causes; such as rancidity, colour loss/change, nutrient losses, dehydration, microbial proliferation, senescence, gas build-up, and off-odours [37]. Up to now, butylated hydroxyanisole (BHA) and butylated hydroxytoluene (BHT) are the most widely used synthetic antioxidants for preventing oxidation in food products [38-39]. The use of both compounds in food packaging formulations is currently under discussion due to toxicological concerns.

The interest in the use of natural antioxidants based on natural extracts obtained from plants, essential oils or agricultural waste products in polymer formulations in the food packaging area is continuously increasing, not only by their harmless character to humans but also by their good performance in limiting oxidation processes in the material and/or food; as well as the good acceptance by consumers of the use of natural additives.

In general, nuts and their by-products, such as almond hulls, hazelnut shell and kernels, hazel leaves or chestnut fruit by-products, have been studied as natural antioxidant sources [40]. *Almond* hulls have been proved to act as natural preservatives for meat products, such as round beef and minced chicken breasts, reducing the lipid oxidation until seven and twelve days in refrigeration [41, 42].

In this context, new possibilities to give added value to nut residues as sources of natural antioxidants to be used to control the oxidative process in the food industry have been recently opened. Their use as natural additives in direct contact with food or incorporated into the packaging material as an active additive is currently under evaluation and it is one of the main objectives of food packaging research. In this sense, Siriwardhana and Shahidi [43] reported highest phenolic content for almond skin extracts (87.8 mg of catechin equivalents per gram of extract) compared to extracts from almond seeds and green shells (leafy cover) with 8.1 and 71.1 mg of catequin equivalents per gram of extract, respectively. A number of flavonoids and phenolic acids, including flavanols, flavanones and simple phenolic acids, identified in blanched almond skin and other almond residues may play a role in reducing risk factors against chronic inflammatory diseases and ageing disorders [44-46]. Bartolomé et al. [46] used a commercial almond skin (AS) extract (Amanda®) obtained from the Spanish Marcona variety in studies in human health. Results suggested that AS polyphenols are bioavailable in humans since they were detected as phase-II and microbial-derived metabolites in plasma and urine samples, but further investigations on the potential of these phenolic metabolites as biomarkers of AS consumption are currently under development by using metabolomic approaches. This study suggested that AS can be used in the elaboration of dietary ingredients, since the total polyphenols content was found to be higher than 10 mg per gram; which is higher than other well-known natural extracts, such as red grape skin. Liu et al. [44] reported that AS could serve as a candidate food for potential prebiotic effects, since the ingestion of 10 g of AS during 4 weeks increased the beneficial bacterial populations, such as *Bifidobacterium spp.* and *Lactobacillus spp.*, whereas the growth of the pathogen *Clostridum perfringens* was significantly repressed. As a consequence, modification of the intestinal bacterial activities was observed, which would induce the promotion of health beneficial features and the inhibition of harmful factors. It was also reported that AS possess potential prebiotic properties due to its abundance of dietary fibre and polyphenols.

Conventional antioxidant extraction methods are time-consuming, eventually lead to thermal degradation of the relatively volatile antioxidants and usually require large quantities of solvents, raising process costs and reducing the environmental sustainability. The exposure to oxygen and radiation during extraction could contribute to the additional degradation of the target compounds [47].

Therefore, alternative methods to lower costs and to reduce sampling time are being developed. These alternative techniques include: supercritical fluid extraction (SFE), pressurized liquid extraction (PLE), microwave-assisted extraction (MAE) and ultrasound-assisted extraction (UAE) [48]. Regarding nuts, MAE was used to extract phenolic antioxidants from *peanut* skins.

The effect of microwave power, irradiation time and sample mass on the total phenolic content (TPC) and oxygen radical absorbance capacity (ORAC) of peanut skin extracts were investigated [49]. Optimal extraction conditions were obtained with 1.5 grams of sample in 30 % ethanol: water solution, 90 % microwave power, and 30 seconds for ORAC and 150 seconds for TPC. The obtained results indicated the potential of revalorisation of nut residues by the development of active packaging systems for food preservation against lipid oxidation during storage. The high polyphenols content in peels, skin and seeds of nuts support the use of these agricultural by-products as sources of natural antioxidants.

The peanut skin residue has been shown to be a potential source of natural antioxidants also by Rosales et al. [50]; even though it is currently considered a by-product of the blanching process of peanuts and only used in the production of animal feed. On the other hand, the antimicrobial and antioxidant activities of *pistachio* hull extracts in relation with the structure of their phenolic compounds were reported by Goli, Barzegar and Sahari [51].

The effect of natural *chestnut* leaves extract on the physico-chemical, lipid oxidation, microbial and sensory characteristics of dry-fermented sausage were investigated by Lorenzo et al. [52] during the ripening period in contrast to natural grape seed extract and synthetic antioxidants (BHT). Although the results of this study indicated that grape seed is the most effective antioxidant against lipid oxidation, both grape seed and chestnut extracts were more effective than BHT. Also, pH, colour, and overall sensory quality were affected by the addition of these antioxidants. According to those analyses, results proved that both natural extracts could be used to improve the safety and quality of dry-fermented sausages and also to obtain more attractive products for the consumers. Lorenzo, Sineiro, Amado and Franco [53] studied the ability of different compounds including chestnut leave extracts to inhibit microbial spoilage. Colour deterioration and lipid oxidation were evaluated, being promising natural compounds towards replacing the use of commercial antioxidants for extending the shelf-life of porcine patties packaged in modified atmosphere under refrigerated conditions.

Hazelnut shells were reported to be a suitable source of antioxidants with total phenol content (TPC) of 181 ± 9 mg gallic acid per litre of extract and antioxidant activity measured by FRAP assay of 21 ± 1 mmol Fe^{2+} per gram of gallic acid [54]. These results suggest that hazelnut shell extract could be further exploited for its application in the development of innovative and sustainable food packaging materials.

4. Recent Uses of Nut By-Products

The growing accumulation of synthetic plastic wastes, together with the difficulty of recycling most packaging, has stimulated food and packaging industries to explore new biodegradable packaging materials. In recent years, there has been a growing interest in edible films and coatings, which offer several advantages over synthetic materials, such as being biodegradable and environmentally friendly. Materials available for forming films and coatings to be used as packaging materials (or to complement packaging materials) for food generally fall into the categories of polysaccharides, proteins and lipids [55].

In the last decade, different nanoparticles based on polysaccharides were produced for drug delivery applications. These systems can also be applied in the agricultural field, where pesticides can be entrapped in the polymer matrix, maximizing their effect, at low concentration. Cashew nut, one of the most important edible nuts in international trade, is the fruit of cashew trees (Anacardium occidentale L.), and its kernel is the most widely industrialized cashew product [56]. Cashew gum is a water soluble heteropolysaccharide exudated from *Anacardium occidentale* tree and has similar properties to those of Arabic gum; whereby their structures have a main chain of galactose units, having branches of arabinose, glucose and rhamnose. Uronic acid units were also found to be present in side chains. Abreu, Oliveira, and de Paula [57] reported the preparation by spray drying of a new matrix composed of chitosan and cashew gum as an encapsulating agent for *L. sidoides* to improve essential oil loading and release profiles. The effects of polymer concentration and chitosan–gum relative ratio were also investigated on the nanoparticles size and encapsulation efficiency. In vitro release studies and in vivo experiments were also carried out.

The loading and encapsulation efficiency of *L. sidoides* oil were optimized for samples produced with matrix:oil 10:2 using a 5 % cashew gum concentration. Particularly, sample gum:chitosan 1:1 presented the highest loading (11.8 %) and encapsulation efficiency (70 %). In vitro release profiles revealed markedly Fickian behaviour; a prolonged release being obtained for the sample with larger chitosan proportion, i.e., gum:chitosan 1:10. All the nanogels produced presented efficacy against *St. aegypti larvae*, where the mortality rate was related to the loading values and gum:chitosan ratios. In particular, samples gum:chitosan 1:1 and gum:chitosan 1:10 showed, respectively, 87 % and 75 % of mortality after 48 h; reaching over 90 % of mortality at 72 h. These results showed that the gum–chitosan nanoparticles were designed and present sustained release features.

Since lipids constitute more than 40 g/100 g of cashew nut kernels, lipid oxidation is their main cause of deterioration due to the formation of off-flavours (oxidative rancidity), impairing their acceptability. Moreover, the kernels are also susceptible to moisture absorption, resulting in texture changes and loss of their characteristic crispness. Then, the application of coatings to cashew nut kernels could significantly increase their stability, avoiding the necessity of using high barrier packaging materials (more expensive and usually thicker). However, polysaccharide-based films and coatings usually have poor mechanical and barrier properties when compared to petroleum-based polymers.

The use of nanoreinforcements, such as montmorillonite-type nanoclays, has been extensively studied to improve the physico-mechanical performance of biopolymer materials. Recently, Pinto et al. [56] studied the development and characterization of films obtained from starch and cashew tree gum added with a montmorillonite nanoparticle in order to test their application as coatings to increase stability of cashew nut kernels. Results suggested that tensile properties of films from starch and cashew tree gum were favoured by intermediate nanoparticle concentrations with improvement of water vapour barrier of the films. Coatings from starch-cashew tree gum with or without montmorillonite were similarly effective to reduce oxidation rates of the kernels.

The development of electrochemical biosensors has been a focal subject in biosciences and biotechnologies. Different kinds of materials have been used as transducer elements in biosensors, and many of them have been built from the layer-bylayer (LbL) technique that was originally developed by Decher and Hong [58].

This is an assembly method based on electrostatic alternated adsorption of oppositely charged polyions. Due to its versatility and simplicity, it has been applicable to assemble a wide variety of materials including carbohydrates, proteins, nanoparticles, dyes, DNA and natural gums. Araújo et al. [58] studied the contribution of a cashew tree gum for the development of LbL films with potential application in nanobiomedical devices such as electrochemical sensors for the dopamine neurotransmitter. This gum has low viscosity comparable in many aspects to gum Arabic. The technological interest in the cashew gum and other natural gums, which has been proved to present similar rheological characteristics and industrial applications to many synthetic polymers, comes mainly from its biodegradability and mechanical properties. It was observed that cashew tree gum contributed favourably to obtain stable films with well-defined redox processes which can detect dopamine with a detection limit of relevance to the pharmaceutical industry.

Conclusion

The currently research situation highlights the potential use of nut agricultural by-products, mainly nut shells, as reinforcement agents in several polymer matrices improving physical and mechanical properties. Nut agricultural residues are rich in natural active compounds making them attractive alternatives to synthetic additives in antioxidant and/or antimicrobial packaging materials in contact with food. Main nut residues obtained from food processing are the outer hull and inner shells, skin or endocarp and leafy covers; showing polyphenols and phenolic acids with antioxidant, antifungal and antimicrobial activity. These materials could be effective for food preservation against lipid oxidation during storage. The developed materials could be also more competitive than the pure ones in the marketplace, reducing costs for the final packaging and the food industry. On the other hand, the utilization of gums is now being taken advantage in research and development, being used in different applications including edible films and electrochemical biosensors.

References

[1] Ros, E. Health benefits of nut consumption. *Nutrients*, 2010, 2, 652-682.

[2] Phenol-Explorer database. http://www.phenol-explorer.eu/ (accessed January 2015).

[3] Coenders, A. Química culinaria: estudio de lo que le sucede a los alimentos antes, durante y después de cocinados. Ed. ACRIBIA, S.A. Zaragoza (Spain), 2007; 88-90.

[4] Vadivel, V.; Kunyanga, C. N.; Biesalski, H. K. Health benefits of nut consumption with special reference to body weight control. *Nutrition*, 2012, 28, 1089-1097.

[5] Food and Agriculture Organization of the United Nations. (FAO Database). http://faostat.fao.org/ (accessed January 2015).

[6] Zajdenwerg, C.; Branco, G.; Alamed, J.; Decker, E.; Castro, I. Correlation between sensory and chemical markers in the evaluation of Brazil nut oxidative shelf-life. *Eur. Food Res. Technol.*, 2011, 233, 109-116.

[7] Biesaga, M. Influence of extraction methods on stability of flavonoids. *J. Chromatogr. A.*, 2011, 1218, 2505-2512.

[8] Laceta, I.; Etxabide, A.; Cabezudo, S.; de la Caba, K.; Guerrero, P. Bio-based films prepared with by-products and wastes: environmental assessment. *J. Clean. Prod.,* 2014, 64, 218-227.

[9] Valdés, A.; Mellinas, C.; Ramos, M.; Garrigós, M.C.; Jiménez, A. Natural additives and agricultural wastes in biopolymer formulations for food packaging. *Front. Chem.*, 2014, 2: 1-10.

[10] Beltrán, A.; Ramos, M.; Valdés, A.; Garrigós, M. C. Linoleic acids: sources, biochemical properties and health effects. Ed. Nova Science Publishers, New York, 2012, 83-96.

[11] Jacobsen, C. Enrichment of foods with omega-3 fatty acids: a multidisciplinary challenge. *Ann. N.Y. Acad. Sci.,* 2010, 1190, 141-150.

[12] Carlsen, M. H.; Halvorsen, B. L.; Holte, K.; Bohn, S. K.; Dragland, S.; Sampson, L.; Willey, C.; Senoo, H.; Umezono, Y.; Sanada, C.; Barikmo, I.; Berhe, N.; Willett, W. C.; Phillips, K. M.; Jacobs, D. R.; Blomhoff, R. The total antioxidant content of more than 3100 foods, beverages, spices, herbs and supplements used worldwide. *J. Nutr.*, 2010, 9, 1-11.

[13] Cordain, L.; Eaton, S. B.; Sebastian, A.; Mann, N.; Lindeberg, S.; Watkins, B. A.; O'Keefe, J. H.; Brand-Miller, J. Origins and evolution of the Western diet: health implications for the 21st century. *Am. J. Clin. Nutr.,* 2005, 81, 341-354.

[14] U.S. Food and Drug Administration (FDA). Qualified health claims: letter of enforcement discretion-nuts and coronary heart disease (Docket No 02P-0505). (July 14, 2003). FDA website: http://www.fda.gov/ (accessed January 2015).

[15] Alasalvar, C.; Shahidi, F. Natural antioxidants in tree nuts. *Eur. J. Lipid Sci. Technol.*, 2009, 111, 1056-1062.

[16] Santana, O.; González, A.; Sánchez, R. Agricultural residues as a source of bioactive natural products. *Phytochem. Rev.*, 2012, 11, 447–466.

[17] Faruk, O.; Bledzki, A. K.; Fink, H. P.; Sain, M. Biocomposites reinforced with natural fibres: 2000–2010. *Prog. Pol. Sci.*, 2012, 37, 1552–1596.

[18] Mande, S. Biomass gasifier-based power plants: potential, problems, and research needs for decentralized rural electrification. In: Lal B, Reddy MRVP (Eds). Wealth from waste: trends and technologies. Ed. TERY: The Energy and Resources Institute, New Delhi, India, 2005.

[19] Bonelli, P.; Della Rocca, P.; Cerrella, E.; Cukierman, A. Effect of pyrolysis temperature on composition, surface properties and thermal degradation rates of Brazil nut shells. *Bioresour. Technol.,* 2001, 76:15–22.

[20] Cardozo, E.; Erlich, C.; Alejo, L.; Fransson, T. Combustion of agricultural residues: An experimental study for small-scale applications. *Fuel*, 2014, 115: 778–787.

[21] Diedhiou, A.; Bensakhria, A.; Ndiaye, L.; Khelfac, A.; Sock, O. Study of Cashew Nut Shells Valorisation by Gasification. *Chem. Eng. Trans.*, 2014, 39, 1171-1176.

[22] Melzer, M.; Blina, J.; Bensakhria, A.; Valette, J.; Broust, F. Pyrolysis of extractive rich agroindustrial residues. *J. Anal. Appl. Pyrol.*, 2013, 104, 448–460.

[23] Johansson, C.; Bras, J.; Mondragon, I.; Nechita, P.; Plackett, D.; Simon, P.; Aucejo, S. Renewable fibers and bio-based materials for packaging applications - A review of recent developments. *BioResources*, 2012, 7, 2506-2552.

[24] Santos, J. C. O.; Dos Santos, I. M. G.; De Souza, A. G.; Prasad, S.; Dos Santos, A. V. Thermal stability and kinetic study on thermal decomposition of commercial edible oils by thermogravimetry. *J.Food Sci.*, 2002, 67, 1393-98.

[25] Kumar, V.; Kumari, M. Processing and characterization of natural cellulose fibres/thermoset polymer composites. *Carbohyd. Polym.*, 2014, 109, 102–117.

[26] Mitra, B.C. Environment friendly composite materials: biocomposites and green composites. *Defence Sci. J.*, 2014, 64, 244-261.

[27] Dong, C.; Davies, I. J. Flexural properties of macadamia nutshell particle reinforced polyester composites. *Composites: Part B*, 2012, 43, 2751–2756.

[28] Pirayesh, H.; Khazaeian, A.; Tabarsa, T. The potential for using walnut (Juglans regia L.) shell as a raw material for wood-based particleboard manufacturing. *Composites: Part B*, 2012, 43, 3276–3280.

[29] Salasinska, K.; Ryszkowska, J. Physico-mechanical properties and dimensional stability of natural fibre composites fabricated from polyethylene waste and walnut shells. ECCM 2012 - Composites at Venice, Proceedings of the 15th European: Composites at Venice, ECCM 2012; Venice; Italy.

[30] Pirayesh, H.; Khanjanzadeh, H.; Salari, A. Effect of using walnut/almond shells on the physical, mechanical properties and formaldehyde emission of particleboard. *Composites: Part B*, 2013, 45, 858–863

[31] Essabir, H.; Nekhlaoui, S.; Malta, M.; Bensalah, M.O.; Arrakhiz, F.Z.; Qaiss, A.; Bouhfid, R. Bio-composites based on polypropylene reinforced with Almond Shells particles: Mechanical and thermal properties. *Mater. Des.*, 2013, 51, 225–230.

[32] Mandalari, G.; Tomaino, A.; Arcoraci, T.; Martorana, M.; Turco, V. L.; Cacciola, F.; Rich, G. T.; Bisignano, C.; Saija, A.; Dugo, P.; Cross, K. L.; Parker, M. L.; Waldron, K. W.; Wickham, M. S. J. Characterization of polyphenols, lipids and dietary fibre from almond skins (Amygdalus communis L.). *J. Food Comp. Anal.*, 2010, 23, 166-174.

[33] Valdés, A.; Ramos, M.; Sanahuja, A.; Garrigós, M.C. Characterization and degradation characteristics of poly(ε-caprolactone)-based composites reinforced with almond skin residues. *Polym. Degrad. Stab.*, 2014, 108, 269-279.

[34] Caraschi, J.; Leão, A.; Chamma, P. Avaliação de Painéis Produzidos a partir de Resíduos Sólidos para Aplicação na Arquitetura. *Polímeros: Ciência e Tecnologia*, 2009, 19, 47-53.

[35] Prado, J. G.; Porto, E.; Bani, C.; Matías de Aléncar, S.; Micotida, E.; Ribeiro, I. S.; de Aquino, L. M. Antimicrobial protential and chemical composition of agroindustrial wastes. *J. Nat. Prod.*, 2012, 5, 27–36.

[36] Osorio, E.; Flores, M.; Hernández, D.; Ventura, J.; Rodríguez, R., Aguilar, C. N. Biological efficiency of polyphenolic extracts from pecan nuts shell (Carya Illinoensis), pomegranate husk (Punica granatum) and creosote bush leaves (Larrea tridentata Cov.) against plant pathogenic fungi. *Ind. Crops. Prod.*, 2010, 1, 155–157.

[37] Realini, C. E.; Marcos, B. Active and intelligent packaging systems for a modern society. *Meat Sci.*, 2014, 98, 404–419.

[38] Singh, P.; Wani, A. A.; Saengerlaub, S. Active packaging of food products: recent trends. *Nut. Food Sci.*, 2011, 41, 249-260.

[39] Ramos, M.; Valdés, A.; Beltrán, A. Desarrollo de biopelículas activas para envasado de alimentos. Aplicación en materiales para envasado de alimentos. Ed. Académica Española, 2012.

[40] Fernández, A. A.; Pereira, E.; Freire, M. S.; Valentão, P.; Andrade, P. B.; González, J. A.; Pereira, J. A. Influence of solvent on the antioxidant and antimicrobial properties of walnut (Juglans regia L.) green husk extracts. *Ind. Crop. Prod.*, 2013, 42, 126-132.

[41] Teets, A. S.; Were, L. M. Inhibition of lipid oxidation in refrigerated and frozen salted raw minced chicken breasts with electron beam irradiated almond skin powder. *Meat Sci.,* 2008, 80, 1326–1332.

[42] Teets, A. S.; Sundararaman, M.; Were, L. M. Electron beam irradiated almond skin powder inhibition of lipid oxidation in cooked salted ground chicken breast. *Food Chem.,* 2008, 111, 934–941.

[43] Siriwardhana, S. S. K. W.; Shahidi, F. Antiradical activity of extracts of almond and its by-products. *J. Amer. Oil Chem. Soc.,* 2002, 79, 903–908.

[44] Liu, Z.; Lin, X.; Huang, G.; Zhang, W.; Rao, P.; Ni, L. Prebiotic effects of almonds and almond skins on intestinal microbiota in healthy adult humans. *Anaerobe,* 2014, 26, 1-6.

[45] Tsujita, T.; Shintani, T.; Sato, H. α-Amylase inhibitory activity from nut seed skin polyphenols. 1. Purification and characterization of almond seed skin polyphenols. *J. Agric. Food. Chem.*, 2013, 61, 4570-4576.

[46] Bartolomé, B.; Monagas, M.; Garrido, I.; Gómez, C.; Martín, P.J.; Lebrón, R.; Urpí, M.; Llorach, R.; Andrés, C. Almond (Prunus dulcis (Mill.) D.A. Webb) polyphenols: from chemical characterization to targeted analysis of phenolic metabolites in humans. *Biochem. Biophys.,* 2010, 501, 124-33.

[47] Pirayesh H.; Khazaeian A. Using almond (Prunus amygdalus L.) shell as a bio-waste resource in wood based composite. *Composites: Part B,* 2012, 43, 1475–1479.

[48] Michel, T.; Halabalaki, M.; Skaltsounis, A. L. New concepts, experimental approaches, and dereplication strategies for the discovery of novel phytoestrogens from natural sources. *Planta Med.,* 2013, 79, 514–532.

[49] Ballard, T. S.; Mallikarjunan, P.; Zhou, K.; O´Keefe, S. Microwave-assisted extraction of phenolic antioxidant compounds from peanut skins. *Food Chem.,* 2010, 120, 1185-1192.

[50] Rosales, P.; Arellano, S.; Dorantes, L.; García, F.; López, M. Comparison between Antioxidant Activities of Phenolic Extracts from Mexican Peanuts, Peanuts Skins, Nuts and Pistachios. *J. Mex. Chem. Soc.*, 2014, 58, 185-193.

[51] Goli, A. H.; Barzegar, M.; Sahari, M. A. Antioxidant activity and total phenolic compounds of pistachio (Pistachia vera) hull extracts. *Food Chem.,* 2005, 92, 521–525.

[52] Lorenzo, J.; González, R.; Sánchez, M.; Amado, I.; Franco, D. Effects of natural (grape seed and chestnut extract) and synthetic antioxidants (buthylatedhydroxytoluene, BHT) on the physical, chemical, microbiological and sensory characteristics of dry cured sausage "chorizo". *Food. Res. Int.*, 2013, 54, 611–620.

[53] Lorenzo, J.; Sineiro, J.; Amado, I.; Franco, D. Influence of natural extracts on the shelf life of modified atmosphere-packaged pork patties. *Meat. Sci.*, 2014, 96, 526–534.

[54] Vellingiri, V.; Amendola, D.; Spigno, G. Screening of Four Different Agro-Food By-Products for the Recovery of Antioxidants and Cellulose. *Chem. Eng. Trans.*, 2014, 37, 757-762.

[55] Salarbashi, D.; Tajik, S.; Shojaee-Aliabadi, S.; Ghasemlou, M.; Moayyed, H.; Khaksar, R.; Shahidi, M. Development of new active packaging film made from a soluble soybean polysaccharide incorporated Zataria multiflora Boiss and Mentha pulegium essential oils. *Food. Chem.*, 2014, 146: 614–622.

[56] Pinto, A.; Santos, T.; Caceres, C.; Lima, J.; Ito, E.; Azeredo, H. Starch-cashew tree gum nanocomposite films and their application for coating cashew nuts. *LWT – Food. Sci. Technol.,* 2014, article in press.

[57] Abreu, F.; Oliveira, E.; de Paula, H.; de Paula, R. Chitosan/cashew gum nanogels for essential oil encapsulation. *Carbohydr. Polym.*, 2012, 89: 1277– 1282.

[58] Araújo, I.; Zampa, M.; Moura, J.; dos Santos, J.; Eaton, P.; Zucolotto, V.; Veras, L.; de Paula, R.; Feitosa, J.; Leite, J.; Eiras, C. Contribution of the cashew gum (Anacardium occidentale L.) for development of layer-by-layer films with potential application in nanobiomedical devices. *Mat. Sci. Eng*., 2012, 32: 1588–1593.

In: Agricultural Wastes
Editor: Camille N. Foster

ISBN: 978-1-63482-359-3

Chapter 5

PERSPECTIVES ON THE UTILIZATION OF RICE HULL IN PRODUCTIVE PROCESSES

Mariane Silva de Miranda, Alexandre Correia Lima, Tatiane Faustino de Moraes and Flávio Aparecido Rodrigues
Laboratório de Materiais e Superfícies, Universidade de Mogi das Cruzes (UMC), Brazil

ABSTRACT

Rice (*Oryza sativa L.*) is a very important component of human diet for many people around the globe. Rice world production is approximately 680 million tons year and Asia leads world harvesting. This is an important source of biomass, especially because there is a tendency to rationalize the use of crude oil and derivatives. Enhancing the utilization of biomass may help to avoid climate and environmental problems.

The industrial processing of rice generates some byproducts, such as rice bran and broken rice. Both components can be used as nutritional constituents and they will not be discussed in this work. On the other hand, agricultural residues are relevant in the process, especially rice hull, which accounts for about 20% of the rice crop.

This work presents some relevant aspects about the utilization of rice hull. There are many possible applications in different areas, including fermentation and production of ethanol, preparation of cellulose, synthesis of inorganic materials, such as pigments, zeolites, cements, composites, fillers, among others.

INTRODUCTION

Rice is one of the most produced and consumed agricultural product. The world production, according to FAO (Food and Agricultural Organization) will reach approximately 680 million tons. [1]

The demand for rice as food and nutritional component is still growing. In the period between 2000 and 2010, rice paddy production has raised in all continents, except Oceania. Asia is the region where most of rice is produced. Figure 1 shows the rice production in this continent, between 2000 and 2012. [2] It can be seen that a considerable growth in production has occurred.

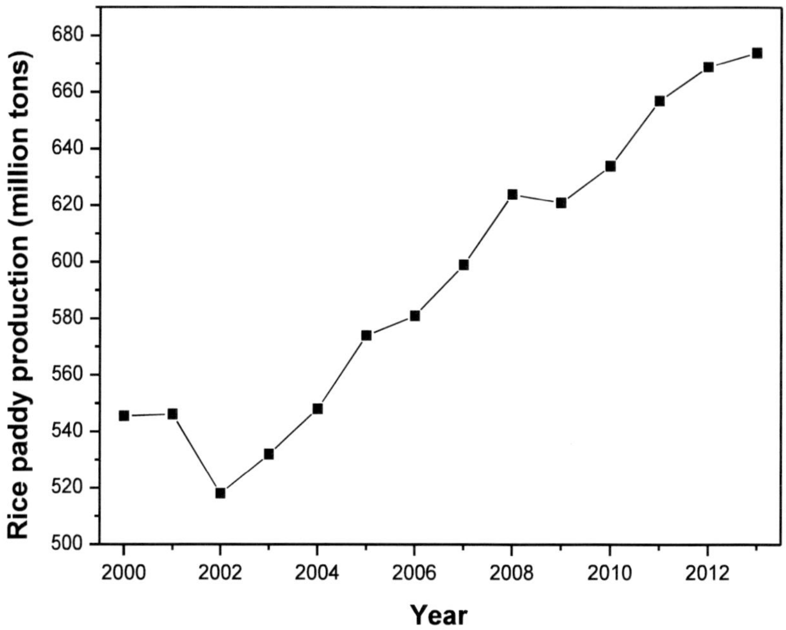

Figure 1. Variation of rice production in Asia (2000-2012).

Some interesting aspects should be highlighted; the production 1 kg of harvested rice paddy renders about 1.35 kg of straws. The hull is a protective coat to the rice grain and it corresponds to about 20% of total mass. It means that almost 140 million tons of rice hulls are produced yearly. Roughly, this value accounts for twice the world annual production of primary aluminum. [3] It is evident that such amount of biomass represents a valuable source of raw-materials and that its adequate utilization could be very productive in terms of economic and environmental aspects.

The average chemical composition of rice hull is presented in Table 1. [4] These values are not absolute and show some variability according to region, type of production, and the availability of nutrients.

Table 1. Average chemical composition of rice hull

Component	% (w/w)
Cellulose	31
Hemicellulose	23
Lignin	22
Mineral ash	14
Moisture	8
Extractives	2

The mineral ash is composed mainly by silica (above 93%) and minor components such as potassium oxide, magnesium oxide, calcium oxide and aluminum oxide. [5]

In this work, the perspectives of rice hull applications will be presented considering the organic (cellulose, hemicellulose and lignin) and the inorganic fraction (mineral ash). There are many methods employed to separate or eliminate these components. For instance, pyrolysis leads to elimination of organic fraction. On the other hand, sometimes the inorganic component is undesirable and may be discarded during processing. Ideally both components should be recycled.

2. Silica Extraction from Rice Hull

Combustion process is the conventional method applied to prepare silica from biogenic resources, due to its simplicity and low-cost. Usually, in open fields, rice hulls are disposed and burnt under uncontrolled conditions. The ash produced contains high carbon content (up to 30%) and silica can be either amorphous or crystalline. However, this material has no market value.

Silica with very small particle size and high purity can be obtained after burning under controlled conditions. The combustion process can be conducted at temperature range between 300 to700 °C, and time varying from few minutes up to 6-7 hours. The conventional laboratory furnace is a very simple method to burn organic materials. However the poor contact between rice hull and oxygen limits the overall process efficiency. [6] At high temperatures textural structure of hull is preserved during burning [7] and under these conditions, silica might contain spots with unburnt carbon. Moreover, high energy consumption and variable chemical composition are some drawbacks associated to this kind of processing. Thus some strategies are used to overcome thermal treatment of rice hull and achieve silica with high quality.

Pre-pyrolysis treatment combined with calcination can be applied to obtain nanosilica particles with controllable surface areas and volume pore. Gu and coworkers [8] prepared nanosilica particles under CO_2 and N_2 pyrolysis atmosphere and achieved silica with different levels of purity and texture. The pyrolysis processes were conducted in temperature range of 300–800°C and post calcination was performed at 610 °C for 2/3 hours under O_2 atmosphere. Silica prepared under CO_2 atmosphere shows a decrease in surface area (from 323.1 to 237.2 $cm^2.g^{-1}$), pore volume (0.5228 – 0.4108 $cm^3.g^{-1}$) and purity (99.56% - 97.54%) with the increasing of pyrolysis temperature. This trend was also observed for pyrolysis under N_2 atmosphere. At 800 °C surface area, pore volume and purity decreased drastically to 129 $cm^2.g^{-1}$, 0.3148 $cm^3.g^{-1}$ and 89,6%, respectively. According to researchers, N_2 fill the pores enhancing inner pressure, so the pores eventually collapses leading to decrease of pore volume and surface area.

Previous treatment of rice hulls with acid solutions has been proven effective to eliminate traces of K^+, Na^+ Ca^{+2} and others elements; also acid-leaching treatment provides a better powder dispersion of the final product. The rice hull sample was pretreated by soaking in deionized water to remove most of the alkali metals and partial fixed carbon. Soaking pretreatment may remove the hydrogen bonds or extractives between hemi-cellulose and cellulose, which affects the thermal stability of both molecules [9]. Della and coworkers [7] compared silica obtained from acid leaching of rice hull (RH) and silica obtained from thermal treatment of rice hull. After leaching RH with 10% hydrochloric acid and calcination at 600 °C for 3 h it was possible to achieve silica with high purity (97%) and a specific surface area of 296 $m^2.g^{-1}$. On the other hand, the calcination of RHA was more effective at 700 °C for 6 h followed by grinding for 80 min; in this case silica content was about 95% with a specific surface area of 81 m^2/g. [7]

Wang and coworkers [10] also used acid-leaching process in rice hulls. In this treatment rice hulls were soaked in aqueous boiling 10% HCl solution for 2 hours. After drying, rice hulls were heated at 700 °C and silica nanoparticles (25-30 nm in diameter) were obtained.

Zhang and coworkers [11] developed a methodology to treat rice hull. First, cellulose was extracted by a hydrolyzation process with sulfuric acid, followed by ethanol washing. D-xyllose and residue 1 was achieved. Lignin was extracted from residue 1 with ethanediol by hydrothermal process, and then a second residue was generated by the end of hydrothermal process. Residue 2 was dried and put into a tube furnace, which was heated to 550°C under the flowing of carbon dioxide gas; when a black powder was achieved (a mixture of silica and carbon), carbon dioxide gas flow is changed to oxygen flow at 550 °C. The black powder turns into white silica in 30 minutes. Silica reached 99.92% of purity and specific surface 225.2 $m^2.g^{-1}$.

Chen and coworkers [12] developed a different approach. In this case, lignocellulose was dissolved with an ionic liquid (butyl-3-methylimidazolium chloride). After the lignocellulose regeneration from solution, the residue was ready for thermal treatment. The residue containing silica was dried and heated at 700 °C for 2 hours. SEM images show that diameter of silica nanoparticles is ca. 70 nm having a surface area 241.1 $m^2.g^{-1}$.

2.1. Chemical Processing Silica from Rice Hull: Alkaline Leaching

Hiesh and coworkers [13] removed silica from rice hull by alkaline treatment using different NaOH concentrations and temperatures. The authors observed that temperature increase had minimal effect on silica removal, but changes in NaOH concentration (up to 1 $mol.L^{-1}$) had a pronounced influence on the efficiency of silica separation. When 1M NaOH was used almost all silica was removed at 50 and 85 °C. After precipitation using 1M HCl, silica powders were characterized by photon correlation spectroscopy and exhibited diameters of 90% particles are smaller than 12.7 μm.

Nanosilica particles are required for advanced materials manufacturing. Usually it can be prepared from tetraalkoxysilanes, $Si(OR)_4$, such as tetraethoxysilane (TEOS) and tetramethoxysilane (TMOS). Despite its high homogeneity and surface area the chemical precursors are expensive and highly toxic. Adam and coworkers [14] used rice hull as source of silica to obtain nanospheres. In this process rice hulls were allowed to react with 1M NaOH for 24 h under continuous stirring at room temperature. The sodium silicate solution was titrated against an aqueous 3.0 $mol.L^{-1}$ HCl solution, until pH 9.0; the gel was aged for 2 days. Afterwards the gel was dried and grounded. Silica nanospheres were obtained with an average diameter of 50.9 nm and high surface area (245 $m^2.g^{-1}$). The pore size distribution was narrow with variation between 5.6 and 9.6 nm.

A different approach is the use of organic solvents to control particle shape and size distribution. Zulkifli and coworkers [15] have prepared spherical silica particles from sodium silicate solution derived from rice hull. They have used phosphoric acid, ethanol and water. Controlling reaction conditions, amounts of ethanol and water and pH, silica particles with high specific surface area of 364 $m^2.g^{-1}$ and spherical morphology with pore size of 8 nm were obtained. Spherical silica is suitable for use as filler in dental restoration, as researchers suggest themselves. [16] Besides ethanol addition Lin and coworkers [17] also have used polyethylene glycol (PEG) to obtain silica microspheres from rice hull ash.

Alkaline extraction method is also used as post-treatment in RHA to enhance silica purity, particle size. In this work, RHA was reflux boiled in NaOH solution in three different periods (3. 5 and 9 hours), although XRD analyses shown no difference, silica with higher specific surface were obtained in 5 hours reflux experiments, achieving specific surface area of 353 m2/g while silica prepared by 9 hours reflux had smaller particle size but specific surface area was lower than 5 hours reflux experiments, according to authors 9 hours refluxed samples were dense due to more agglomerated particles, reducing surface area. [18]

Le and coworkers [19] produce homogenous silica particles by addition of a water/butanol mixture with CTAB. In optimized conditions silica was amorphous with an average diameter of 3 nm and specific surface area of $340m^2/g$.

2.2. NH_4F Dissolution Process

Ma et al., [20] proposed an innovative and recyclable method to prepare silica from RHA. In this approach silica was firstly dissolved into NH_4F solution producing $(NH_4)_2SiF_6$ and NH_3. Ammonia was recycled by a receiver to produce $NH_3.H_2O$ which was used in the precipitation silica from $(NH_4)_2SiF_6$. The process can be resumed as following reactions:

$$6NH_4F + SiO_2 \rightarrow (NH_4)_2SiF_6 + 4NH_3 + 2H_2O \quad (1)$$

$$(NH_4)_2SiF_6 + 4NH_3 + (n+2)H_2O \rightarrow 6NH_4F + SiO_2\downarrow + nH_2O \quad (2)$$

Silica dissolution and precipitation procedure are described in reactions (1) e (2), respectively. Silica obtained by this procedure was spherical with a diameter of 50-60 nm and a recovery yield up to 94.6%. The authors claim this ecological procedure since no residues was created and both reactants NH_4F and NH_3 returned to the process.

2.3. Biological Approach

Biological route is also a green alternative procedure to obtain nano sized silica from RH/RHA at room temperature. *Fusarium oxysporum* is a fungus capable of convert metallic salts into metals or metal oxide form. This fungus was applied to convert silica present in RH to quasi-spherical and nano-sized silica particle [21, 22]. Fusarium Oxysporum was capable to biotransformated amorphous silica to a crystalline state with spherical shape and particle size of 2-6 nm. [23]

3. Applications of Silica Derived from Rice Hull

3.1. Synthesis of Cements and Zeolites from Rice Hull Ash

Portland cement is one of the most consumed materials in the world. The traditional method used for the production of cement is based reactions at temperatures around 1450 °C. [24] Portland cement is a complex material, composed basically by calcium silicates, calcium aluminates and calcium aluminoferrites, among others. The calcium silicates, Ca_3SiO_5 and β-Ca_2SiO_4, determine most of the adhesive properties of concrete. The cement production releases huge amounts of CO_2, about 6% of total CO_2 emission caused by human activities. [25]

Rodrigues and coworkers [26, 27] developed a synthetic method for the synthesis of β-Ca_2SiO_4 using silica from rice hull. Silica was obtained from rice hull by heating this material at 600°C in an open furnace. [28] The silica was mixed with other solids (calcium oxide and barium chloride) in stoichiometric proportions. It was found that addition of small amounts of barium chloride (2%, molar ratio) facilitates the synthesis of desired material, having the β-structure. After hand-mix, water was added to solids and a suspension having water: solid ratio of about 20:1 was obtained. The suspension was submitted to an ultrasonic cleaner bath (Thornton, 25 kHz) for 60 minutes. After this procedure a mixture of solids was obtained: an intermediate silicate and calcium hydroxide. Heating of intermediate calcium silicate renders β-Ca_2SiO_4. This material shows potential to be used as a cement.

Many papaers deal with the synthesis of zeolites from rice hull ash. Cheng and co-workers [29] synthesized MCM-22 by hydrothermal method and studied the adsorption of red 5GN, a dye used in textile industry. The preparation of MVM-22 was carried put after 3 days of hydrothermal processing. NaA zeolite [30], ZSM5. [31]

3.2. Silica as Adsorbent and Support for Catalysis

There are several forms of treatment in rice hulls that lead to adsorbents products with environmental applications. Soltani and coworkers describe in detail different methods of preparation of adsorbents products from the hull of rice that are less costly and efficient in adsorption of ions of environmental interest. [32]

Silica extracted from rice hull has been extensively used mainly as support for heterogeneous catalysts. [33] Gan and Li [34] studied the degradation of rhodamine B, using iron as catalyst, supported on silica (RH); they studied several variables, such as the presence of H_2O_2, reactants concentration, pH, ionic strength and temperature. This system was highly efficient. process. Ahmed and Adam [35] studied the incorporation of aluminum, gallium and iron onto silica (RH) and the catalytic effect on Friedel-Crafts benzylation of benzene. Once again several variables were evaluated in the process. Overall, iron-supported catalyst showed higher catalytic activity and gallium-supported material presented higher selectivity.

Adsorption from solutions is also a very active research area. [36] Manique and co-workers [37] have used silica (RH) for the purification of biodiesel from waste frying oil. They studied the effect of concentration of silica on the removal of impurities, using several concentrations of this support and compared to magnesol, a commercial product. They have found that both supports show about the same efficiency. The adsorption of methylene blue, a classical dye, has also been investigated. [38, 39]

There many papers dealing with other applications of rice hull as the removal of heavy metals, [40] cooper [41, 42] nickel [43], humic acids [44] and hydrogen storage capacity. [45]

4. RICE HULL FERMENTATION

The term fermentation can be broadly defined as all processes in which microorganisms catalyze the conversion of a given substance in a given product [46]. Some bacteria and other micro-organisms use the fermentation process in order to obtain energy. It occurs with breaking of glucose (or other substrates such as starch) to pyruvate, which is then converted to other products, such as ethanol, lactic acid, acetic acid, among others, by the action of enzymes. The production and use of microbial enzymes, in a controlled manner, are relatively recent, but currently constitute the largest sector of the biotechnology industry. The production of chemical inputs from the biotechnology industry adheres to the principles of green chemistry, that is, uses materials - renewable materials (biomass) to replace the fossil fuels to produce biodegradable compounds into carbon dioxide and water [47, 48] and the reduction of energy consumption and waste generation.

As Genovese and coworkers stated [49] biomass is organic matter in soil, especially plant waste. In the field of energy, biomass is the term used to describe all forms and derivatives of plants that can be converted into useful energy such as, wood, forest and municipal wastes, grain stalks, vegetable oils and sludge biological treatment of effluents. The energy produced from biomass is also known as "green energy" or "bioenergy". The lignocellulosic biomass to be extremely abundant and high processing capacity, renewable and hardly expensive, and do not compete for food level, it becomes more attractive for the production of bioproducts, biofuels and energy [50, 51, 52].

Rice husk is one of lignocellulosic residues that has attracted the attention of researchers due to its cellulose and hemicellulose content and its potential for the production of bioproducts and biofuels. The rice hull can be used as a substrate for fermentation processes, since 70% of waste is constituted by carbohydrates which has attracted attention for being an underdeveloped source of fermentable sugars for industrial use [54]. Thus these facts make it interesting to study the use of biomass, widely available, renewable and virtually no commercial value. [54, 55]

However, it is important to note that the lignocellulosic components which form part of the rice husk need to be previously separated so that the cellulose and hemicellulose fractions are available in the form of fermentable saccharides. For this purpose various pre-treatment have been proposed in the literature for the separation of such lignocellulosic complex and are based on biological processes or physical-chemical, depending on the intended purpose [57].

According to Sarkar et al., [57] because the lignocellulosic complex is composed of a cellulose matrix of hemicellulose and lignin chains linked, pre-treatment of the material is an essential step because objectively separate the lignin matrix, reducing the crystallinity of cellulose, increasing the amorphous fraction of it and solubilize the hemicellulose, separating the hydrolyzed cellulose so that it becomes more accessible for enzyme attack. Table 1 is a summary of various processes used as pre-treatment of lignocellulosic biomass.

Table 2. Summary of the various processes used for pre-treatment of lignocellulosic biomass

Process pre-treatment	Advantages	**Limitations / Disadvantages**
Mechanical trituration	• Reduces the cellulose crystallinity.	• Energy consumption usually higher than the energy inherent biomass.
Steam explosion (with / without the addition of acid as catalyst)	• causes degradation of hemicellulose and lignin.	• Destruction a portion of the xylan • Incomplete rupture lignin matrix in carbohydrates • Formation of degradation products
Expansion of fiber with Ammonia (AFEX)	• Low energy expenditure • Low formation of inhibitors.	• High cost: large amount of ammonia needed • The fraction of hemicellulose must be enzymatically hydrolyzed • Despicable lignin removal.
Explosion with CO_2	• Increase the accessible surface area • Cost-effective • It does not cause the formation of inhibitory compounds.	• Do not modify the lignin or hemicellulose.
Ozonolysis	• Reduces the lignin content; does not produce toxic waste.	• Requires large amount of ozone; cost high.
Acid Hydrolysis	• High throughput of glucose • Low formation of inhibitors • Carried out at ambient temperature	• High cost of acid or acid recovery used • Expensive equipment due to corrosion problem
Alkaline hydrolysis	• Remove the contents of hemicellulose and lignin • Increase the area of the accessible surface.	• Long residence time is required • There is formation of salts that are incorporated into the biomass.
Process Organosolv	• Hydrolyzed hemicellulose and lignin (depending on the solvent used) • Increased digestibility of cellulosic biomass.	• High cost of organic solvents as well as in its recovery • High cost recovery of by-products • Rate cost liquid-solid must be fixed carefully.
Ionic Liquids	• Reuse of the ionic liquid after the process • Final substrate with digestibility > 90%.	• The ionic liquid must be completely removed before hydrolysis (interferes with the hydrolytic activity of enzymes) • High cost of liquids.
Pyrolysis	• Production of gaseous and liquid compounds.	• Elevated temperature and ash production.
Pulsed Electric Field	• Favorable environmental conditions • Disruption of cells • Requires simple equipment.	• The process needs more research.
Biological	• Degradation of lignin and hemicellulose • Low power requirement.	• Bacteria / fungi consume some of available carbohydrate, so lower sugar yield.

Font: [58, 59].

These treatments may be used alone or combined to obtain large quantities of fermentable carbohydrates. The work Megawati and et al., [60] studied the kinetics of the acid hydrolysis of rice husk using sulfuric acid as a catalyst to produce fermentable carbohydrates, moreover, has been observed experimentally that the carbohydrates produced by acid hydrolysis can be converted into ethanol by fermentation using micro-organisms such as yeast. The experiments performed in this study were conducted using various concentrations of the catalyst and it was found that during the hydrolysis, there was degradation of carbohydrates.

Aguiar and coworkers [61] used the acid catalyst in the acid hydrolysis processes that releases protons to break the glycosidic bonds between the sugar monomers in the polymer chains. The breaking of these bonds principally liberates sugars as xylose, glucose and arabinose. These carbohydrates were investigated in Cunha Pereira and coworkers[62]. The hydrolyzed rice hull was used as a substrate to produce ethanol and production xylitol by conversion of glucose, xylose and arabinose, by fermentation processes using *Spathaspora arborariae* cultures (a new yeast strain isolated from Atlantic Forest in Brazil) *Saccharomyces cerevisiae* and cultures combined these stems.

The use of co-cultures for the production of ethanol appears to have advantages over the use of a single culture provided that there a synergistic action of the metabolic pathways of the strains [63]. The central objective when using co-culture for ethanol production is to combine a glucose fermenter microorganism and another pentose fermenter microorganism simultaneously. The work involving the production of ethanol by the hydrolyzed rice hulls in an acid medium, was developed by Moon et al., [64] which conclude that the rice hulls hydrolyzed in an acid medium is a promising alternative for the production ethanol via fermentation processes using *Saccharomyces cerevisiae* as microorganism. Under optimum conditions for pretreatment acid (2: 1 of acid to the biomass, 5 hours of impregnation and 2 hours of hydrolysis), it was possible to obtain high conversion yields of glucose and xylose. Extreme acidity and increased in the retention time under high temperature were the most important factors influencing the breakdown of glucose and xylose. The presence of inhibitors in the hydrolyzed rice hull decreased the growth rate of the microorganisms and yield of ethanol, which was 0.47 g ethanol / g glucose was slightly lower than the reference medium which was 0.49 g ethanol / g glucose.

Hickert and coworkers [65] developed a method that showed good prospects for the production of xylitol and ethanol from the hydrolyzed rice husk using two types of micro-organisms in the same culture medium. The micro-organism *Saccharomyces cerevisiae* was able to metabolize glucose from the hydrolyzed rice hulls, which contained small amounts of acetic acid, furfural and hydroxymethylfurfural, reaching good yields of ethanol. When there was the mixture of microorganisms *Saccharomyces cerevisiae* and *Spathaspora arborariae* in the same culture medium, hexoses and pentoses obtained from hydrolyzed rice hulls were converted into ethanol and xylitol giving satisfactory yields.

Nichols and coworkers [66] reported in their study that the hydrolyzed rice hull is an agricultural available with high glucose content (from 0.32 to 0.33 g glucose / g shells) and xylose (15-19g xylan / g shells) and thus can be used as substrates for fermentative processes These results indicated that glucose respondents showed a good conversion to ethanol but the metabolism xylose was strongly affected by pH and the concentration acetate of derived groups acetyl and degradation products lignin.

Besides the production of ethanol obtained by rice husk, other works are also available in the literature aiming at the production of other chemicals as from the rice husk as furfural by solid acid catalyst, [53] levulinic acid and 5-hydroxymethyl-furfural, xylitol [67], via dilute acid self-hydrolysis; lactic acid by hydrolysis in the production of xylo-oligosaccharides and subsequent fermentation. [68]

5. Miscellaneous Applications

The use of $MnFe_2O_4$ dispersed in ash of rice hull has been used for removal of Chemical Demand of Oxygen (COD) in the treatment of organic sewage. The material was prepared by calcination leading to a product with a large amount of silica and activated carbon. To these it was dispersed $MnFe_2O_4$ with an average diameter of 59 nm and using microwave radiation for 6 minutes, it was obtained a 73.5% removal in COD. [69]

In the polymer industry there is great interest in using this agricultural waste incorporating its ashes to change some property in some polymers.

The pyrolysis in a fluidized bed under nitrogen atmosphere and with different temperature controls results in white or black ash, the product characteristic is temperature dependent. White ash contains about 95% silica with high porosity and reactive OH groups. The black ash contains both silica and carbon and have as high porosity and specific surface area. These pyrolysis products have been used in composites with polymers: polypropylene and tetrafluoroethylene-etileno. [

In bio fibers composites it is common the use of coupling agents such as, for example, maleic acid to improve adhesion between the fibers resulting in better results of tensile strength and flexural strength. The reason for this is that maleic acid interacts with the OH group of fibers and it is also present in the ashes of rice hull. [71] With similar strategies they obtained similar results for polymers: polyester, epoxy, phenolic and polystyrene. [72]

Rice hull has also been used for high density polyethylene composites with the incorporation of maleic acid to improve adhesion between the polyethylene and rice hull. Flexibility and resistance to traction were increased with the increase in ratio of rice hull, but only with the presence of maleic acid, about 5% in weight. Prolonged exposure to UV light led to the bleaching and the water absorption was low since the maleic acid was present. [73]

The lignin present in the rice hull was used in the preparation of phenol resin. This lignin has undergone fenolation generating ligninafenol product thus enhancing the reactivity of lignin. This ligninafenol was used in the preparation of a polymer: phenol / phenol formaldehyde. The formulated resin exhibited water resistance characteristics, good stability, and resistance to UV radiation. These qualities have been attributed to the covalent interaction between the resin and lignin. [74]

Among the various proposals of reuse of rice hull is the microcrystalline cellulose production. The rice hulls together with the beans and potatoes hulls were employed for performing such preparation of different compositions of these hulls. The obtained product was silica microcrystalline cellulose. The produced compound was tested in hyperlipidemic rats and the results were compared with the commercial compound orlistat. The result with the proposed compound demonstrated antilipidemic action with decreased insulin. [75]

Another mixture of cellulose fibers was used for the preparation of microcrystalline cellulose. The fibers were obtained from rice hulls and bananas peel. These were chemically treated: acid treatment - alkali or alkaline - acid and bleached with hypochlorite. The post chemical treatment product was taken to enzymatic treatment to obtain the crystalline microcellulose. The microcrystalline cellulose obtained from rice hull showed better result. [76]

There is interest in preparing cellulose nanofibers from rice hulls, these feature sizes in the range of 100-250 nm. The refining of the fibers used to prepare nanofibers was described as extremely important. [77, 78]

REFERENCES

[1] Friedman, M. "Rice Brans, Rice Bran Oils, and Rice Hulls: Composition, Food and Industrial Uses, and Bioactivities in Humans, Animals, and Cells*", J. Agricultural and Food Chem.* 61, 10626−10641 (2013).

[2] *Food and Agricultural Organization,* site: http://faostat3.fao.org/ browse/Q/QC/E, accessed 01/07/2015.

[3] U.S. Department of the Interior, U.S. Geological Survey, *Mineral Commodity Summaries* 2013, http://minerals.usgs.gov/ minerals/pubs/mcs/2013/mcs2013.pdf, accessed 12/20/2014.

[4] Kumar, P. S., Ramakrishnan, K., Kirupha, S. D. and S. Sivanesan, "Thermodynamic and kinetic studies of cadmium adsorption from aqueous solution onto rice hull," *Brazilian Journal of Chemical Engineering,* 27, 347 – 355 (2010).

[5] Rodrigues, F. A. and Joekes, I. "Cement Industry: Sustainability, Challenges and Perspectives", *Environmental Chemistry Letters,* 9, 2, 151-166 (2011).

[6] Sugita, S. "*Method of producing active rice hull ash"* US patent 5329867 A, 1994.

[7] Possamai, V., Hotza, D. D., Accordi, J., Oliveira, A. P. N. "Estudo comparativo entre sílica obtida por lixívia ácida da casca de arroz e sílica obtida por tratamento térmico da cinza de casca de arroz". *Quimica Nova,* 29, 1175-1179 (2006).

[8] Gu, S., Zhou, J., Yu, C., Luo, Z., Wang, Q. and Shi, Z. "A novel two-staged thermal synthesis method of generating nanosilica from rice hull via pre-pyrolysis combined with calcination" *Industrial Crops and Products,* 65, 1–6, (2015).

[9] Gu, S., Zhou, J., Luo, Z., Wang, Q. and Ni, M. "A detailed study of the effects of pyrolysis temperature and feedstock particle size on the preparation of nanosilica from rice hull". *Industrial Crops and Products,* 50, 540–549 (2013).

[10] Wang, W., Martin, J. C., Fan, X., Han, A., Luo, Z. and Sun, L. "Silica Nanoparticles and Frameworks from Rice Hull Biomass." *ACS Applied Materials & Interfaces, 4,* 977−981, (2012).

[11] Zhang, H., Ding, X., Wang, Z. and Zhao, X. "Consecutively preparing d-xylose, organosolv lignin, and amorphous ultrafine silica from rice hull." *Bioinorganic Chemistry and Applications,* Article ID 603481 (2014).

[12] Chen, H., Wang, W., Martin, J. C., Oliphant, A. J., Doerr, P. A., Xu, J. F., DeBorn, K. M., Chen, C. and Sun, L. "Extraction of Lignocellulose and Synthesis of Porous Silica Nanoparticles from Rice Hulls: A Comprehensive Utilization of Rice Hull Biomass." *ACS Sustainable Chemistry & Engineering,* 1, 254−259 (2013).

[13] Hsieh, Y., Du, Y., Jin, F., Zhou, Z. and Enomoto, H. "Chemical Alkaline pre-treatment of rice hulls for hydrothermal production of acetic acid." *Chemical Engineering Research and Design,* 87, 13-18 (2009).

[14] Adam, F., Seng, T. S. Chew and Andas, J. "A simple template-free sol–gel synthesis of spherical nanosilica from agricultural biomass." *Journal of Sol-Gel Science and Technology,* 59, 580–583 (2011).

[15] Noushad, M., Rahman, I. A., Sheeraz, N. Zulkifli, C., Husein, A. and Mohamad, D. "Low surface area nanosilica from an agricultural biomass for fabrication of dental nanocomposites.*" Ceramics International,* 40, 4163–4171 (2014).

[16] Zulkifli, N. S. C., Rahmann, I. A., Mohamad, D. and Husein, A. "A green sol–gel route for the synthesis of structurally controlled silica particles from rice hull for dental composite filler." *Ceramics International,* 39, 4559–4567 (2013).

[17] Lin, L., Zhai, S. R., Xiao, Z. Y., Liu, N., Song, Y., Zhai, B. and Na, Q. D. "Cooperative effect of polyethylene glycol and lignin on SiO_2 microsphere production from rice hulls.*" Bioresource Technology,* 125,172–174 (2012).

[18] Pijarn, N., Jaroenworaluck, A., Sunsaneeyametha, W. and Stevens, R. "Synthesis and characterization of nanosized-silica gels formed under controlled conditions." *Powder Technology,* 203, 462–468 (2010).

[19] Le, V. H., Thuc, C. N. H. and Thuc, H. H. "Synthesis of silica nanoparticles from Vietnamese rice hull by sol–gel method." *Nanoscale Research Letters,* 8:58 (2013).

[20] Ma, X., Zhou, B., Gao, W., Qu, Y., Wang, L., Wang, Z. and Zhu, Y. "A recyclable method for production of pure silica from rice hull ash." *Powder Technology,* 217, 497–501 (2012).

[21] Vásquez, T. G. P., Casas-Botero, A. E., Ramírez-Carmona, M. E., Torres-Taborda, M. M., Soares, C. H. L. and Hotza, D. "Biogeneration of silica nanoparticles from rice hull ash using *Fusarium Oxysporum* in two different growth media." *Industrial & Engineering Chemistry Research,* 53, 6959−6965 (2014).

[22] Pineda, T., Soares, C. H. L., Hotza, D., Casas-Botero, A. E., Ramírez-Carmona, M., Torres-Taborda, M. "Extracellular synthesis of silica oxide particles by *Fusarium oxysporum* from rice hull ash." *Materials Science Forum,* 727-728, 1153-1157 (2012).

[23] Bansal, V., Ahmad, A. and Sastry, M. "Fungus-mediated biotransformation of amorphous silica in rice hull to nanocrystalline silica." *Journal of the American Chemical Society,* 128, 14059-14066 (2006).

[24] Young, J. F. and Mindess, S., *Concrete*, Prentice-Hall, New Jersey, USA 1981

[25] J. Davidovits, "Global warming impact on the cement and aggregates industries", *World Resour. Rev.* 6 (2) (1994) 263–278.

[26] Rodrigues, F. A. and Monteiro, P. J. M. "Hydrothermal synthesis of cements from rice hull ash," *J. Mat. Sci. Lett.,* 18 (19) 1551-1552 (1999).

[27] Romano, J. S., Miranda, M. S., Oliveira, M. B. R. and Rodrigues, F. A. "Biogenic cements and encapsulation of zinc." *J. Clean Production,* 19, 1222-1228 (2011).

[28] Amorim, R. P., Miranda, M. S., Oliveira, M. B. R. and Rodrigues, F. A. "Synthesis, hydration and durability of rice hull cements doped with chromium." *J. Haz. Mat.,* 186 (1) 497-501 (2011).

[29] Cheng, Y., Lu, M., Li, J. Su, X. Pan, S., Jiao, C. and Feng, M. "Synthesis of MCM-22 zeolite using rice hull as a silica source under varying-temperature conditions", *Journal of Colloid and Interface Science* 369, 388–394 (2012).

[30] Ghasemi, Z., and Younesi, H. "Preparation and Characterization of Nanozeolite NaA from Rice Hull at Room Temperature without Organic Additives", *Journal of Nanomaterials,* Article ID 858961, (2011).

[31] Kordatos, K., Gavela, S., Ntziouni, A., Pistiolas, K.N. Kyritsi, A., Rigopoulou, V. K. "Synthesis of highly siliceous ZSM-5 zeolite using silica from rice hull ash." *Microporous and Mesoporous Materials,* 115 189–196 (2008).

[32] Loganathan, P., Vigneswaran, S., and Kandasamy, J. "Enhanced removal of nitrate from water using surface modification of adsorbents A review." *Journal of Environmental Management,* 131, 363-374 (2013).

[33] Liu, H., Chen, T., Chang, D., Chen, D. and Frost, L. R. "Catalytic cracking of tars derived from rice hull gasification over goethite and palygorskite" *Applied Clay Science* 70 51–57 (2012).

[34] Gan, P. P., Fong, S. and Li, Y. "Efficient removal of Rhodamine B using a rice hull-based silica supported iron catalyst by Fenton-like process." *Chemical Engineering Journal* 229 351–363 (2013).

[35] Elhag, A. and Adam, F. "The benzylation of benzene using aluminium, gallium and iron incorporated silica from rice hull ash", *Microporous and Mesoporous Materials* 118 35–43 (2009).

[36] Foo, K.Y. and Hameed, B.H. "Utilization of rice hull ash as novel adsorbent: A judicious recycling of the colloidal agricultural waste", *Advances in Colloid and Interface Science* 152, 39–47 (2009).

[37] Manique, M. C., Faccini, C. C., Onorevoli, B., Benvenutti, E. V. and Caramão, E. B. "Rice hull ash as an adsorbent for purifying biodiesel from waste frying oil." *Fuel,* 92 56–61 (2012).

[38] Chen, X. G., Lv, S. S., Liu, S. T., Zhang, P. P. Zhang, A. B., Sun, J. and Ye, Y. "Adsorption of Methylene Blue by Rice Hull Ash," *Separation Science and Technology,* 47, 1, (2012).

[39] Moreno, A., Figueroa, D. and Hormaza, A. "Adsorción de azul de metileno sobre cascarilla de arroz", *Produccion + Limpia,* 7, 1 (2012).

[40] Hua, L. H. and Wang, C. L. "The removal of heavy metal ions from spiked aqueous solutions using solid wastes—Comparison of sorption capability", *Journal of the Taiwan Institute of Chemical Engineers* 41, 585–590 (2010).

[41] Jeon, C. "Removal of copper ion using rice hulls, *Journal of Industrial and Engineering Chemistry.*" 17, 517–520 (2011).

[42] Wang, L. H., Lin, C. I. and Wu, F. C. "Kinetic study of adsorption of copper (II) ion from aqueous solution using rice hull ash" *Journal of the Taiwan Institute of Chemical Engineers* 41 599–605 (2010).

[43] Lin, C. I. and Wang, L. H. "Adsorption of Nickel(II) Ion from Aqueous Solution Using Rice Hull Ash", *Journal Chemical Eng. of Japan,* 44, 278-285 (2011).

[44] Imyim, A. and Prapalimrungsi, E., "Humic acids removal from water by aminopropyl functionalized rice hull ash", *Journal of Hazardous Materials* 184, 775–781 (2010).

[45] Chen, H., Wang, H., Xue, Z. Yang, L., Xiao, Y., Zheng, M., Lei, B., Liu, Y. and Sun, L. "High hydrogen storage capacity of rice hull based porous carbon", *International J. Hydrogen Energy,* 37, 18888-18894 (2012).

[46] Tortora, G. J.; Funke, B. R. and Case, A. L. Microbiologia. 8ª edição, 1ª reimpressão, Artmed, Porto Alegre (2006).

[47] John, R. P., Anisha, G. S.; Nampoothiri, K. M. and Pandey, A. "Direct lactic acid fermentation: Focus on simultaneous saccharification and lactic acid production." *Biotechnology Advances,* 27, 145-153 (2009).

[48] Hatti-Kaul, R., Tornvall, U., Gustafsson, L. and Borjesson, P. "Industrial biotechnology for the production of bio-based chemicals – A cradle-to-grave perspective." *Trends in Biotechnology,* 25, 3, 119-124 (2007).

[49] Genovese, A. L., Udaeta, M. E. M. and Galvão, L. C. R. "Aspectos energéticos da biomassa como recurso no Brasil e no mundo. In: *Encontro energia no meio rural,* 6 (2006), <http://www. proceedings.scielo.br/scielo.php?script=sci_arttext&pid= MSC0000000022006000100021&lng=en&nrm=abn>. Acess on: 28 Dec. 2014.

[50] Damartziz, T. and Zabaniotou, A. "Thermochemical conversion of biomass to second generation biofuels through integrated process design – A review." *Renewable and Sustainable Energy Reviews,* 15, 366-378 (2011).

[51] Demirbas, A. "A. Competitive liquid biofuels from biomass. *Applied Energy,* 88, 17-28 (2011).

[52] Wang, L., Zhao, B.; Liu, B., Yu, B.; Ma, C., Su, F., Hua, D., Li, Q., Ma, Y., Xu, P. "Efficient production of L-lactic acid from corncob molasses, a waste by-product in xylitol production, by a newly isolated xylose utilizing Bacillus sp. strain." *Bioresource Technology,* 101, 7908-7915 (2010).

[53] Saha,, B. C. and Cotta, M. A. "Lime pretreatment, enzymatic saccharification and fermentation of rice hulls to ethanol. *Biomass and Bioenergy."* 32, 971-977 (2008).

[54] Suxia, R., Haiyan, X., Jinling, Z., Shunqing, L., Xiaofeng, H. and Tingzhou, L. "Furfural production from rice husk using sulfuric acid and a solid acid catalyst through a two-stage process." *Carbohydrate Research,* 359, 1-6 (2012).

[55] Suxia, R., Haiyan, X., Jinling, Z., Shunqing, L., Xiaofeng, H. and Tingzhou, L.. Furfural production from rice husk using sulfuric acid and a solid acid catalyst through a two-stage process." *Carbohydrate Research,* 359, 1-6 (2012).

[56] Moiser, N.; Wyman, C., Dale, B., Elander, R., Lee, Y. Y., Holtzapple, M. and Ladisch, M. "Features of promising technologies for pretreatment of lignocellulosic biomass." *Bioresource Technology,* 96, 673-686 (2005).

[57] Sarkar, N. "Bioethanol production from agricultural wastes: An overview." *Renewable Energy,* 19-27 (2012).

[58] Kumar, P., Barrett, D. M., Delwiche, M. J., Stroeve, P. "Methods for Pretreatment of Lignocellulosic Biomass for Efficient Hydrolysis and Biofuel Production." *Industrial & Engineering Chemistry Research, 48,* 8, 3713-3729 (2009).

[59] Oggeda, T. L. and Petri, D. F. S. "Hidrólise Enzimática de Biomassa." *Química Nova,* 33, 1549 – 1558(2010).

[60] Megawati, S., Sulistyo, H. and Hidayat, M. "Kinetics of sequential reaction of hydrolysis and sugar degradation of rice husk in ethanol production: Effect of catalyst concentration." *Bioresource Technology*, 102, 2062-2067 (2010).

[61] Aguiar, R., Ramirez, J. A., Garrote, G. and Vazques, M. "Kinetic study of the acid hydrolysis of sugar cane bagasse." *Journal of Food Enginnering*, 5, 309-318 (2002).

[62] Cunha-Pereira, F., Hickert, L. R., Sehnem, N. T., Souza-Cruz P. B.; Rosa, C. A., and Ayub, M. A. "Conversion of sugars present in rice hull hydrolysates into ethanol by Spathaspora arborariae, Saccharomyces cerevisiae, and their cofermentations." *Bioresource Technology*, 102, 4218-4225 (2011).

[63] Chen, Y. "Development and application of co-culture for rthanol production by co-fermentation of glucose and xylose: a systematic revew." J. Ind, Microbiol. Biotechnol., 38, 581-597 (2011).

[64] Moon, H. C., Jeong, H. R. and Kim, D. H. "Bioethanol production from acid-pretreated rice hull." *Asia-Pacific Journal of Chemical Engineering*, 7, 206-211 (2012).

[65] Hicker, R. L., Souza-Cruz, P. B., Rosa, C. A., Ayub, M. A. Z. "Simultaneous saccharification and co-fermentation of un-detoxified rice hull hydrolysate by *Saccharomyces cerevisiae* ICV D254 and *Spathaspora arborariae* NRRL Y-48658 for the production of ethanol and xylitol." Bioresource Technology, 143, 112-116 (2013).

[66] Nichols, N. N., Hector, R. E., Saha, B. C., Frazer, S. E. and Kennedy, G. J. "Biological abatement of inhibitors in rice hull hydrolyzate and fermentation to ethanol using conventional and engineered microbes." *Biomass and Bioenergy*, 67, 79-88 (2014).

[67] Rambo, M. D. *"Aproveitamento da casca de arroz para produção de xilitol e sílica xerogel."* Dissertação (Mestrado em Química) - Universidade Federal de Santa Maria, (2009).

[68] Gullón, P., Moura, P., Estevea, M. P., Girio, F. M., Dominguez, H. and Parajó, J. C. "Assessment on the fermentability of xylooligosaccharides from rice husks by probiotic bacteria." *Journal of Agricultural and Food Chemistry*, 56, 7482-7487 (2008).

[69] Foo K.Y. and Hameed B.H "The environmental applications of activated carbon/zeolite composite materials." *Advances in Colloid and Interface Science*, 162, 22–28 (2011).

[70] Turmanova, S. Genieva, S. and Vlaev, L. "Obtaining Some Polymer Composites Filled with Rice Husks Ash-A Review." *International Journal of Chemistry*, 4, 4 (20120.

[71] Aggarwal1, P. K., Chauhan1, S., Raghu1, N., Karmarkar, S. and Shashidhar, G. M. "Mechanical properties of bio-fibers-reinforced high-density polyethylene composites: effect of coupling agents and bio-fillers." *Journal of Reinforced Plastics and Composites* 32, 22, 1722–1732 (2013).

[72] Khalil, H. P. S A., Tehrani, M. A., Davoudpour, Y., Bhat, A. H., Jawaid, M. M. and Hassan, A. "Natural fiber reinforced poly(vinyl chloride) composites: A review." *Journal of Reinforced Plastics and Composites* 32, 330–356 (2013).

[73] Petchwattana, N., Covavisaruch, S. and Chanakul, S. "Mechanical properties, thermal degradation and natural weathering of high density polyethylene/rice hull composites compatibilized with maleic anhydride grafted polyethylene." *Polym Reseasrch*, 19, 9921-9925 (2012).

[74] Ma, Y., Zhao, X., Chen, X. and Wang, Z. "An approach to improve the application of acid-insoluble lignin from rice hull in phenol–formaldehyde resin." *Colloids and Surfaces A: Physicochem, Eng. Aspects*, 377, 284–289 (2011).

[75] Adela, A. M. and El-shinnawy, N. A. "Hypolipidemic applications of microcrystalline cellulose composite synthesized from different agricultural residues." *International Journal of Biological Macromolecules,* 51, 1091– 1102 (2012).

[76] Ibrahim, M. M., El-Zawawy, W. K., Juttke, Y., Koschella, A. and Heinze, T. " Cellulose and microcrystalline cellulose from rice straw and banana plant waste: preparation and characterization." *Cellulose,* 20, 2403–2416 (2013).

[77] Brinchia, L., Cotanaa, F., Fortunatib and Kenny, E. M. "Production of nanocrystalline cellulose from lignocellulosic biomass: Technology and applications." *Carbohydrate Polymers,* 94, 154– 169 (2013).

[78] Khalil, H.P.S. A., Davoudpoura, Y., Islama, M. N., Mustaphaa, A., Sudeshd, K., Dungania, R. and Jawaidb, M. "Production and modification of nanofibrillated cellulose using various mechanical processes: A review." *Carbohydrate Polymers,* 99, 649– 665 (2014).

In: Agricultural Wastes
Editor: Camille N. Foster
ISBN: 978-1-63482-359-3

Chapter 6

THE USE OF AGRICULTURAL RESIDUES: A TECHNICAL AND SOCIOECONOMIC CHALLENGE FOR THE BIOREFINERY

Araceli García*[1,2,*]*, Cristina Sánchez*[1]*, Itziar Egüés*[1]*,
***Luis Serrano*[1] *and Jalel Labidi*[1]**
[1]Department of Chemical and Environmental Engineering, University of the Basque Country UPV-EHU, Donostia-San Sebastián, Spain
[2]Laboratoire de Génie des Procédés Papetiers (LGP2), UMR CNRS 5518, Grenoble INP-Pagora – 461, Saint-Martin-d'Hères, France

ABSTRACT

Biorefining has been defined by the International Energy Agency as the sustainable processing of biomass into a spectrum of marketable products and energy. The concept of sustainability, defined by the World Commission on Environment and Development, arose as consequence, among other reasons, of the energy crisis derived from the imminent depletion of the fossil resources. Thus, a global mindset change is required to face the present scenario, through the reconcilement of three important pillars: economical, societal and environmental issues. Since the current energy system results unsustainable because of imbalance concerns that will have environmental, economical and geopolitical implications far into the future, the sustainable development should be achieved by learning how to use/reuse our resources.

In this sense, the use of biomass as a source of products and energy is not new, but its use under a sustainable perspective may imply interesting novelties. With this aim, the Biorefinery outlook should be constructed fulfilling some requirements such as the responsible and optimal exploitation of resources, the application of energy efficient processes and the accessibility of the resulting energy and products, i.e., a compliance in terms of a viable, bearable and equitable development.

The available source of biomass (the biorefinery feedstock) determines not only the range of products obtained in the Biorefinery, but also the more or less specific technology and the optimal conversion pathway required for its transformation. These

[*] Corresponding author: araceli.garcia@ehu.es.

three parameters (feedstock, technology and conversion pathway) allow classifying the biorefining processes, and their combination offers a huge range of possibilities for the biomass exploitation.

The use of agricultural wastes in a biorefinery concept offers a promising perspective of sustainable development. The agricultural activity generates significant quantities of lignocellulosic residues (over 60% of the total crop) that are usually left on the cropland or incinerated to prevent the spread of pests and uncontrolled fires. Against the high availability of this biorefinery feedstock, some other issues appear concerning the use of agricultural residues as bioproducts source: volume variability, crop seasonality, low density, heterogeneous chemical composition, localized generation … These factors are negatively considered when agricultural wastes are proposed as biorefinery feedstock.

In the present work, several crop residues (woody and non-woody wastes) were chemically characterized for determining their contents of the main lignocellulosic components (cellulose, hemicelluloses and lignin). Other biomass components, such as moisture and ash, were also determined. In addition, hot water and weak soda solubilities were measured in order to establish the treatability of these agricultural wastes in a biorefinery concept. According to the results, and after an exhaustive crop production assessment, some biorefinery scenarios were proposed considering different worldwide agricultural productions.

The Agricultural Residues

Properly speaking, the term "agricultural waste" describes the fraction of the crop that does not constitute the harvest itself. The residues derived from the forests maintaining and from the industrial processing of both agricultural and forestry resources should be also considered under this classification. These materials show a diverse composition (moisture, minerals, carbohydrates and lignin contents) depending on the crop source, growing and harvesting conditions.

Currently, the main uses of the agricultural and forestry residues are low value-added, mainly meeting needs that concern farming activities (bed and feed for livestock), soil fertilization and compensation (composting) or energetic requirements (pellets for combustion). These uses do not cover the real potential of this feedstock from technologic and profitability points of view. Considering this fact, these lignocellulosic residues should be considered optimal sustainable and renewable source of products and fuels within the Biorefinery concept [1] due its relatively high availability and potentiality.

Figure 1 displays the worldwide production of total cereals, fruits, vegetables or oil crops and their distribution over different areas. As noted, the agricultural activities are globally common and could supply enough renewable feedstock to biorefinery mills of different size and production capacities, being able to limit the environmental impact of these activities. This issue is clearly dependent of the specific residue volume generated by each particular crop [2]–[4] and of the harvesting seasonality, both factors could lead to a discontinuous or insufficient fed operation in a biorefining plant. Lal (1995) [5] estimated that 400×10^6 Mg of crop residue was produced annually during the early 1990s by 18 crops comprising of cereals, legumes and oil crops grown on 95 Mha of cropland. Most of grain production crops generate huge amounts of residues after their harvesting (in a variable residue/grain ratio of 1 – 3) [2]–[4], [6]. The production of most of vegetables, tubers and roots leads in low wastes (of approx. 0.25 of residue/product), whereas other crops as beans, peas and different oil crops (sunflower or linseed) generate almost the same amount of product and waste per harvested area.

In this sense, under baseline assumptions and considering the future necessity of increase the agricultural production to meet the needs of world population, in the coming years large amounts of lignocellulosic wastes will be generated. While it is true that cropland residues serve as soil remediation and auto-fertilization of farmland, most of the agricultural residues do not represent a profitable management, and they are burned, with consequential environmental problems. Indeed, due to the potential excess of crop residues and processing wastes, there should be other interesting and profitable applications for them. This amount of agricultural residue that can potentially be removed from agricultural cropland [7] could represent a potential and profitable source for the world development and to avoid greenhouse emissions.

Chemical Composition of the Lignocellulosic Biomass

Lignocellulose is the term used to describe the three-dimensional polymeric composites formed by plants as structural material. It consists of variable amounts of cellulose, hemicellulose and lignin [8] besides other minor compounds (**Figure 2**). The high heterogeneity of this feedstock mainly depends on its origin (**Table 1**) but also on other less manageable factors related to growing, harvesting and storage conditions.

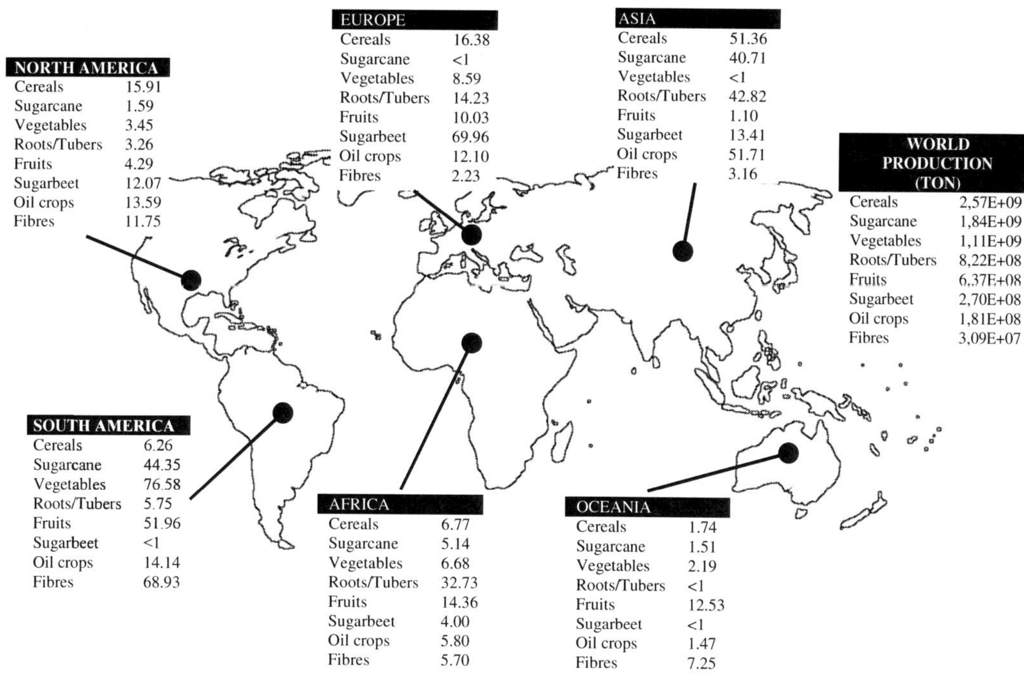

Figure 1. World production (Ton) of some crops and their distribution (% of total production) in different areas (data source FAO, 2012).

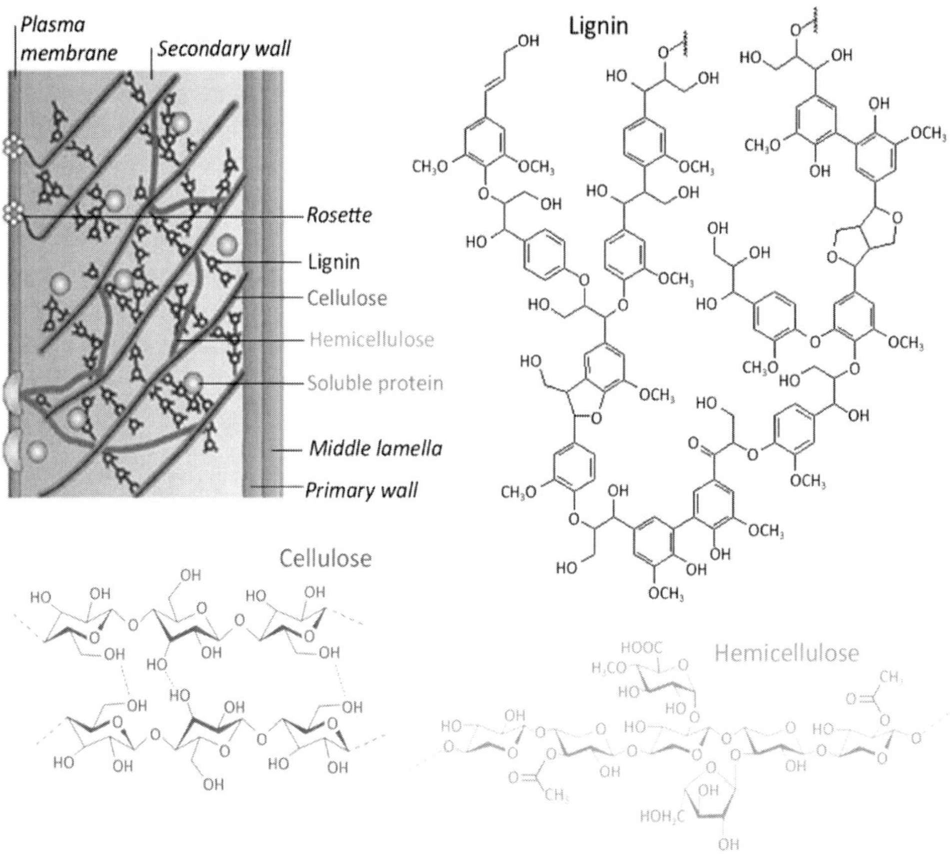

Figure 2. Structure and main constituents of lignocellulosic biomass (adapted from [9]).

Carbohydrates

The carbohydrate constituents of the lignocellulosic materials are cellulose and hemicelluloses. Other forms of carbohydrates present in biomass are starch, pectin and chitin.

The cellulose constitutes the main support of the plant cell wall and provides strength to the lignocellulosic structure. It consists on long linear straight chains of glucose units, bound together by β-1,4-glycoside linkages, leading to highly polymerized crystalline fibres packed so tightly that not only enzymes, even water cannot enter the complex framework [9].

Hemicelluloses are an amorphous and heterogeneous group of branched polysaccharides constituted by pentoses (xylose and arabinose), hexoses (mannose, glucose and galactose) and sugar acids (glucuronic and galacturonic acids) which act as linkage between cellulose and lignin. Xylans are the most abundant hemicelluloses, consisting in homopolymeric backbone chains of 1,4-linked xylose units from that ramify secondary heterogeneous chains formed by different monomeric sugars and acids. The degree of polymerization of xylans (150-200) and its composition differs significantly between different lignocellulosic feedstocks.

Lignin

Lignin is a highly complex hydrophobic, cross-linked aromatic polymer synthesized by plants from three major phenylpropane units (syringyl, guaiacyl and p-hydroxyphenylpropane) linked together by ether and carbon-carbon bonds. It is tremendously complex and condensed structure together with high functionality and molecular weight, make almost impossible to certainly know how lignin is composed. Its main function in the cell wall is to achieve structural rigidity, holding the polysaccharides fibres together. It also serves as barrier against moisture, insects and diseases, protecting the more important carbohydrate structure of the plant.

Minor organic and inorganic components

Besides the structural constituents described above, the lignocellulosic biomass also contents minor organic and inorganic components. The content of these compounds can significantly vary among the lignocellulose source due genetic and environmental factors, as well as physiological and morphological differences between crops [9], [10]. The minor organic oligo-constituents, usually so-called extractives, are formed by small size carbohydrates, proteins, hydrocarbons, oils, aromatics, lipids, fats, starches, phenols, waxes... synthesized by the plant itself with specific purposes (elasticity or permeability supplying, antifungal or disease against protection...).

The inorganic minerals (silicates, carbonates, sulphates, sulphites, nitrates...) are low molecular weight and simple structure compounds that can be found in soils. These components are absorbed by plants and fixed in their cell wall. The amount of inorganic matter varies not only among lignocellulose specimens (from 0.1 to 46%) but also within different parts of the plant, being usually more abundant in bark.

The Heterogeneous Composition of the Agricultural Wastes

The main drawback in the use of agricultural wastes in biorefineries lies in their heterogeneous chemical composition. An evidence of this is shown in Table 2, that displays the chemical composition of 30 different agricultural residues, among which different crop stems (pepper [11], okra and bean [12]), stalks (corn [13], cotton [14] and grapevine [15]), trimmings and pruning wastes (apple and olive trees), leaves, grasses and other crop subproducts (hesperaloe [16], corn cobs [17], palm oil empty fruit bunch [18]) are included.

Table 1. Main composition (cellulose, hemicelluloses and lignin in % of raw material, w/w dry basis of some lignocellulosic feedstocks (from [9], [10])

Feedstock	Cellulose	Hemicelluloses	Lignin
Barley straw	48.6	29.7	21.7
Coconut coirs	52.2	28.4	19.4
Corn cob	48.1	37.2	14.7
Corn stover	47.4	30.3	22.3
Cotton	87.5	17.1	0.0
Cotton stalk	66.2	18.4	15.4
Elephant grass	31.5	34.3	34.2
Eucalyptus	52.7	15.4	31.9
Flax fibres	75.9	20.7	3.4

Table 1. (Continued)

Feedstock	Cellulose	Hemicelluloses	Lignin
Flax shives	39.9	26.8	24.2
Legume straw	29.2	35.5	35.3
Newspaper	45.6	31.3	23.1
Oilseed rape	27.3	20.5	14.2
Olive husks	25.0	24.6	50.4
Pine	48.1	23.5	28.4
Rice straw	52.3	32.8	14.9
Sugarcane bagasse	47.4	29.1	23.5
Sunflower shells	56.5	28.0	15.5
Switchgrass	48.7	38.4	12.9
Tea residue	33.3	23.2	43.5
Tobacco stalks	44.6	30.2	25.2
Wheat straw	44.5	33.2	22.3
Wood bark	25.2	30.3	44.5

Table 2. Chemical composition (in % wt dry basis) of some agricultural residues

Raw Material	Standard/Method						
	Inorganics	**Hot water solubles**	**1% NaOH solubles**	**Extractives**	**Lignin**	**Cellulose**	**Hemicelluloses**
	TAPPI T211 om-93	***TAPPI T207 om-93***	***TAPPI T212 om-98***	***TAPPI T204 om-97***	***TAPPI T222 om-98***	***From* [19], [20]**	
 Almond shell	0.9	Nd	Nd	0.1	52.6	41.3	8.6
 Aloe Vera	19.9	Nd	Nd	12.7	4.6	23.8	21.1
 Apple tree pruning	1.2	Nd	Nd	10.7	26.2	27.3	31.1
 Asparagus stalk	8.0	9.4	23.3	4.5	22.9	44.7	37.8

Raw Material	Standard/Method						
	Inorganics	**Hot water solubles**	**1% NaOH solubles**	**Extractives**	**Lignin**	**Cellulose**	**Hemicelluloses**
	TAPPI T211 om-93	***TAPPI T207 om-93***	***TAPPI T212 om-98***	***TAPPI T204 om-97***	***TAPPI T222 om-98***	***From* [19], [20]**	
 Chili stalks	7.9	8.3	24.8	2.2	28.9	62.5	22.4
 Corn cobs	1.0	5.1	30.0	2.4	23.1	36.8	30.0
 Corn stalks	9.3	18.0	49.2	1.1	17.2	49.2	25.6
 Cotton stalks	6.0	10.5	24.0	1.3	23.0	51.2	23.7
 Datepalm trunk (degla)	8.0	12.5	17.5	23.5	13.8	23.9	19.0
 Date palm leafs (degla)	16.3	24.0	36.6	16.3	28.7	27.5	17.8
(ftimi)	14.7	26.6	31.4	18.5	34.9	27.3	18.6
(kinta)	12.9	18.9	36.8	18.5	18.8	26.7	17.9
 Date palm stems (degla)	5.0	10.7	23.3	11.5	21.0	31.2	27.3
(ftimi)	10.6	18.1	30.8	13.7	16.3	31.2	24.1
(kinta)	8.2	17.4	31.2	16.9	19.5	29.7	26.6

Table 2. (Continued)

Raw Material	Standard/Method						
	Inorganics	**Hot water solubles**	**1% NaOH solubles**	**Extractives**	**Lignin**	**Cellulose**	**Hemicelluloses**
	TAPPI T211 om-93	***TAPPI T207 om-93***	***TAPPI T212 om-98***	***TAPPI T204 om-97***	***TAPPI T222 om-98***	***From* [19], [20]**	
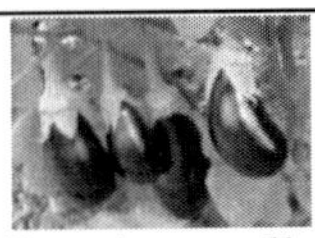 Eggplant stalks	6.8	12.4	27.4	3.1	22.2	42.3	27.8
 Faba bean stalks	7.3	29.7	52.9	10.7	16.6	31.1	25.6
 Field pea stems	12.6	36.5	47.0	13.4	25.6	29.7	38.2
 Grapevine stalks	3.9	15.4	33.0	5.4	34.0	52.0	25.3
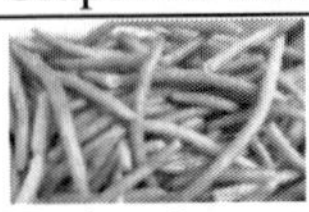 Haricot bean stalks	10.2	Nd	Nd	5.2	11.0	33.8	21.6
 Hesperaloe	5.9	13.5	29.5	4.2	7.9	52.3	21.8
 Okra	3.2	14.4	25.4	2.9	17.7	37.7	30.7
 Olive tree pruning	3.1	23.4	34.4	9.5	21.0	35.4	25.1

Raw Material	Standard/Method						
	Inorganics	**Hot water solubles**	**1% NaOH solubles**	**Extractives**	**Lignin**	**Cellulose**	**Hemicelluloses**
	TAPPI T211 om-93	***TAPPI T207 om-93***	***TAPPI T212 om-98***	***TAPPI T204 om-97***	***TAPPI T222 om-98***	***From [19], [20]***	
 Palm oil EFB	1.7	5.7	30.8	2.8	23.8	38.0	31.0
 Pepper stems	6.0	12.9	20.2	4.1	25.6	41.3	23.3
 Sunflower (head)	7.9	Nd	Nd	10.5	10.5	26.4	11.5
(stalks)	9.4	Nd	Nd	13.8	13.8	29.4	19.5
 Tomato stems	17.5	Nd	Nd	7.8	11.0	38.5	14.0

Nd: not determined.

Generally, the variability in the amount of the main lignocellulosic components (cellulose, hemicelluloses and lignin) should indicate the primary potential application of each agricultural residue. Lignocellulosic wastes with high cellulose content (stalks of pea 38.2% or asparagus 37.8%) could be suitable feedstocks for bioethanol production. However, the biorefinery processes implies different pretreatment steps, after that cellulose-rich solid fractions can be obtained for this purpose [12]. This means that other agricultural residues that contain large amount of hemicelluloses (chilli stalks 62.5%, hesperaloe 52.3%, grapevine 52%, cotton 51.2% and corn stalks 49.2%) should represent an interesting source of carbohydrates-based platform chemicals through fermentation pathways [15]. Likewise, the pretreatment of highly lignin-containing agricultural wastes (almond shell 52.6%) should offer a renewable source for the obtaining of different aromatic chemicals and additives.

However, the key point in the use of agricultural residues for value-added applications lies not only in the individual chemical composition of each waste, but also in the possible challenge of use different lignocellulosic residues together with specific biorefinery technologies, and following the same processing pathways. In this sense, regardless of the lignocellulosic components containing (cellulose, hemicelluloses and lignin) other components of the lignocellulosic biomass (inorganics and extractives) are the drawback that establish how easily treated/transformed will be these agricultural residues into products, and therefore also the investments and cost associated to their management/processing.

The inorganic matter and extractives contained in the lignocellulosic wastes could lead to important problems during their processing, because acidification, foaming and corrosion mechanisms. For this reason, even though two agricultural wastes could be processed together because of similar lignocellulosic components, e.g., corn cobs and tomato stems due to similar cellulose content (36.8% and 38.5%, respectively), they should be separately pretreated to avoid problems derived from inorganic load into the biorefining system (corn cobs 1% and tomato stems 17.5% of inorganics). Analogously, those agricultural wastes with high extractives contents (pea stalks 13.4%, sunflower stalks 13.8% or date palm wastes 11.5 – 23.5%) require of special pretreatment steps (using high temperature hydrolysis or extraction with organic solvents) in order to avoid fouling and foaming problems.

On the other hand, similar biomass treatability ranges should lead to advantages in the use of different agricultural wastes. The hot water and NaOH solubilities of a raw material determine the severity of the fractionation processes in a biorefinery [12] and therefore the cost associated to its transformation into added value products. As shown in **Figure 3**, these two parameters present equal behaviours, allowing to identify those agricultural residues that should be more easily treated (e.g., pea and bean stalks) and which present more recalcitrance, and consequently should require of more energy/chemicals for its suitable fractionation (date palm trunk, pepper stems).

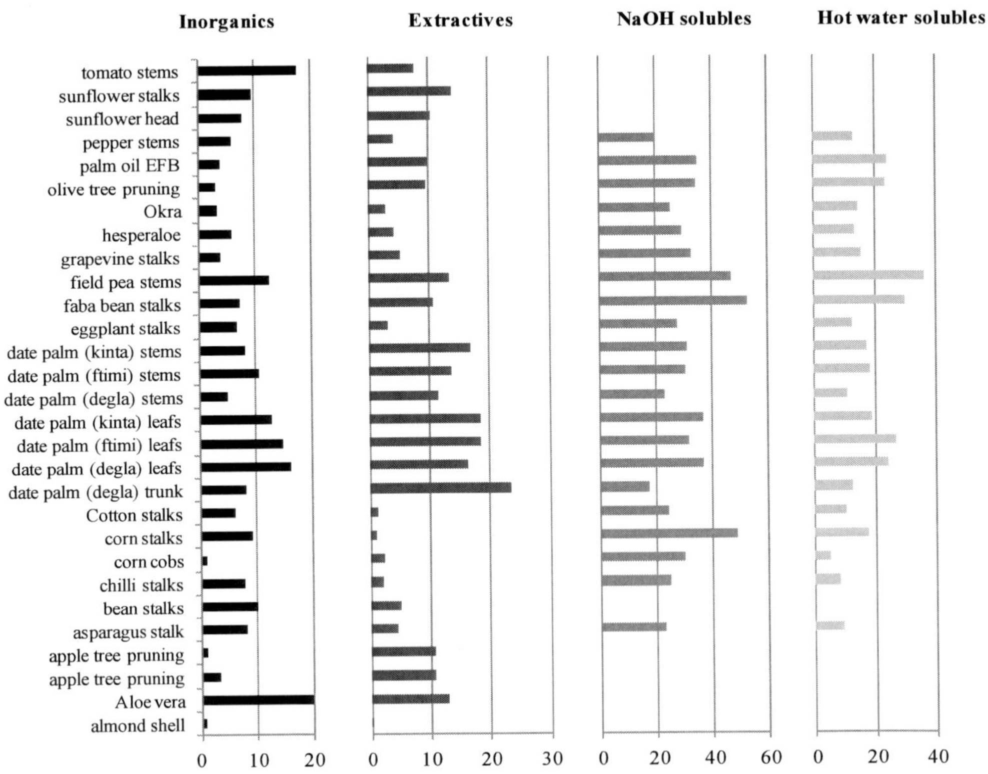

Figure 3. Compositional features determining the treatability of agricultural residues.

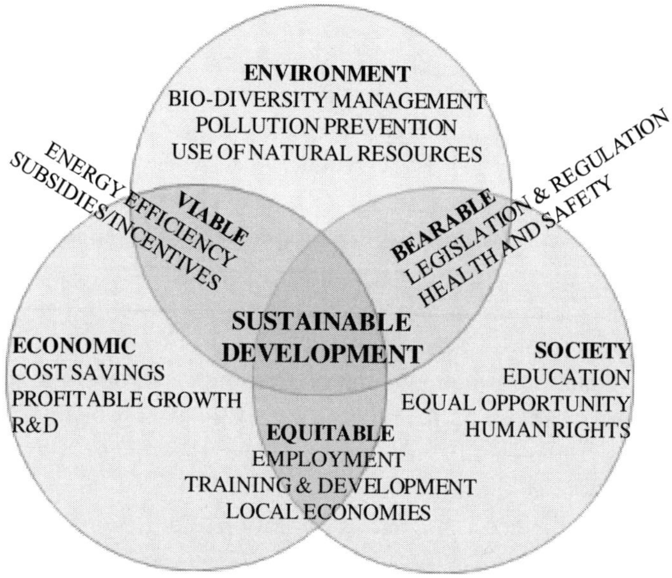

Figure 4. The three aspects on which the sustainable development is based.

THE BIOREFINERY CONCEPT: NEED FOR A CHANGE

Biorefining has been defined by the International Energy Agency as the sustainable processing of biomass into a spectrum of marketable products and energy [21], [22]. The concept of sustainability, defined by the World Commission on Environment and Development [23], arose as consequence, among other reasons, of the energy crisis derived from the imminent depletion of the fossil resources. The massive dependency of these feedstocks, on which our society has been built, is leading to plenty of important implications at economic, social and environmental scale. Thus, a global mindset change is required to face the present scenario, through the reconcilement of these three important pillars (**Figure 4**).

The current energy system results unsustainable because of imbalance concerns that will have environmental, economical and geopolitical implications far into the future [24]. In order to counteract the effects derived from this unsustainable global growth, the energy obtained from naturally renewed resources (sunlight, wind, tides, biomass and geothermal heat) should be exploited in an appropriate, efficient and profitable way. An energetic sustainable development should consider the improvement of the current technologies in order to maximize the raw materials usage yields, as well as to reconsider the commodities consumption and the minimizing of the waste generation. To sum up, the sustainable development should be achieved by learning how to use our resources.

In this sense, the use of biomass as a source of products and energy is not new, but its use under a sustainable perspective may imply interesting novelties [25]. With this aim, the Biorefinery outlook should be constructed fulfilling some requirements such as the responsible and optimal exploitation of resources, the application of energy efficient processes and the accessibility of the resulting energy and products, i.e., a compliance in terms of a viable, bearable and equitable development.

The Replacement of Oil Refineries

The fossil resources (petroleum, coal and natural gas) exploitation technology, the traditional oil refinery, is being questioned from the ecological and economic points of view due to the huge amounts of resources that consumes and the large waste flows that produces [26]. In this sense, the Biorefinery concept represents the path towards the integrated systems technologies in which sustainable resources are used [22]. Analogously to the common petroleum-based refineries, the Biorefinery aims to diversify a feedstock into several profitable products, but starting on the renewability, sustainability and availability that distinguish the biomass from oil.

Regarding the exploitation of biomass and oil, two major differences can be found. The main difference between biomass and petroleum feedstock lies on its composition, in terms of heterogeneity and oxygen content [10]. Biomass contains from 10 to 45 % of oxygen, while petroleum has essentially none, making the chemical properties of biomass derived products very different from those obtained from petroleum. Furthermore, referring to the composition homogeneity, fossil resources present the advantage of being more or less fixed hydrocarbons mixtures, whereas biomass shows a very complex structure formed by different constituents. The second difference, highly dependent on the first one, concerns to the required transformation and products obtaining processes (**Figure 5**).

Initially, the biomass processing technology is similar to the petroleum one, since it must be first fractionated, i.e., separated in its main constituent components. However, as commented above, the biomass presents high heterogeneous composition, entailing more complex and consecutive separation and transformation stages than petroleum for its proper exploitation. The existing petroleum-based infrastructure and the optimized large-scale facilities [27] make it difficult for the biomass to compete with these well established products from an economic point of view.

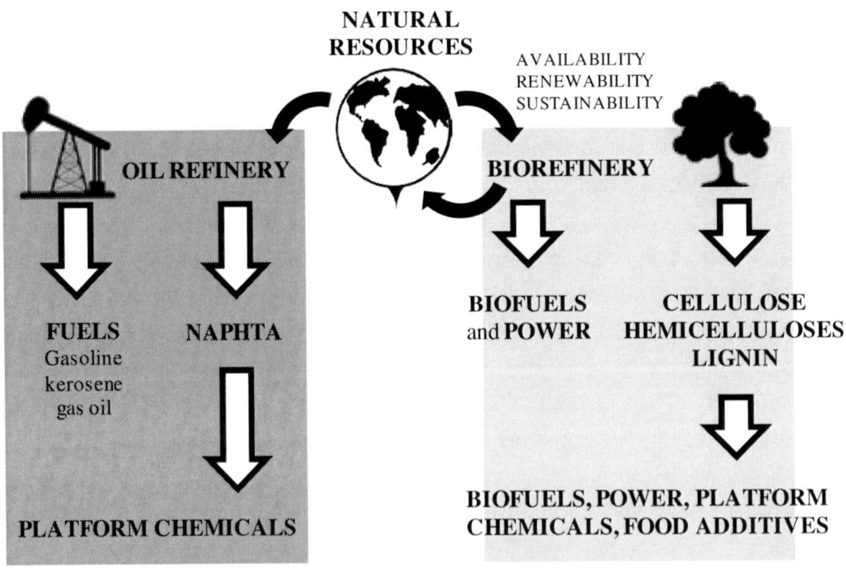

Figure 5. The use of natural resources in oil and biomass-based refineries.

In this sense, for the substitution of petrochemical products by the new bioproducts, the Biorefinery technology should focus on the development of bio-sourced intermediate chemicals with an indirect replacement approach, that is, on generating products that would perform similar functions as the existing chemicals at a lower cost, or products with specific properties different from the oil-based products ones.

The Biorefinery Perspective

Ideally, the Biorefinery concept should integrate biomass conversion processes to produce a range of fuels, power, materials and chemicals from biomass [27] offering competitive performance to traditional petrochemical refineries [28], with significant technological and environmental advantages compared to the petrochemical processes. For example, all the energy requirements of the several biomass conversion processes should be internally supplied by the production of heat and electricity from combustion of residues within a properly designed set of processes and technologies [22]. Furthermore, the reduction and reutilization of production wastes could be linked, not only to one transformation process, but also to the arrangement between several plants with. Optimization of the raw materials-residues-energy flow between different industrial production units would result in a generalization of the Biorefinery concept through an actual "Industrial Metabolism" [29].

Most of the existing biofuels and biochemicals are currently produced in single production chains apart from the Biorefinery concept [22], which entails the diversification of raw materials, processes and products. Considerably more investment in research, development, demonstration and deployment are needed to ensure that the biomass exploitation can be undertaken sustainably [24] and competitive against the oil-refineries.

In the Lignocellulosic Biorefinery the transformation into products/energy from non-food materials (agricultural and forest residues, energy crops) is carried out. The feedstock in this category is mainly integrated by the residues generated from agricultural-forestry activities (straw, shrubs, pruning, and trimmings) and thereof derived industries. The Second generation Biorefinery is designed on the basis of non-food crops, including lignocellulosic material such as agricultural, forestry and industry residues, an unutilized biomass that would not necessarily influence the land use [9]. Moreover, the related biomass processing allows a diversification of products, obtaining not only biofuels but platform chemicals, polymeric precursors and food additives. At present, the proper implementation of the second generation Biorefineries is not cost-effective because of technical barriers related to the suitable and efficient use of all the components of the lignocellulosic material [26].

The development of the Biorefinery processes that may fulfil all the above mentioned features seems to be a complex challenge. The drawbacks derived from the crops seasonality, the heterogeneity of lignocellulose composition, the difficulties due the integration of several technologies... make the implementation of the Biorefinery technologies an arduous task. Moreover, the social and political arrangement that involves the promotion and development of local economies through the biomass exploitation also present some difficulty.

More efforts should be done in the development of the Lignocellulosic Biorefinery, taking advantage of the wide feedstocks availability that offer the agricultural, forestry and industrial crops residues. The processing of this type of biomass could fully supply a sustainable platform of chemicals and also contribute in the replacement of fossil fuels requirements, together with other sources of renewable energy, by providing of some environmentally more accepted biofuels.

The Challenge of the Agricultural Wastes Biorefining

Benefits and Drawbacks of Agricultural Wastes as Biorefinery Feedstock

So far, the benefits from the use of agricultural residues for the production of value-added products result evident (Figure 6). First of all, the sustainable and renewable nature of these feedstocks implies an environmental advantage against the use of other natural resources.

Agricultural activities are worldwide widespread, so the development of local economies based on this valueless feedstock could represent a clear positive milestone. Much of the rural population is employed directly in agriculture, as smallholders or farm labourers. Their income can be strengthened by exploiting this type of waste, improving access to services, including education and health, from less developed areas.

However there are some economic drawbacks related to the development of biorefineries of agricultural wastes. Even though this feedstock is worldwide available, the periodicity or seasonality in the harvesting of most crops could lead to their batch-processing into marketable products. This is the reason why nowadays only high capacity centralized mills for the processing of highly yielding crops are being projected (based on intensive crops such as cereals, oil palm or sugarcane). Also the physicochemical properties of the agricultural residues (heterogeneous composition and low density) seem to set aside the utopia of the processing of different wastes from various crops and from delocalized farms. Solving of the last drawback should lead to larger investments, not only for management and transportation of the agricultural residues to the processing mills, but for the equipments and technology required for their suitable transformation.

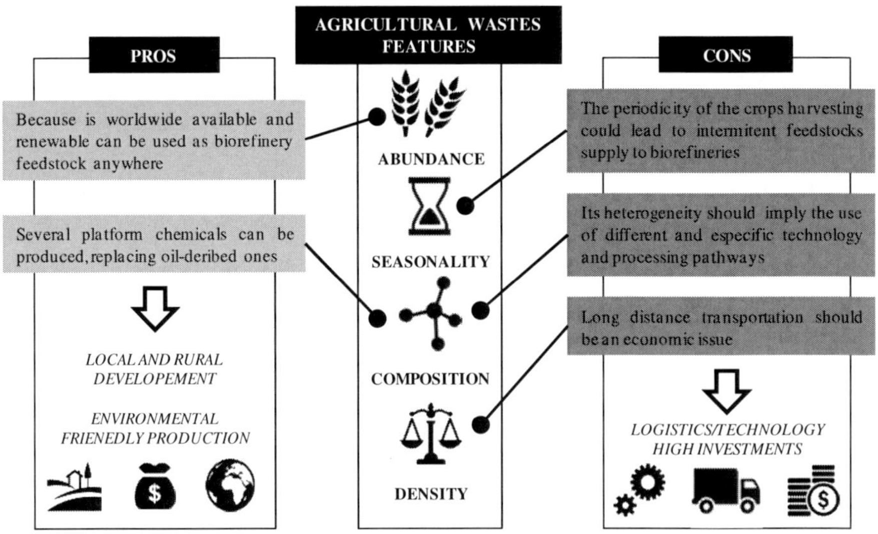

Figure 6. Pros and cons of the use of agricultural wastes in Biorefineries.

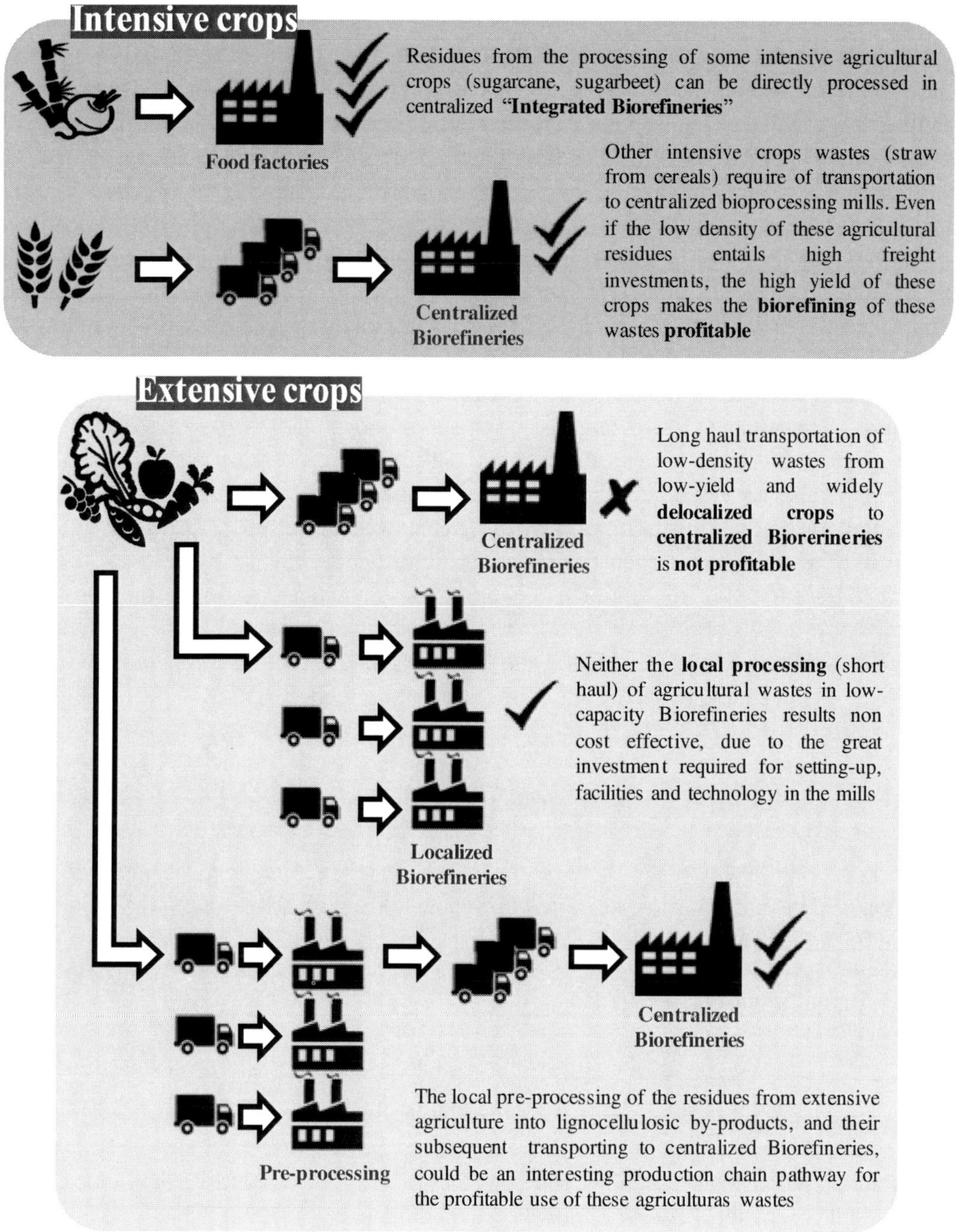

Figure 7. Biorefinery scenarios for intensive and extensive crops.

The Agricultural Wastes – Biorefinery Scenario

Based on the discussed above, hypothetic scenarios for the biorefining of agricultural wastes are proposed (Figure 7).

The most profitable use of agricultural residues implies the exploitation of intensive crops (cereals, oil and sugar crops), which have a high yield and require of extensive cultivation areas. The large generation of these wastes ensures the suitable supply of feedstocks for the biorefineries, which can result in high capacity and productive mills. An interesting chance is the integration of food factories and biorefineries, where the residues generated from the processing of crops into primary products, can be biorefined in-situ for the obtaining of other high value sub-products.

But the real challenge in the use of agricultural wastes resides in the integral use of any type of lignocellulosic residues. This subject involves the suitable exploitation of extensive crops, quite delocalized and with lower residue/crop ratios than intensive ones, but that should lead to the socioeconomic development or rural areas.

Obviously, the use of this kind of agricultural residues in high-capacity Biorefineries should not be profitable due to the costly transportation of large amounts of low density residues from local areas to centralized mills. The in-situ processing of these agricultural wastes should also not result economically feasible, because of low capacity processing mills would entail large investments in equipments, manpower and utilities. However, local pre-processing of residues from extensive agriculture into lignocellulosic by-products, and their subsequent transportation to centralized Biorefineries, could be an interesting production chain pathway for the profitable use of these agricultural wastes. In this way, local/rural development could be ensured as well as the proper use of a residue that otherwise would not provide benefits. It should be called ambitious targets for "modest" gains.

Acknowledgments

Authors would like to thank the Department of Education, Universities and Investigation of the Basque Government (Postdoctoral Development Program fellowship of Ms. García and project IT672-13), the University of the Basque Country (Postdoctoral fellowships of Ms. Egüés, Ref. ESPDOC 14/3) and Spanish Ministry of Economy and Competitiveness (CTQ2013-4126-R) for financially supporting this work.

References

[1] Mabee, WE; McFarlane, PN; Saddler, JN. "Biomass availability for lignocellulosic ethanol production," *Biomass Bioenergy*, vol. 35, no. 11, 4519–4529, Nov. 2011.

[2] Kim, S; Dale, BE. "Global potential bioethanol production from wasted crops and crop residues," *Biomass Bioenergy*, vol. 26, no. 4, 361–375, Apr. 2004.

[3] Lal, R. "World crop residues production and implications of its use as a biofuel," *Environ. Int.*, vol. 31, no. 4, 575–584, May 2005.

[4] Unger, PW. *Managing Agricultural Residues*, 1 edition. Boca Raton: CRC Press, 1994.

[5] Lal, R. "The Role of Residues Management in Sustainable Agricultural Systems," *J. Sustain. Agric.*, vol. 5, no. 4, 51–78, Jun. 1995.

[6] Stout, BA. *Handbook of energy for world agriculture*. London, GB: Elsevier Applied Science, 1990.

[7] Wolf, B; Snyder, GH. "Sustainable soils: the place of organic matter in sustaining soils and their productivity.," p. xx + 352 pp., 2003.

[8] Speight, JG. "Chapter 1 - Feedstocks," in *The Refinery of the Future*, J. G. Speight, Ed. Boston: William andrew Publishing, 2011, 1–37.

[9] Menon, V., Rao, M., "Trends in bioconversion of lignocellulose: Biofuels, platform chemicals & biorefinery concept," *Prog. Energy Combust. Sci.*, vol. 38, no. 4, 522–550, Aug. 2012.

[10] Vassilev, SV; Baxter, D; andersen, LK; Vassileva, CG; Morgan, TJ. "An overview of the organic, and inorganic phase composition of biomass," *Fuel*, vol. 94, 1–33, Apr. 2012.

[11] Serrano, L; Marín, F; Gonzalo, A; Labidi, J. "Integral use of pepper stems," *Ind. Crops Prod.*, vol. 40, 110–115, Nov. 2012.

[12] García, A; González Alriols, M; Labidi, J. "Evaluation of different lignocellulosic raw materials as potential alternative feedstocks in biorefinery processes," *Ind. Crops Prod.*, vol. 53, 102–110, Feb. 2014.

[13] Egüés, I; Sanchez, C; Mondragon, I; Labidi, J. "Antioxidant activity of phenolic compounds obtained by autohydrolysis of corn residues," *Ind. Crops Prod.*, vol. 36, no. 1, 164–171, Mar. 2012.

[14] Egüés, I; Eceiza, A; Labidi, J. "Effect of different hemicelluloses characteristics on film forming properties," *Ind. Crops Prod.*, vol. 47, 331–338, May 2013.

[15] Egüés, I; Serrano, L; Amendola, D; De Faveri, DM; Labidi, JSpigno, G. "Fermentable sugars recovery from grape stalks for bioethanol production," *Renew. Energy*, vol. 60, 553–558, Dec. 2013.

[16] Sánchez, R; Ferrer, A; Serrano, L; Toledano, A; Labidi, J; Rodríguez, A. "Hesperaloe funifera as a raw material for integral utilization of its components.," *BioResources*, vol. 6, no. 1, 3–21, 2011.

[17] Sánchez, C; Serrano, L; Andres, MA; Labidi, J. "Furfural production from corn cobs autohydrolysis liquors by microwave technology," *Ind. Crops Prod.*, vol. 42, 513–519, Mar. 2013.

[18] Alriols, MG; Tejado, A; Blanco, M; Mondragon, I; Labidi, J. "Agricultural palm oil tree residues as raw material for cellulose, lignin and hemicelluloses production by ethylene glycol pulping process," *Chem. Eng. J.*, vol. 148, no. 1, 106–114, May 2009.

[19] Wise, LE; Murphy, M; AD'Addieco, A. "Chlorite holocellulose, its fractionation and bearing on summative wood analysis and on studies on the hemicelluloses," *Pap. Trade J.*, vol. 122, no. 2, 35–42, 1946.

[20] Rowell, RM. *The Chemistry of solid wood.* Washington, D.C: American Chemical Society, 1984.

[21] Moshkelani, M; Marinova, M; Perrier, M; Paris, J. "The forest biorefinery and its implementation in the pulp and paper industry: Energy overview," *Appl. Therm. Eng.*, vol. 50, no. 2, 1427–1436, Feb. 2013.

[22] Cherubini, F. "The biorefinery concept: Using biomass instead of oil for producing energy, and chemicals," *Energy Convers. Manag.*, vol. 51, no. 7, 1412–1421, Jul. 2010.

[23] "Brundtland Commission," *Wikipedia, the free encyclopedia*. 11-Dec-2014.

[24] Demirbas, A. "Biorefineries: Current activities and future developments," *Energy Convers. Manag.*, vol. 50, no. 11, 2782–2801, Nov. 2009.

[25] Koutinas, AA; Du, C; Wang, RH; Webb, C. "Production of Chemicals from Biomass," in *Introduction to Chemicals from Biomass*, J. H. Clark and F. E. I. Deswarte, Eds. John Wiley & Sons, Ltd, 2008, 77–101.
[26] Naik, SN; Goud, VV; Rout, PK; Dalai, AK. "Production of first and second generation biofuels: A comprehensive review," *Renew. Sustain. Energy Rev.*, vol. 14, no. 2, 578–597, Feb. 2010.
[27] FitzPatrick, M; Champagne, P; Cunningham, MF; Whitney, RA. "A biorefinery processing perspective: treatment of lignocellulosic materials for the production of value-added products," *Bioresour. Technol.*, vol. 101, no. 23, 8915–8922, Dec. 2010.
[28] Dorado, MP; Lin, SKC; Koutinas, A; Du, C; Wang, R; Webb, C. "Cereal-based biorefinery development: utilisation of wheat milling by-products for the production of succinic acid," *J. Biotechnol.*, vol. 143, no. 1, 51–59, Aug. 2009.
[29] Octave, S; Thomas, D. "Biorefinery: Toward an industrial metabolism," *Biochimie*, vol. 91, no. 6, 659–664, Jun. 2009.

In: Agricultural Wastes
Editor: Camille N. Foster

ISBN: 978-1-63482-359-3

Chapter 7

VALORIZATION OF WASTES FROM INDUSTRIAL PROCESSING OF AN AGRICULTURAL PRODUCT VIA THERMOCHEMICAL CONVERSION PROCESSES

***Pablo R. Bonelli*[1,2] *and Ana L. Cukierman*[1,2,3,*]**

[1]Programa de Investigación y Desarrollo de Fuentes Alternativas de Materias Primas y Energía (PINMATE) - Depto. de Industrias, Facultad de Ciencias Exactas y Naturales, Universidad de Buenos Aires. Ciudad Universitaria. Buenos Aires, Argentina

[2]Consejo Nacional de Investigaciones Científicas y Técnicas (CONICET), Buenos Aires, Argentina

[3]Cátedra de Tecnología Farmacéutica II. Depto. de Tecnología Farmacéutica, Facultad de Farmacia y Bioquímica, Universidad de Buenos Aires. Buenos Aires, Argentina

ABSTRACT

Large amounts of wastes arising from industrial processing of agricultural products constitute alternative renewable bioresources potentially attractive for bioenergy generation and/or for the manufacture of other useful products. Their conversion additionally contributes to reduce environmental pollution. The present chapter examines thermochemical conversion of the wastes generated from industrialization of an agricultural product into biofuels and/or products potentially applicable for environmental remediation. The selected wastes arise from industrial processing of whole branches (leaves and twigs) from a native evergreen tree *Ilex paraguariensis,* belonging to the Aquifoliaceae family, for the manufacture of yerba mate. It is a widespread product massively consumed in Southern Latin America countries to prepare a popular herbal tea–like beverage. The commercial final product generally contains less than ~ 35% twigs, since they provide an unpleasantly bitter taste to the infusion, and therefore huge quantities of unused twigs emerge as a by-product. Kinetics for the pyrolysis of the twigs is characterized by non-isothermal thermogravimetric analysis from room temperature up

[*] E-mail: analea@di.fcen.uba.ar; anacuki@ffyb.uba.ar.

to 900 °C to obtain information for the proper design of full–scale pyrolyzers. A deactivation model which assumes an overall first-order process and considers the physicochemical changes taking place in the biomass with the pyrolysis course through variations of the reaction rate constant with the temperature and solid conversion enables a proper representation of the experimental data over the whole temperature range, with estimated energy activation values between 49 and 137 kJ mol^{-1}. Likewise, yield and characteristics of the three kinds of pyrolysis products, comprising bio-char, bio-oil, and gases, are examined from experiments conducted in a bench-scale fixed-bed installation at temperatures in the range 400 – 700 °C. Gas yield increases with increasing temperature, attaining 43% at 700 °C, while the biochar yield decreases from 30% to 20% with temperature rise. Yield of the bio-oil attains a maximum (53%) at 500 °C, likely arising from the competition between primary formation of volatiles, at relatively low temperatures, and secondary degradation of the condensable vapors at the higher temperatures. All the pyrolysis products could be used in energy applications. The obtained biochars with higher heating value (HHV) of 23 – 24 MJ kg^{-1} have potential as environmentally friendly solid biofuel and could be employed for the manufacture of briquettes mainly for domestic use. Accounting for their high stability, as judged from the molar O:C ratio, another possible application could be incorporation of the biochars into the soil for the storage of atmospheric carbon. In turn, the bio-oils show organic fractions with HHV between 28 and 33 MJ kg^{-1}. Density values of the as-produced liquids (~1 kg dm^{-3}) are rather higher than those for conventional hydrocarbon fuels due to their higher contents of oxygen and water. The crude bio-oils could be directly burnt or subjected to further upgrading to attain characteristics similar to those of fuel-oil. Pyrolysis of the twigs yields low to medium heating value-gases (5 – 11 MJ m^{-3}), mostly composed by CO_2, CO, CH_4 and H_2. Gas composition depends on the temperature, even though CO_2 is the major generated species, followed by CO. Proportion of CO_2 decreases with temperature, particularly at 700 °C, accompanied by enhancements in the HHV of the gaseous mixtures, as a consequence of compositional variations, attaining a maximum value of 11 MJ m^{-3}. They might contribute to the energy sustainability of the process. Besides, phosphoric acid activation of the yerba mate twigs at pre-established moderate conditions leads to good quality activated carbons with well-developed porous structures characterized by textural parameters (BET surface area of ~ 1000 m^2 g^{-1}; total pore volume of 1cm^3 g^{-1}) comparable to those of commercially available samples.

Keywords: pyrolysis; agro-industrial waste; yerba mate twigs; kinetics; bioenergy; bio-oil; bio-char; activated carbons

1. INTRODUCTION

The progressive depletion of fossil fuels along with the need to mitigate global climate change associated to their use, have stimulated an increasing interest in alternative resources for energy generation and chemicals production (Baggio et al., 2008; Guo et al., 2015). Biomass is a promising, clean source of renewable energy and valuable chemicals primarily due to its abundance, wide distribution, and CO_2 neutrality (Kanaujia et al., 2014; Ma et al., 2015). Biomass comes from a large number of different sources, and in a wide variety of forms. In particular, agricultural residues and food processing wastes from agroindustry represent an important source of biomass having widespread availability, especially in countries where economies are considerably based on activities related to the agroindustry (White et al., 2011; Bonelli et al., 2012; Long et al., 2013). Use of wastes or by-products generated from industrial processing as biomass feedstock is recommended in order to avoid competition with biomass derived from energy crops for the soil, an issue that has engendered controversy.

Thermochemical processing is a recognized route for conversion of biomass feedstock. Pyrolysis, i.e., thermal degradation of biomass in an oxygen-depleted atmosphere, is a key thermochemical process which enables conversion of biomass into value-added products mainly directed to energy applications (González et al., 2008; Cukierman et al., 2012; Shi et al., 2012). They are often lumped into three groups: permanent gases, a pyrolytic liquid (bio-oil/tar), and a carbon enriched solid product (char or bio-char). Biomass pyrolysis has been extensively investigated to optimize the process for the production of bio-oil or charcoal and for the implications in the companion thermochemical processes of gasification and combustion (Di Blasi et al., 2010). Biomass pyrolysis is a complex process. The process kinetics as well as yields and properties of the pyrolysis products depend on various parameters, such us the origin and physico-chemical characteristics of the biomass, and the experimental conditions used (Shi et al., 2012; Guizani et al., 2014).

Besides, the advantage of using agricultural by-products or wastes as raw materials for the manufacture of activated carbons (ACs) has been highlighted. Activated carbons in powder or granular forms are used worldwide in several different applications, which include domestic and industrial uses, such as gas storage and delivery, removal of liquid and gaseous pollutants, purification and separation of gases, metal recovery, in catalysis, as a catalyst or catalytic support, and also in biomedical applications, among others (Kwiatkowski et al., 2011; Mezohegyi et al., 2012). The increasing demand of this adsorbent, primarily due to its growing use in applications pertaining to environmental pollution, has promoted the necessity to develop activated carbons from cheaper and readily available raw materials in order to favor their sustainable large scale production (Valente Nabais et al., 2008; 2013; Nabarlatz et al., 2012; Cukierman, 2013; El Sayed et al., 2014). Wood and coconut shell have been mostly used as conventional lignocellulosic precursors for the production of activated carbons. Agricultural solid wastes also investigated for the production of low cost effective ACs include olive wastes (Baccar et al., 2012), cherry stones (Duran Valle et al., 2005), shells from peach stone, almond and walnut (Girgis et al., 2007; González et al., 2009); sugarcane bagasse (Blanco Castro et al., 2000; Bonelli et al., 2012), corncob (El Sayed et al., 2014), orange peels (Fernandez et al., 2014), rose stems (Cifuentes et al., 2013), among many others. Conversion of these wastes into ACs adds economic value, helps reduce the cost of waste disposal, and provides a potentially less expensive alternative to the existing commercial samples (Fernandez et al., 2014). Thermochemical processes are also used for biomass conversion into activated carbons. The two main processes applied for the development of activated carbons are the so-called physical or thermal activation, and chemical activation. Briefly, physical activation generally involves pyrolysis of the precursor, subsequently followed by gasification with an oxidizing agent, often steam or CO_2, under strictly controlled conditions. On the other hand, chemical activation involves impregnation of the precursor frequently with a Lewis acid, such as zinc chloride or phosphoric acid, followed by pyrolysis of the impregnated precursor. Washing of the so-obtained product to eliminate the remaining impregnating agent leads to complete porous structure development. Advantages and drawbacks of these processes have been reported in detail elsewhere (Marsh and Rodriguez-Reinoso, 2006). Properties of the resulting activated carbons are highly dependent on the precursor, the activation process, and process parameters used (Basso et al., 2002;

Cukierman, 2013). Therefore, research on the production of ACs from different precursors and processes is relevant to the global community (Valente Nabais et al., 2013).

In this context, the aim of the present chapter is to examine the feasibility of converting a native, abundant agro-industrial biomass – yerba mate twigs (YMT) – into useful, value-added products by applying thermochemical processes. YMT arise from industrial processing of whole branches (leaves and twigs) from a native evergreen tree *Ilex paraguariensis* Saint Hilaire, belonging to the Aquifoliaceae family, for the manufacture of yerba mate. It is a widespread product massively consumed in Southern Latin America countries – Argentina, Brazil, Paraguay, and Uruguay – to prepare a popular herbal tea–like beverage appreciated for its flavor, and increasingly recognized worldwide for its nutritional and medicinal value, being included in several food codes (Anesini et al., 2006). Argentina is the largest producer of yerba mate worldwide with a yearly average production of around 260 x 10^3 ton, which is in part exported, followed by Brasil and Paraguay. It is mostly produced in the provinces of Misiones and Corrientes, Argentina (Canitrot et al., 2011). The per capita domestic consumption is estimated in 5-6 kg per year. Yerba mate is a strategic crop in the country due to its relevance related to regional development and workforce occupation. Industrial processing of the commercial product consists of a series of sequential stages that include heat treatment at relatively high temperatures in order to inactive enzymes and avoid enzymatic degradation, drying, grinding, and seasoning (Heck et al., 2008; Scipioni et al., 2010). The final product usually contains more than 65% dried leaves and less than 35% twigs, since the latter provides an unpleasantly bitter taste to the infusion. Therefore, huge quantities of unused twigs are discarded, thus constituting an attractive biomass resource for their conversion into valuable products. An advantage of YMT as biomass feedstock is their low moisture content (< 5 – 6 wt%) because of the thermal processing involved in the manufacture of the commercial product. This characteristic avoids the need of a first drying-conditioning stage often required for the thermochemical conversion of a wide variety of biomasses mostly characterized by high moisture contents (Cukierman et al., 1999: Mohan, 2006).

Pyrolysis of YMT is firstly investigated focusing on the kinetic characterization of the process in order to obtain fundamental information for the proper design of full-scale pyrolyzers. Likewise, yields, physicochemical characteristics, and higher heating value of the three kinds of pyrolysis products, comprising bio-char, bio-oil, and gases, are determined with emphasis on the effect of the temperature. Furthermore, conversion of the yerba mate twigs into activated carbon by the chemical activation process using phosphoric acid solution as activating agent is also examined. In this sense, it must be stressed that although a great deal of research has been published on the subject of producing activated carbons from several different biomasses, use of yerba mate twigs as raw material for this purpose has been unexplored. Potential applications of the different resulting products are analyzed and discussed in terms of their main characteristics.

2. Experimental Section

2.1. Materials

Dry yerba mate twigs (YMT) (0.35 cm average diameter and 1.25 cm length) were supplied by a local manufacture (Mate Larangeira S.A., Argentina). They were milled and sieved. Fractions of average particle size less than 2 mm were reserved for pyrolysis experiments. Proximate and ultimate analyses of the YMT are reported in Table 1. They were carried out by conventional ASTM standards and using a Carlo Erba EA 1108 instrument, respectively. Higher heating value (HHV) was measured using a Parr 1341 oxygen bomb calorimeter.

Table 1. Proximate and elemental analyses of the yerba mate twigs used in pyrolysis experiments

Proximate analysis (wt%, dry basis)	
Volatiles	72.1
Ash	3.1
Fixed Carbon[a]	24.8
Elemental analysis (wt%, dry, ash free-basis)	
C	46.4
H	5.9
N	1.2
O[a]	46.5
HHV (MJ kg^{-1})	17.8

[a] Estimated by difference.

2.2. Kinetic Measurements of the Pyrolysis of Yerba Mate Twigs

Kinetic measurements for the pyrolysis of samples of yerba mate twigs were carried out by non-isothermal thermogravimetry (TG-DTG) from ambient temperature up to 900 °C, operating under flowing nitrogen. A Netzsch STA 409 thermogravimetric balance equipped with a N_2 flow device and a data acquisition system was used to perform the measurements. Preliminary experiments were first conducted in order to assess operating conditions for which diffusional effects were negligible. Accordingly, masses of 10 mg, N_2 flow rates of 400 cm^3 min^{-1}, and a heating rate of 100 °C min^{-1} were employed. Further details of the experimental procedures employed have been depicted in previous own works (Bonelli et al., 2007; González et al., 2008).

2.3. Bench-Scale Pyrolysis Experiments

Pyrolysis assays were conducted in a bench-scale equipment in order to obtain and characterize the three kinds of pyrolysis products using the YMT and different temperatures. They enabled to calculate yields as well as to examine the effect of the pyrolysis temperature on yields and characteristics of the pyrolysis products.

The equipment and procedure employed have been already reported elsewhere (Cukierman et al., 2012). Basically, the equipment consisted of a stainless steel fixed-bed reactor (2.6 cm outer diameter, 110 cm total length). At the top of the reactor, a special device enabled to support a basket built in stainless steel mesh. It was used as a container of the biomass sample. The bottom part of the reactor was externally heated by an electric furnace driven by a temperature controller. The basket with the sample, constituting the solid fixed-bed, was centrally placed in the heated bottom zone of the reactor. In this way, the reduced length of the heated reactor zone enabled to attain short residence times of the volatiles evolved with the process course, thus minimizing possible extra-bed secondary reactions.

A system located at the reactor outlet allowed for condensation and collection of the condensable volatiles generated with the pyrolysis course. It consisted of a series of four sequential flasks (250 cm^3), acting as cold tramps, immersed in a cooling bath. The first flask was empty, while the two latter were partially filled with acetone in order to ensure completely retention of the condensable vapors. Non-condensable vapors, after passing through the condensation system, were sampled periodically using Teflon gas bags for further analysis by gas chromatography. A tubular cotton filter located before the main ventilation system was used to capture any residual liquid droplets formed downstream the condensation system. Finally, all the off-gases were sent to a main ventilation system.

All the assays were performed under isothermal conditions, according to the following protocol. The basket was first loaded with the pre-weighed sample. A slip thermocouple was located inside the solid bed. Then, the loaded basket was introduced into the upper part of the reactor by means of the support device. Before heating the reactor, all the installation was purged by flowing N_2 (200 cm^3 min^{-1}) for 1 h. Afterwards, the heating system was connected and the desired temperature was set. Once the pre-established temperature was attained, the basket containing the sample was displaced to the heated zone of the reactor. As determined from temperature measurements by the thermocouple inserted in the solid bed and monitoring the time required to attain the pre-established temperature, heating the sample in this way may be considered to take place almost instantaneously in all cases.

After the pre-established reaction time was achieved, heating was cut-off and the basket was immediately shifted towards the upper (non-heated) part of the reactor, keeping the N_2 stream. Once at ambient temperature, the basket was removed from the reactor. The residual solid and the accumulated liquid products contained in the flasks were weighted to determine product yields. These products were then carefully stored in closed containers for further characterization. Gas yields were obtained by difference from overall mass balances. Each experiment was repeated three times and averaged values are reported.

Preliminary assays were performed at the highest temperature of the investigated range (700 °C) in order to examine the influence of the reaction time, sample mass (bed depth), particle diameter, and gas flow rate on process characteristics. The conditions used for which variations were found almost negligible are listed in Table 2.

Table 2. Experimental conditions used for YMT pyrolysis

Temperature range	400 – 700 °C
Reaction time	30 min
Total gas flow rate	200 cm^3 min^{-1}
Sample mass	6 – 7 g
Particle diameter	0.5 – 1 mm

2.4. Characterization of the Pyrolysis Products Generated at the Different Temperatures

Characterization of the bio-char: As for the YMT, proximate analysis of the biochar obtained at temperatures in the range 400 – 700 °C was carried out by thermogravimetric analysis following conventional ASTM standards (D5142) using the instrument already detailed in section 2.2. Biochar samples were first heated at 105 °C under N_2 to determine moisture content; the temperature was then raised at 25 °C min^{-1} to 900 °C and kept for further 10 min to eliminate volatile matter. Finally, air was introduced to the system and the sample was combusted, maintaining the temperature at 900 °C for 20 min, in order to determine ash content. Fixed carbon was calculated on a weight per cent basis by subtracting moisture, volatile and ash values from the original mass.

Likewise, elemental compositions (C, H, N, S) were assessed in triplicate using an elemental analyser (Carlo Erba EA 1108). The O content was determined by difference. Higher heating value (HHV) was measured using the Parr 1341 oxygen bomb calorimeter.

Characterization of the liquid products: Density of the liquid products collected at the end of each experiment was determined in triplicate by pycnometry at room temperature. To determine HHV and elemental composition, the liquid products resulting from each experiment were separated into two phases by addition of methylene chloride. In this way, an aqueous phase and an organic phase were obtained. HHV of the latter was determined after evaporation of the solvent under vacuum. For this purpose, the oxygen bomb calorimeter earlier described was used following ASTM D240-02, for assessment of the heating value of liquid hydrocarbons. Elemental composition of the organic phase was determined with the aforementioned elemental analyzer.

Characterization of the gaseous products: Gas fractions (mainly composed of H_2, CO, CO_2, and CH_4) evolved at the different temperatures of the investigated range, were analyzed by gas chromatography. A Shimadzu GC-8A instrument equipped with a thermal conductivity detector and an Altech CTR I concentric packed column was employed. Argon as carrier and a temperature of 25 °C were used. For quantification, calibration curves were previously assessed using a standard gas mixture (Scott Specialty Gases).

2.5. Experiments for Conversion of the Yerba Mate Twigs into Activated Carbons

To examine the feasibility of developing activated carbons using the yerba mate twigs as a precursor, 10 g of sample (0.75 mm average particle diameter) was impregnated with 30

cm^{-3} of H_3PO_4 acid solution (50 wt%). The impregnated samples were first dried in an oven at 110 °C for 2 h. Afterward, they were placed in a horizontal, fixed-bed reactor externally heated by an electric furnace, and thermally treated at a heating rate of 3 °C min^{-1} up to 500 °C. Thermal treatment of the impregnated samples was carried out under a self-generated atmosphere. Once the selected temperature was reached, it was held for 0.5 h. Temperature selection was based on previous findings which demonstrate that maximum porosity development for ACs obtained by phosphoric acid activation from different lignocellulosic precursors is attained at 400-500 °C (Blanco Castro et al., 2000; Vernersson et al., 2002; Puziy et al., 2007).

To remove the excess of acid, the ACs were first contacted with a NaOH solution (0.1 N) and then extensively rinsed with distilled hot water until neutral pH in the wash water was attained. In addition, the lack of phosphates in the samples was verified following a reported procedure (Basso et al., 2002). Afterward, the samples were dried in an oven to constant weight. Yields were evaluated from weight differences. Further details of the equipment and procedure used in our laboratory have been earlier reported for other precursors (Blanco Castro et al., 2000; Basso et al., 2002; De Celis et al., 2009; Nunell et al., 2012; Fernandez et al., 2014).

On the other hand, for a better understanding about the behaviour of the twigs impregnated with the phosphoric acid solution during the thermal treatment stage, measurements by non-isothermal thermogravimetric analysis were conducted using acid-treated samples. They were carried out on the thermal analyzer earlier detailed in 2.2. The samples (~10 mg) were placed in the thermobalance, and heated from room temperature up to 500 ºC operating under a flow of nitrogen.

The developed activated carbons were examined by scanning electronic microscopy (SEM) in a Zeiss Supra 40 microscope equipped with a field emission gun. The images were taken with an in-lens detector and 3 kV acceleration voltages. The samples were placed on an aluminum holder, supported on conductive carbon tape, dried under vacuum, and sputter coated with Au-Pd.

Textural characterization of the ACs developed from the YMT was conducted. N_2 adsorption-desorption isotherms at (-196 °C) were determined with an automatic Micromeritics ASAP-2020 HV volumetric sorption analyzer. Prior to gas adsorption measurements, the samples were outgassed at 120ºC overnight. Textural properties were assessed from the isotherms, according to conventional procedures earlier depicted in own previous studies (Ramos et al., 2011; Nunell et al., 2012).

The Brunauer-Emmett-Teller (BET) surface area (S_{BET}) was determined by the standard BET procedure. Total pore volumes (V_T) were estimated from the amount of nitrogen adsorbed at the relative pressure of 0.95 (P/P_0 = 0.95).Micropore volume (V_{micro}) was estimated applying Dubinin Radushkevich equation (Bonelli et al., 2001b). The mean pore radius (R_m) was calculated from $R_m = 4\ V_T / S_{BET}$. For comparison, the isotherms for the yerba mate twigs used as precursor and for a commercial activated carbon were also determined following the procedure described above.

3. Results and Discussion

3.1. Kinetic Characterization of the Pyrolysis of the Yerba Mate Twigs

Typical non-isothermal TG-DTG curves for the pyrolysis of the yerba mate twigs are shown in Figure 1. Instantaneous weight fractions (w), namely instantaneous weight losses normalized by the initial mass of the sample, represented as a function of the temperature (T) are illustrated in the figure. Likewise, the reaction rates, obtained by differentiation of the normalized weight losses-time curve, versus T can also be appreciated in the same figure.

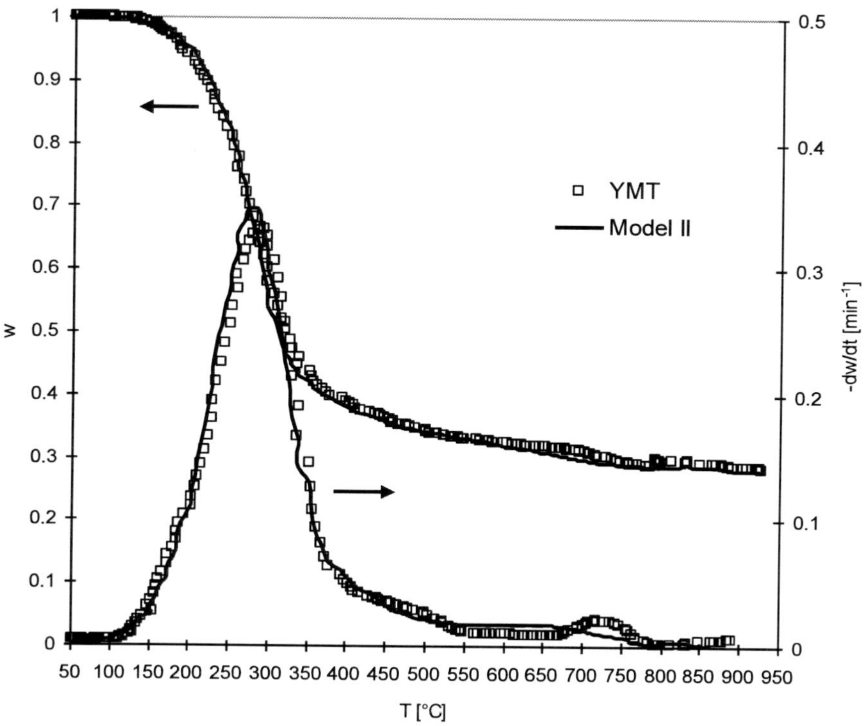

Figure 1. TG – DTG curves for the pyrolysis of yerba mate twigs. Comparison between experimental data (symbols) and predictions of the deactivation model (Model II, lines).

As seen, a noticeable decrease in weight fractions, associated with the release of volatiles present in the twigs (Table 1), takes place between 150 – 350 °C. At higher temperatures, only slight variations occur accompanied by a pronounced decrease in reaction rates. The latter attained a sharp maximum at 300 °C. It is generally accepted that biomass pyrolysis proceeds through a sequence of primary transformations. Initially release of free moisture from the biomass takes place, followed by degradation of the more unstable biopolymers, namely cellulose and hemicellulose possessing a polysaccharide structure relatively easy to breakdown, predominantly at the lower temperatures. With increasing temperature the more refractory components, basically lignin, which is more resistant to degrade due to its cross-linked aromatic structure, begin to decompose and volatiles are released from the substrate matrix. Solid char residue that is formed during the primary decomposition phase, taking place in the range 200 – 400 °C, apparently undergoes slow aromatization in a secondary pyrolysis stage that occurs at temperatures higher than 400 °C (White et al., 2011; Bonelli and Cukierman, 2012).

To evaluate the kinetic parameters, modeling of the TG curves was carried out. Kinetic modeling of pyrolysis of carbonaceous materials, such as coal and biomass, is complex because of numerous chemical reactions taking place simultaneously. To describe kinetic data, the process was assumed as a single first-order overall decomposition reaction. The reaction rate is given by the following equation:

$$-dw/dt = k_{app}(w - w_r) \tag{1}$$

w and w_r in Eq.(1) are the instantaneous and residual weight fractions, respectively, and k_{app}, the apparent reaction rate constant. Two different models were adopted to account for the dependence of k_{app} on the temperature. On one side, the conventional Arrhenius-type expression (Model I) was considered:

$$k_{app} = k_{0I}\, exp\ (-\ E_A\ /\ RT) \tag{2}$$

where k_{0I} is the pre-exponential factor, E_A, the activation energy, R, the universal gas constant, and T, the absolute temperature.

On the other hand, the second model applied supposes that the significant physicochemical changes which take place within the YMT as pyrolysis proceeds cause their deactivation (Model II). This fact may affect the reaction rate constant (k_{app}) and is taken into account through an increase of the activation energy with the temperature and the solid conversion according to the following expression (Balci et al., 1993; Della Rocca et al., 1999; Bonelli et al., 2003):

$$k_{app} = k_{0II}\, exp\ [-E_{Ai}\ (1+\beta T X^{\gamma})\ /\ RT] \tag{3}$$

X in Eq. (3) is the normalized fractional conversion of the solid defined as:

$$X = (1 - w)\ /\ (1 - w_r) \tag{4}$$

whereas E_{Ai} is the initial activation energy for X=0; β and γ are fitting parameters (β, the deactivation rate, and γ, the order with respect to X).

Characteristic parameters of both models are reported in Table 3 along with standard deviations (s) and the range of temperatures for which each model enabled to properly describe the experimental data. They were evaluated by non-linear regression analysis, by minimizing the sum of squares of the differences between the experimental and predicted model values for weight losses curves (w vs T) and considering the measured linear relationship between temperature and time. Thus, both models were adjusted to the data minimizing the following objective function (O.F.):

$$O.F. = \sum \left(w_{exp} - w_{mod}\right)^2 \tag{5}$$

where w_{exp} and w_{mod} represent, respectively, the experimental data and the value predicted by the model.

The capability of each applied model to represent the experimental data was compared by estimating the standard deviation (s) as:

$$s = \sqrt{\frac{\sum\left(w_{exp} - w_{mod}\right)^2}{N - p}} \qquad (6)$$

being N, the number of data, and p, the number of fitted parameters.

Table 3. Model characteristic parameters estimated for the pyrolysis of yerba mate twigs and range of temperature

Model I	
k_{0I} [min^{-1}]	2.8 x 10^4
E_A [kJ mol^{-1}]	46
w_r	0.37
s [%]	1.5
Temperature range [ºC]	25 – 450
Model II	
k_{0II} [min^{-1}]	3.5 x 10^4
E_{Ai} [kJ mol^{-1}]	49
β [K^{-1}]	1.5 x 10^{-3}
γ	5.5
s [%]	1.3
Temperature range [ºC]	25 – 900

Although both models led to acceptable s values (Table 3), Model I only provided a proper fit up to 450 °C. Instead, Model II succeeded in representing the experimental curves over the whole temperature range up to 900 ºC, as shown in Figure 1. As seen in Table 3, the estimated values for the pre-exponential factor and the initial activation energy of Model II were quite similar to those of Model I.

The initial activation energy is within values obtained for the pyrolysis of other biomasses under similar conditions (Bonelli et al., 2001b). Estimated values of the activation energy varied from 49 to 137 kJ mol^{-1} for the temperature range 25 – 900 °C.

3.2. Effect of the Temperature on Yields and Characteristics of the Pyrolysis Products

The effect of the temperature on yields of the bio-char (YS), bio-oil (YL), and the gases (YG) generated from the pyrolysis of the yerba mate twigs, on a dry-basis, is illustrated in Figure 2. As may be noticed, the results show that the temperature markedly affected the distribution of the pyrolysis products, especially at temperatures higher than 500 °C, in agreement with other reported data (García Perez et al., 2008; Goyal et al., 2008).

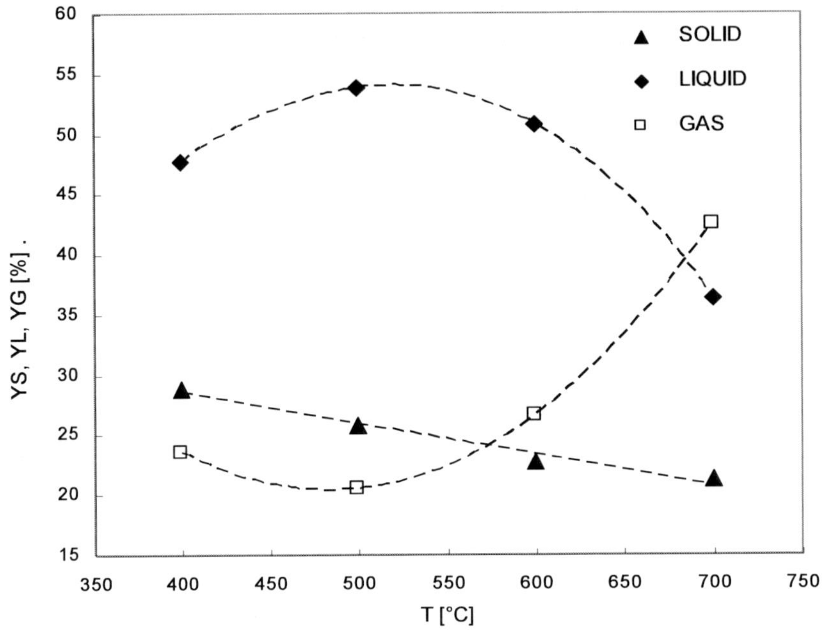

Figure 2. Effect of the temperature on yields of the three kinds of pyrolysis products: solid (YS), liquid (YL), and gas (YG).

Increasing the temperature led to enhance gas formation and to reduce the solid fraction, even though the effect for the latter was less pronounced. Maximum yield for the gaseous products was attained at the highest temperature within the investigated range, whereas formation of the bio-char achieved the minimum value at 700 °C. These trends may be due to a competition between devolatilization reactions promoting gas production, which are progressively favored with temperature increase, and reactions which induce char formation (Di Blasi, 2008). Dehydration, slow decarboxylation, depolymerization, and recombination of the decomposition products taking place during pyrolysis at relatively low temperatures have been ascribed as the main reactions responsible for the formation and higher yield of the bio-char. Instead, at higher temperatures, cleavage of macromolecules and subsequent degradation to volatiles of low-molecular weight would take place.

In turn, bio-oil yield increased with temperature, attaining a maximum at around 500 ºC, and decreased at temperatures higher than 500 °C. It agreed with results reported in the literature for bio-oils derived from other biomasses (Encinar et al., 2000; Kim et al., 2013). The maximum yield of bio-oil would be due to the competition between primary formation of volatiles, taking predominantly place at relatively low temperatures, and secondary degradation of the condensable vapors, that should be favored at higher temperatures (Jand and Foscolo, 2005). Consequently, increase in gas yield at temperatures higher than 500 ºC would result from contributions of bio-char degradation and secondary reactions of the condensable vapors. The pronounced decrease in bio-oil yield accompanied by the steep rise of the gas fraction at 700 ºC suggests that secondary reactions were predominant.

Chemical characteristics and HHV for the bio-char produced at the different pyrolysis temperatures are reported in Table 4. As expected, the results in Table 4 show that increasing the temperature led to a bio-char with higher fixed carbon content, slight increase in ash, and gradual decrease in volatiles. The increase in ash content should result from progressive concentration of minerals and destructive volatilization of lignocellulosic matter as temperature increased (Zhang et al., 2015). Likewise, enhanced %C and reductions in %H and %O, as a consequence of volatiles release, may be noticed. The temperature had only a weak effect on the HHV of the resulting bio-char. Its influence on both chemical characteristics and HHV was similar to trends reported for bio-char samples derived from other biomass resources (Claoston et al., 2014).

Table 4. Effect of the temperature on bio-char features

BIO-CHAR	400°C	500°C	600°C	700°C
Proximate analysis (dry basis, wt%)				
Volatiles	19.1	17.1	14.7	12.4
Ash	9.7	10.3	10.5	10.7
Fixed carbon[a]	71.2	72.6	74.7	76.6
Elemental analysis (dry, ash-free basis, wt%)				
%C	74.2	75.9	77.7	82.6
%H	2.9	2.7	2.0	1.0
%N	1.3	0.9	0.6	0.9
%O[(a)]	21.6	20.5	19.7	15.5
HHV (MJ kg^{-1})	23.4	23.5	23.4	24.1

[a] Estimated by difference.

Concerning the potential use of the bio-char as biofuel and the manufacture of briquettes, international standards establish contents of fixed carbon higher than 76 wt% to obtain high quality briquettes (Encinar et al., 2000; González et al., 2005; Basso et al., 2005; Cukierman et al., 2012). Accordingly, the bio-char obtained at the highest temperature, possessing a fixed carbon content of ~ 77wt%, fulfils the requirement for this purpose. Another criterion adopted as an index of good quality takes into account the volatile matter content of the char together with the intended application. For domestic use, volatile matter contents ranging between 20% and 30 % are acceptable, whereas these contents are restricted to 10 – 15% for use in the steel industry. Since the biochars derived at 400 – 700 °C have volatile matter contents ranging between ~ 12 and 19 wt % (Table 4), they could be employed for domestic use.

Apart from its use as biofuel, other possible application increasingly reported in the last years, is incorporation of the biochar into the soil in order to increase the long-term storage of carbon, simultaneously providing soil amendment benefits and reduction of greenhouse gases (Lehmann et al., 2006; Woolf et al., 2010; Zimmerman, 2010; Bruun et al., 2012; Creamer et al., 2014; Windeatt et al., 2014; Schimmelpfennig et al., 2014). Although the mechanisms by which the biochar favors soil quality are still not fully understood, it is reported that it positively impacts on soil organic carbon, water holding capacity, cation exchange capacity, pH, and soil microbial ecology (Crombie and Mašek, 2014). The stability in soil has been pointed out as a paradigmatic attribute of the biochar when selecting its application in environmental or agricultural applications, and potential indicators for stability prediction have been proposed (Fabbri et al., 2012). Among them, an indicator of the bio-char stability to store atmospheric carbon in soil considers the molar oxygen to carbon ratio (O:C) of the samples. In general, a ratio of O:C lower than 0.2 appears to provide a very prolonged biochar half-life, since it is linked with slower biochar mineralization rates (Spokas, 2010).

Accordingly, based on this indicator, present values of the molar O:C ratios for the biochar obtained at the different temperatures, varying within the range $1 - 1.5 \times 10^{-3}$, point to highly stable samples.

In Table 5, characteristics of the bio-oil generated at the different temperatures are reported. Bio-oils are multi-component mixtures of different size molecules possessing appreciable proportions of water, usually ~ 15 – 30 wt%, derived both from the original moisture in the biomass feedstock and as a product of dehydration with the pyrolysis course and storage. The distribution of chemicals in bio-oils in terms of polarity is quite complex including the presence of highly non-polar to intermediate polarity substituents, such as normal and branched aliphatics, alicyclics, olefins, and aromatics. Polar and highly polar chemicals mainly include oxygenates like carboxylates, carbonyls, aryl ethers, phenols, and alcohols (Kanaujia et al., 2014). Physico-chemical and energy properties of bio-oil differ significantly from petrol distillate fuels (Qi et al., 2007; Guo et al., 2015). As shown in Table 5, the organic fraction of the bio-oils showed carbon contents ranging between 61% and 67%. Instead, %C for conventional fuels is comprised between 83% and 89% (Ghazi, 2013). Values of the HHV of the YMT-derived bio-oils were relatively higher than those reported for bio-oils generated from other biomasses because the former were determined for the organic fraction. However, if water content of the bio-oils is considered, lower values, comparable to the reported ones (Mohan et al., 2006; Kim et al., 2013), are obtained. Density of the bio-oils as produced, i.e., including both the aqueous and organic fractions, was rather higher (~ 1 kg dm^{-3}) than that for conventional hydrocarbon fuels, attributable to their higher oxygen content and presence of water. The results also indicated that the temperature did not affect drastically the characteristics of the resulting bio-oils. The derived crude bio-oils could be directly burnt by standard atomization techniques in industrial scale combustion systems, using a proper burner to account for bio-oils particular characteristics. Atomization and combustion of bio-oils in engines are favored by their content of water since it reduces viscosity and enhances fluidity (Kanauija et al., 2014). However, upgrading of the bio-oils should be necessary in order to employ them as a petrol distillate fuel alternative. The upgrading of bio-oils has been subject of intense research in the last years with the aim of reducing their moisture content and acidity, as well as of improving the heating value and storage ability. It may include hydrogenation, hydrodeoxygenation, esterification, catalytic cracking, molecular distillation, supercritical fluidization, emulsification, steam reforming, and blending, among other methods. For instance, moderate upgrading has been applied to substitute for heavy fuel oils to power static appliances including boilers, furnaces, engines, and electric generators. Nevertheless, large-scale application of bio-oils as transportation fuels requires sustainable upgrading techniques that have not been still achieved. Hydrodeoxygenation, which is a high pressure process involving H_2 to exclude oxygen from the bio-oil to yield a high grade oil product equivalent to crude oil, appears to have the greatest potential, even though several issues involved in this process still have to be solved (Mortensen et al., 2011). Other possible applications of bio-oils concern their potentialities to obtain valuable chemicals, preservatives, lubricants, binders, paints, thickeners, stabilizers (Qi et al., 2007; García Perez et al., 2008; Butler et al., 2011; Bridgwater, 2012).

On the other hand, main gaseous species evolved with pyrolysis course comprise CO_2, CO, CH_4, H_2 and, to a lesser extent, C_2H_6, and C_2H_4. Typical concentration-time profiles for the main species are illustrated in Figure 3 (a-d) for the different temperatures investigated. Furthermore, Figure 4 shows the molar fraction of the gaseous species produced from YMT pyrolysis and the HHV of the gases mixture, as a function of the pyrolysis temperature. HHV was calculated considering the amount of moles produced per mass unit of biomass (G_i) and the heat of combustion for each of the species, according to the following equation:

$$HHV\ [MJ\ kg^{-1}] = 0.802\ G_{CH4} + 0.286\ G_{H2} + 0.283\ G_{CO} \quad (7)$$

G_{CH4}, G_{H2}, and G_{CO} in Equation (7) are the total amount of each of the gaseous species produced per mass unit of biomass at each temperature.

Table 5. Effect of the temperature on bio-oil characteristics

BIO-OIL	400°C	500°C	600°C	700°C
Elemental analysis* (dry, ash-free basis, wt%)				
%C	66.6	63.0	61.0	66.5
%H	6.5	5.3	5.5	8.5
%N	4.5	3.3	3.5	3.0
%O[a]	22.4	28.4	30.0	22.0
HHV* ($MJ\ kg^{-1}$)	28.5	25.2	24.6	30.9
Density ($kg\ dm^{-3}$)	1.2	1.2	1.0	0.9

[a] Estimated by difference; * corresponding to the organic fraction.

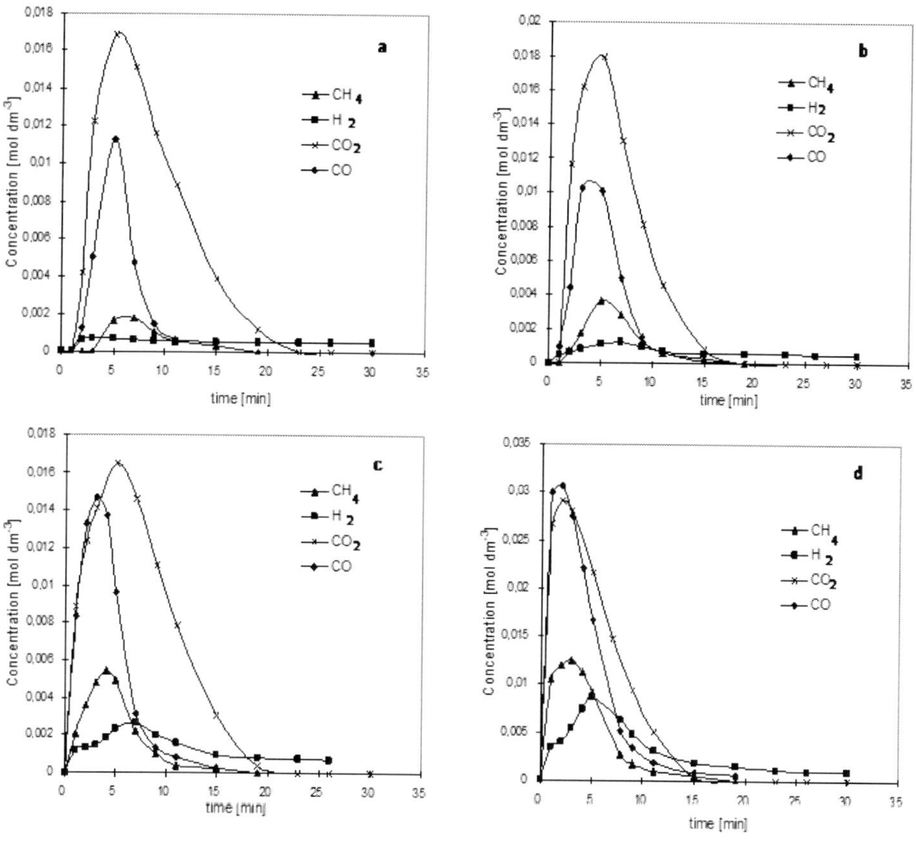

Figure 3. Effect of the temperature on concentration profiles of the main gaseous species produced at: a) T = 400 ºC, b) T = 500 ºC, c) T = 600 ºC, d) T = 700 ºC.

As may be appreciated in Figure 3, composition of the gaseous mixtures depends on the pyrolysis temperature. For all the temperatures investigated, the mixtures were mostly composed of CO_2, followed by CO, and in less proportion, by CH_4 and H_2. Seemingly, CO_2 results mainly from reforming and cracking of carbonyl and carboxyl functional groups, whereas CO is generated from the rupture of C–O–C and C=O. In turn, CH_4 should be produced from decomposition of methoxyl–O–CH_3. Cracking and deformation of C=C and C–H groups could explain H_2 generation (Qu et al., 2011; Park et al., 2012; Cukierman et al., 2012).

Proportions of CO_2 in the mixtures progressively decreased with temperature rise, whereas those corresponding mainly to CO and H_2 increased, thus leading to HHV enhancement. As seen in Figure 4, HHV of the generated gas attained values from 1 to 4 MJ kg^{-1} of biomass in the range of investigated temperatures, namely from 5 to 11 MJ m^{-3}, indicating that pyrolysis of the twigs yields low to medium heating value-gases. It agrees with some data reported in the literature for gases arising from the pyrolysis of other biomassic feedstocks for similar conditions (Encinar et al., 2009). The gas produced could contribute to provide heat, which may help to the sustainability of the pyrolysis process (Strezov et al. 2008; Bridgwater, 2012).

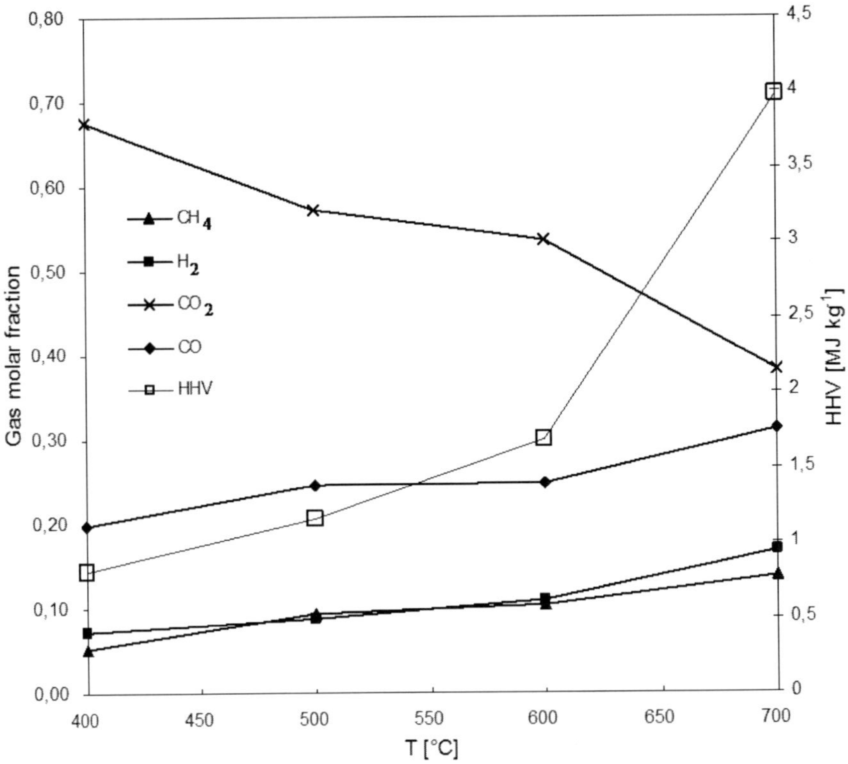

Figure 4. Effect of the temperature on the composition and HHV of the gaseous mixtures produced from pyrolysis of the yerba mate twigs.

3.3. Conversion of the Yerba Mate Twigs into Activated Carbons

In Figure 5, thermal degradation of the yerba mate twigs impregnated with the phosphoric acid solution is illustrated. For the sake of comparison, the results for the degradation of the untreated twigs are also included in the figure.

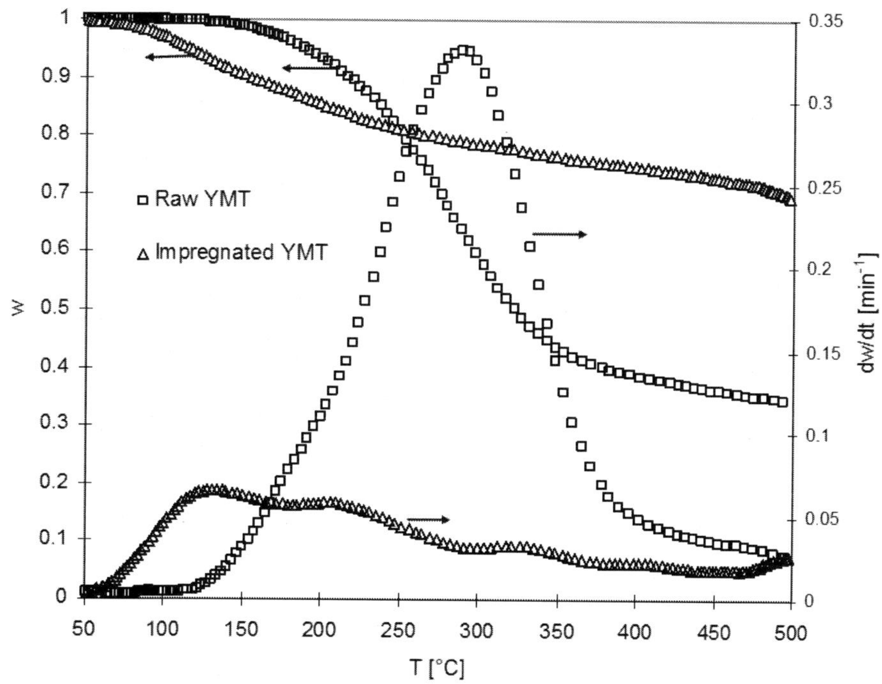

Figure 5. Effect of the phosphoric acid impregnation on the thermal degradation behavior of the yerba mate twigs (YMT). TG and DTG curves for the raw and acid impregnated YMT.

The results in Figure 5 show that impregnation of the yerba mate twigs with the phosphoric acid solution promoted noticeable modifications in the thermal degradation behavior in comparison with that determined for the untreated sample. As seen, the acid-treated sample started to degrade at lower temperatures than the untreated one, degradation proceeding steadily throughout the whole range of temperatures, including the low temperature region where no weight losses for the untreated sample were detected. Apparently, phosphoric acid facilitates the reactions in the lignocellulosic sample that lead to weight loss at the initial stages of carbonization. For temperatures above 250 °C, values of the recorded weight fractions for the degradation of the acid-treated twigs were higher than those for the untreated ones. At the highest temperature of the investigated range (500 °C), the residual weight fraction for the former (w = 0.7) almost duplicated the value determined for the latter (w = 0.3), pointing to a higher solid yield. Besides, the reaction rates for the impregnated YMT were markedly lower than those determined for the untreated one for almost all the temperature range (T > 150 °C). While a sharp maximum at 300 °C was observed for the untreated YMT, it was shifted to lower temperatures and attained a lower value for the acid-treated sample. The behavior is in general similar to those reported for the thermal degradation of other phosphoric acid-treated lignocellulosic biomasses (Vernersson et al., 2002; Basso et al., 2005a; 2005b; Bonelli et al., 2012; Cukierman et al., 2012; Nunell et al., 2015). It may be attributed to modifications in physicochemical properties of the sample induced by the phosphoric acid, markedly affecting its thermal behavior. Seemingly, the acid acts as a catalyst promoting bond-cleavage reactions and redistribution of fragments of the constituent biopolymers of the YMT, i.e., cellulose, hemicellulose, and lignin, that favor conversion of aliphatic to aromatic compounds at lower temperatures in comparison with the degradation of the sample in the absence of the reagent (Jagtoyen and Derbyshire, 1998). Alterations in the sample's structure caused by incorporation of the acid could also favor cross-linking reactions, decreasing losses of volatile compounds and leading to the higher weight fraction for the acid-treated twigs (Puziy et al., 2002; Ramos et al., 2007). In turn, phosphoric acid activation of YMT led to ACs yield of 45%. In general, phosphoric acid besides contributing to develop ACs with a suitable porous structure, promotes the insertion of different P- and O-containing functional groups in the resulting ACs, that favor some applications, such as the removal of metal ions from polluted water. The direct introduction of heteroatoms in the ACs that takes place upon chemical activation with H_3PO_4 has the potential advantage of avoiding the need for a further functionalization treatment, which often requires the use of oxidizing agents (Basso and Cukierman, 2006; Vázquez-Santos et al., 2012).

Representative SEM micrographs of the surface of the developed activated carbons at two different magnifications are illustrated in Figure 6 (a-b). At the lower magnification, the typical honeycomb-like structure, characteristic of activated carbons usually obtained from lignocellulosic precursors by H_3PO_4 acid activation (Vernersson et al., 2002, De Celis et al., 2009; Cukierman et al., 2012; Nunell et al., 2012), may be observed (Figure 6a). In turn, the image of the surface obtained at a higher magnification evidences a spongiform appearance, with numerous, diverse tiny pores (Figure 6b).

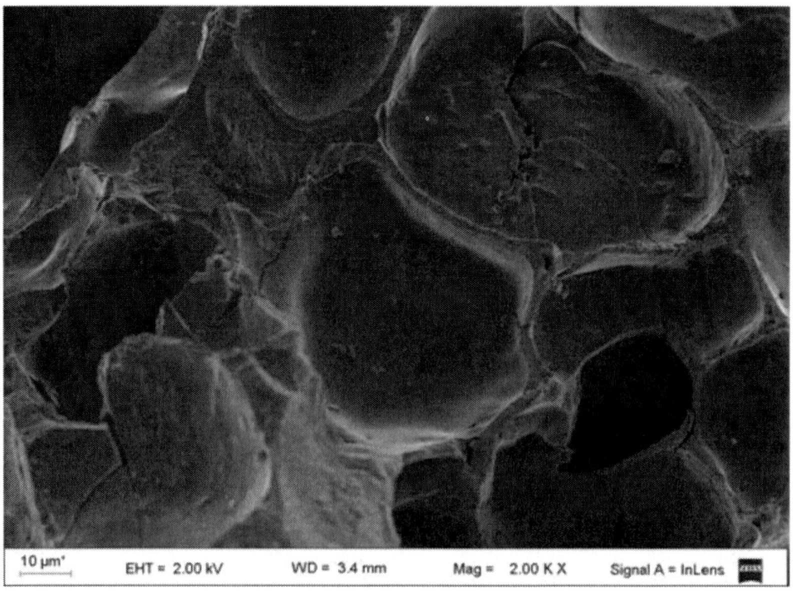

a

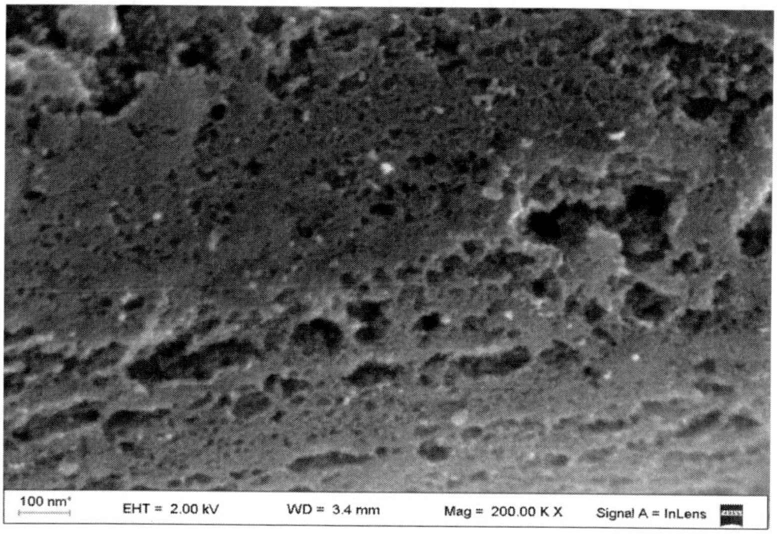

b

Figure 6. SEM micrographs of the activated carbons developed by phosphoric acid activation of the yerba mate twigs at magnifications of: a) 2000x; b) 200000x.

Figure 7 illustrates N_2 (-196 °C) adsorption isotherm for the activated carbons developed from the yerba mate twigs. The isotherms determined for the untreated twigs and the commercial activated carbon, used as a reference, are also included in the figure for comparison. N_2 volumes adsorbed onto the samples, at standard temperature and pressure conditions per sample mass unit (Vads), are represented as a function of the relative pressure (P/P_0), where P is the equilibrium pressure and P_0, the saturation pressure of the adsorbate at (-196 °C). The shape of the adsorption isotherms for the developed activated carbon and the commercial sample were quite similar, and showed features between those of Type I and II, according to IUPAC classification. They pointed to the presence of micropores (< 2 nm) and mesopores (2 - 50 nm). In turn, the pronouncedly higher N_2 volumes adsorbed on the prepared activated carbon compared to those for the raw sample attested the porosity development in YMT as a consequence of the activation process.

Textural parameters evaluated from the isotherms are listed in Table 6. Close similarities between the textural parameters for the developed and commercial activated carbon samples indicate that phosphoric acid activation of the yerba mate twigs used as a precursor led to a good quality adsorbent, at least with textural characteristics comparable to some others commercially available.

Table 6. Textural parameters for the raw yerba mate twigs (YMT), the activated carbon derived from the twigs (AC-YMT) and a commercial activated carbon (CAC)

Sample	YMT	AC-YMT	CAC
S_{BET} (m^2 g^{-1})	1.3	1040	1090
V_T (cm^3 g^{-1})	2 x 10^{-3}	1.0	1.1
V_{micro} (cm^3 g^{-1})	1 x 10^{-3}	0.55	0.53
R_m (nm)	3.1	1.9	2.0

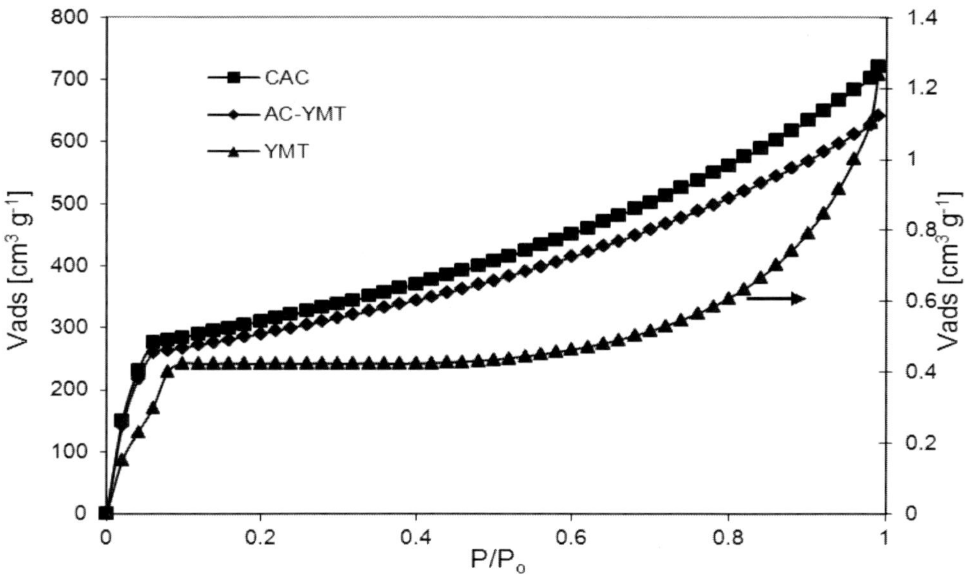

Figure 7. N_2 adsorption isotherms for the yerba mate twigs (YMT), the activated carbons developed from the yerba mate twigs (AC-YMT), and the commercial sample used as a reference (CAC).

Conclusion

The feasibility of converting the discarded twigs, which emerge in large quantity from the industrial processing of *Ilex paraguariensis* St Hilaire for the manufacture of yerba mate, into valuable energy products has been investigated by applying the pyrolysis process. The kinetics of the process was characterized by non-isothermal TG analysis, from room temperature up to 900 °C, and modeling of the experimental information. A deactivation model which accounts for the physicochemical changes taking place in the biomass with the process course by assuming variations of the reaction rate constant with the temperature and solid conversion enabled a proper representation of the data over the whole temperature range. Values of the activation energy between 49 and 137 kJ mol^{-1} were obtained.

Yields and characteristics of the three types of products arising from the pyrolysis of the yerba mate twigs have been also examined from experiments conducted in a fixed-bed reactor at temperatures in the range 400 ºC – 700 ºC. The temperature exerted a noticeable influence on the products distribution. Whereas gas yield increased with increasing temperature, attaining 43% at 700 °C, yield of the solid product decreased from 30% to 20% with temperature rise. The trends may be respectively attributed to devolatilization of the twigs and secondary reactions of the condensable vapors promoting gas production, which are favored at the higher temperatures, and to competing reactions that lead to biochar formation, with predominance of those related to the cleavage of the constituent biopolymers and their further degradation to volatiles with temperature increase. In turn, yield of the bio-oils showed a maximum (53%) at 500 °C, likely arising from the competition between primary formation of volatiles, at relatively low temperatures, and secondary degradation of the condensable vapors at the higher temperatures.

Characterization of the biochar obtained at the temperatures investigated indicated that it has potential as environmentally friendly solid biofuel (HHV: 23 – 24 MJ kg^{-1}) and could be employed for the manufacture of briquettes mainly for domestic use. Accounting for the high stability of the biochar, as judged from the molar O:C ratio of the samples, another possible application could be its incorporation into the soil for the storage of atmospheric carbon. Nevertheless, further characterization for this purpose and for its use as soil amendment should be required. The temperature did not affect drastically the characteristics of the resulting bio-oils. They showed organic fractions characterized by HHV between 28 and 33 MJ kg^{-1} for the range 400-700 °C, while density values of the as-produced liquids (~ 1 kg dm^{-3}) were rather higher than those for conventional hydrocarbon fuels due to the higher contents of oxygen and water of the bio-oils. The crude bio-oils derived from the yerba mate twigs could be directly burnt by standard atomization in industrial scale combustion systems. Likewise, they could be used as a fuel with characteristics similar to those of fuel-oil by further upgrading or as feedstock for extraction of other valuable chemicals. Pyrolysis of the twigs also yielded low to medium heating value-gaseous products, mostly composed by CO_2, CO, CH_4 and H_2. Gas composition depended on the temperature, even though CO_2 was the major generated species, followed by CO. Proportion of CO_2 decreased with temperature, particularly at 700 °C, accompanied by enhancements in the HHV of the gaseous mixtures, as a consequence of compositional variations, attaining a maximum value of 4 MJ kg^{-1} (11 MJ m^{-3}). The gases generated might contribute to the energy sustainability of the process.

On the other hand, the unused yerba mate twigs also showed potential as precursor for their transformation into activated carbon by phosphoric acid activation at moderate conditions. The developed activated carbons showed honeycomb-like porous structures, BET surface area of ~ 1000 m^2 g^{-1}, and total pore volume of 1 cm^3 g^{-1}. These characteristics are comparable to those of other commercially available activated carbon samples, and point to a good quality adsorbent.

Accordingly, present results show that the unused yerba mate twigs could be transformed into valuable products with potential applicability to bioenergy generation and/or environmental remediation, through their proper transformation *via* thermochemical processes.

ACKNOWLEDGMENTS

Grants from Consejo Nacional de Investigaciones Científicas y Técnicas (CONICET), Universidad de Buenos Aires (UBA), and Agencia Nacional de Promoción Científica y Tecnológica (ANPCYT – FONCYT) from Argentina, are gratefully acknowledged.

REFERENCES

Anesini, C., Ferraro, G., Filip, R. 2006. Peroxidase-like activity of *Ilex paraguariensis*. *Food Chemistry* 97, 459-464.

Baccar, R., Sarrà, M., Bouzid, J., Feki, M., Blánquez, P. 2012. Removal of pharmaceutical compounds by activated carbon prepared from agricultural by-product. *Chemical Engineering Journal* 211-212, 310-317.

Baggio, P., Baratieri, M., Gasparella, A., Longo, G.A. 2008. Energy and Environmental Analysis of an Innovative System Based on Municipal Solid Waste (MSW) Pyrolysis and Combined Cycle, *Applied Thermal Engineering,* 28, 136-144.

Balci, S., Dogu, T., Yucel, H. 1993. Pyrolysis kinetics of lignocellulosic materials. *Industrial Engineering Chemistry Research,* 32, 2573-2579.

Basso, M.C., Cerrella, E.G., Cukierman, A.L. 2002. Activated carbons developed from a rapidly renewable biosource for removal of cadmium (II) and nickel (II) ions from dilute aqueous solutions. *Industrial and Engineering Chemistry Research,* 41, 180-189.

Basso, M. C., Cukierman, A. L. 2005a. *Arundo donax*-based activated carbons for aqueous-hase adsorption of volatile organic compounds. Industrial and Engineering Chemistry Research, 44, 2091-2100.

Basso, M.C., Cerrella, E.G., Buonomo, E.L., Bonelli, P.R., Cukierman, A.L. 2005b. Thermochemical conversion of *Arundo donax* into useful solid products. *Energy Sources,* 27, 1429-1438.

Basso, M.C., Cukierman, A.L. 2006. Wastewater treatment by chemically activated carbons from giant reed: effect of the activation atmosphere on properties and adsorptive behavior. *Separation Science & Technology*, 41, 149-156.

Blanco Castro, J., Bonelli, P.R., Cerrella, E.G., Cukierman, A.L. 2000. Phosphoric acid activation of agricultural residues and bagasse from sugar cane: influence of the experimental conditions on adsorption characteristics of activated carbons. *Industrial and Engineering Chemistry Research,* 39, 4166-4172.

Bonelli, P.R., Della Rocca, P.A., Cerrella, E.G., Cukierman, A.L. 2001a. Effect of pyrolysis temperature on composition, surface properties and thermal degradation rates of Brazil nut shells. *Bioresource Technology,* Volume 76, 15-22.

Bonelli, P.R., Della Rocca, P.A., Cerrella, E.G., Cukierman, A.L. 2001b. Comparative study on char properties and pyrolysis kinetics of different lignocellulosic wastes. In "Progress in Thermochemical Biomass Conversion". Editor A.V. Bridgewater. *Blackwell Science,* London, UK. Volume 2, 1116 - 1128.

Bonelli, P.R., Cerrella, E.G., Cukierman, A.L. 2003. Slow pyrolysis of nutshells: characterization of derived chars and of process kinetics. *Energy Sources,* 25, 8, 767-778.

Bonelli, P.R., Buonomo, E.L., Cukierman, A.L. 2007. Pyrolysis of sugarcane bagasse and co-pyrolysis with an Argentinean subbituminous coal. Energy Sources. Part A: *Recovery, Utilization, and Environmental Effects*, 29, 731-740.

Bonelli, P.R., Cukierman, A.L. 2012. Pyrolysis characteristics of different kinds of lignins. In "Lignin: Properties and Applications in Biotechnology and Bioenergy". Editor Ryan J. Paterson. *Nova Science Publishers Inc.,* N.Y., USA. Chapter 11, 355-380.

Bonelli, P.R., Nunell, G.V., Fernández, M.E., Buonomo, E.L., Cukierman, A.L. 2012. The potential applications of the bio-char derived from the pyrolysis of an agro-industrial waste. Effects of temperature and acid-pretreatment. Energy Sources. Part A: *Recovery, Utilization, and Environmental Effects,* 34, 746-755.

Bridgwater, A.V. 2012. Review of fast pyrolysis of biomass and product upgrading. *Biomass and Bioenergy,* 38, 68-94.

Bruun, E.W., Ambus, P., Egsgaard, H., Hauggaard-Nielsen, H. 2012. Effects of slow and fast pyrolysis biochar on soil C and N turnover dynamics. *Soil Biology & Biochemistry,* 46, 73-79.

Butler, E.; Devlin, G.; Meier, D.; McDonnell, K. 2011. A review of recent laboratory research and commercial developments in fast pyrolysis and upgrading. *Renewable and Sustainable Energy Reviews,* 15, 4171-4186.

Canitrot L, Grosso MJ, Méndez A. 2011. Available: http://www.mecon.gov.ar /peconomica/docs/Complejo_Yerbatero.pdf.

Cifuentes, A.R., Avila, K., García, J.C., Daza, C.E. 2013. The pyrolysis of rose stems to obtain activated carbons: A study on the adsorption of Ni(II). *Ind. Eng. Chem. Res.* 52, 16197-16205.

Claoston, N., Samsuri, A.W., Ahmad Husni, M.H., Mohd Amran, M.S. 2014. Effects of pyrolysis temperature on the physicochemical properties of empty fruit bunch and rice husk biochars. *Waste Management & Research,* 32, 331-339.

Creamer, A.E., Gao, B., Zhang, M. 2014. Carbon dioxide capture using biochar produced from sugarcane bagasse and hickory wood. *Chemical Engineering Journal,* 249, 174-179.

Crombie, K., Mašek, O. 2014. Pyrolysis biochar systems, balance between bioenergy and carbon sequestration. Global Change Biology Bioenergy Published by John Wiley & Sons Ltd., *GCB Bioenergy,* doi: 10.1111/gcbb.12137

Cukierman, A.L., Della Rocca, P.A., Bonelli, P.R., Cerrella, E.G. 1999. Pyrolysis of an agricultural by-product: a characterization study. In "Biomass: A Growth Opportunity in Green Energy and Value-Added Products". Editors R.P. Overend and E. Chornet. *Pergamon Elsevier Science.* Volume 2, 1201-1208.

Cukierman, A.L., Nunell, G.V., Fernández, M.E., De Celis, J., Kim, M.R., Gurevich Messina, L., Bonelli, P.R. 2012. Thermochemical processing of wood from invasive arboreal species for sustainable bioenergy generation and activated carbons production. In: "Invasive Species: Threats, Ecological Impact and Control Methods". Eds Blanco JJ, Fernandes A. Chapter 1, 1-46. *Nova Science Publishers Inc.* New York, USA.

Cukierman, A.L. 2013. Development and environmental applications of activated carbon cloths. Review. ISRN Chemical Engineering. Volume 2013. Article ID 261523, 31 pages. http://dx.doi.org./10.1155 /2013/261523

De Celis, J., Amadeo, N.E., Cukierman, A.L. 2009. In situ modification of activated carbons developed from a native invasive wood on removal of trace toxic metals from wastewater. *Journal of Hazardous Materials*, 161, 217-223.

Della Rocca, P.A., Cerrella, E.G., Bonelli, P.R., Cukierman, A.L. 1999. Pyrolysis of hardwood residues: on kinetics and chars characterization. *Biomass & Bioenergy,* 16, 79-88.

Di Blasi, C. 2008. Modeling Chemical and Physical Processes of Wood and Biomass pyrolysis, *Progress in Energy and Combustion Science,* 34, 47-90.

Di Blasi, C., Branca, C., Galgano, A. 2010. Biomass screening for the production of furfural via thermal decomposition. *Industrial and Engineering Chemistry Research,* 49, 2658-2671.

Durán-Valle, C.J., Gómez-Corzo, M., Pastor-Villegas, J. & Gómez-Serrano, V. 2005. Study of cherry stones as raw material in preparation of carbonaceous adsorbents. *Journal of Analytical and Applied Pyrolysis*, 73, 59-67.

El-Sayed, G.O., Yehia, M.M., Asaad, A.A. 2014. Assessment of activated carbon prepared from corncob by chemical activation with phosphoric acid. *Water Resources and Industry*, 7-8, 66-75.

Encinar, J.M., González, J.F., González, J. 2000. Fixed-bed pyrolysis of Cynara cardunculus L. Product yields and compositions. *Fuel Processing Technology*, 68, 209-222.

Encinar, J.M., Gónzalez, J.F., Martínez, G., Romám, S. 2009. Jerusalem artichoke pyrolysis: Energetic evaluation. *Journal of Analytical and Applied Pyrolysis*, 85, 294-300.

Fabbri, D., Torri, C., Spokas, K.A. 2012. Analytical pyrolysis of synthetic chars derived from biomass with potential agronomic application (biochar). Relationships with impacts on microbial carbon dioxide production. *Journal of Analytical and Applied Pyrolysis*, 93, 77–84.

Fernandez, M.E., Nunell, G.V., Bonelli, P.R., Cukierman A.L. 2014. Activated carbon developed from orange peels: Batch and dynamic competitive adsorption of basic dyes. *Industrial Crops and Products*, 62, 437-445.

Ghazi, K.A. 2013. Fuels, Energy, and the Environment. CRC Press, Taylon and Francis Group, *Boca Raton*, USA.

García-Perez, M., Wang, X.S., Shen, J., Rhodes, M.J., Tian, F., Lee, W-J., Wu, H., Li, C. 2008. Fast pyrolysis of oil mallee woody biomass: Effect of temperature on the yield and quality of pyrolysis products. *Industrial and Engineering Chemistry Research*, 47, 1846-1854.

Girgis, B.S., Attia, A.A., Fathy, N.A. 2007. Modification in adsorption characteristics of activated carbon produced by H_3PO_4 under flowing gases. Colloids and Surfaces A: *Physicochemical and Engineering Aspects*, 299, 79-87.

González, J.F.; Ramiro, A.; González García, C.M.; Gañán, J.; Encinar, J.M.; Sabio, E.; Rubiales J. 2005. Pyrolysis of Almond Shells. *Energy Applications of Fractions, Industrial and Engineering Chemistry Research*, 44, 3003-3012.

González, J., Buonomo, E., Bonelli, P., Cukierman, A.L. 2008. Pyrolysis of Biomass from Sustainable Energy Plantations: Effect of Mineral Matter Reduction on Kinetics and Charcoal Pore Structure. *Energy Sources Part A: Recovery, Utilization, and Environmental Effects*, 30, 809-817.

González, J.F., Román, S., Encinar, J.M., Martínez, G. 2009. Pyrolysis of various biomass residues and char utilization for the production of activated carbons. *Journal of Analytical and Applied Pyrolysis*, 85, 134-141.

Goyal, H.B., Seal, D., Saxena, R.C. 2008. Bio-fuels from thermochemical conversion of renewable resources: A review. Renewable and Sustainable Energy Reviews, 12, 504-517.

Guizani, C., Escudero Sanz, F.J., Salvador, S. 2014. Effects of CO_2 on biomass fast pyrolysis: Reaction rate, gas yields and char reactive properties. *Fuel*, 116, 310-320.

Guo, M., Song, W., Buhain J. 2015. Bioenergy and biofuels: History, status, and perspective. *Renewable and Sustainable Energy Reviews*, 42, 712-725.

Heck, C.I., Schmalko, M., Gonzalez de Mejia E. 2008. Effect of growing and drying conditions on the phenolic composition of mate teas (*Ilex paraguariensis*). *Journal of Agricultural and Food Chemistry*, 56, 8394-8403.

Jand, N.; Foscolo, P.U. 2005. Decomposition of wood particles in fluidized beds. *Industrial and Engineering Chemistry Research*, 44, 5079-5089.

Jagtoyen, M., Derbyshire, F. 1998. Activated carbons from yellow poplar and white oak by H_3PO_4 activation. *Carbon* 36, 1085-1097.

Kanaujia, P.K., Sharma, Y.K., Garg, M.O., Tripathi, D. Singh, R. 2014. Review of analytical strategies in the production and upgrading of bio-oils derived from lignocellulosic biomass. *Journal of Analytical and Applied Pyrolysis* 105, 55-74.

Kim, K.H., Kim, T.S., Lee, S.M., Choi, D., Yeo, H., Choi, I.G., Choi, J.W. 2013. Comparison of physicochemical features of biooils and biochars produced from various woody biomasses by fast pyrolysis. *Renewable Energy,* 50, 188-195.

Kwiatkowski, J.F. (Ed.). 2011. Activated Carbon: Classifications, Properties and Applications. *Nova Science Publishers Inc.,* N.Y., USA.

Lehmann, J., Gaunt, J., Rondon, M. 2006. Bio-char sequestration in terrestrial ecosystems - A review. *Mitigation and Adaptation Strategies for Global Change*, 11, 403-427.

Long, H., Li, X., Wang, H., Jia, J. 2013. Biomass resources and their bioenergy potential estimation: A review, *Renewable and Sustainable Energy Reviews,* 26, 344-352.

Ma, Z., Chen, D., Gu, J., Bao, B., Zhang, Q. 2015. Determination of pyrolysis characteristics and kinetics of palm kernel shell using TGA–FTIR and model-free integral methods. *Energy Conversion and Management* 89, 251-259.

Marsh H., Rodriguez-Reinoso F. 2006. Activated Carbon. *Elsevier B.V,* Amsterdam, The Netherlands.

Mezohegyi, G., van der Zee, F.P., Font, J., Fortuny, A., Fabregat, A. 2012. Towards advanced aqueous dye removal processes: A short review on the versatile role of activated carbon. *Journal of Environmental Management,* 102, 148-164.

Mohan, D., Pittman, C.U., Steele, P.H. 2006. Pyrolysis of Wood/Biomass for Bio-oil: A Critical Review. *Energy & Fuels,* 20, 848-889.

Mortensen, P.M., Grunwaldt, J.D., Jensen, P.A., Knudsen, K.G., Jensen, A.D. 2011. A review of catalytic upgrading of bio-oil to engine fuels. *Applied Catalysis A: General* 407, 1-19.

Nabarlatz, D.A., De Celis, J., Bonelli, P.R., Cukierman, A.L. 2012. Batch and dynamic sorption of Ni(II) ions by activated carbon based on a native lignocellulosic precursor. *Journal of Environmental Management*, 97, 109-115.

Nunell, G.V., Fernández, M.E., Bonelli, P.R., Cukierman, A.L. 2012. Removal of nitrate from wastewater by activated carbons developed from sawdust of an invasive wood. *Biomass and Bioenergy*, 44, 87-95.

Nunell, G.V., Fernández, M.E., Bonelli, P.R., Cukierman, A.L. 2015. Nitrate uptake improvement by modified activated carbons developed from two species of pine cones. *Journal of Colloids and Interface Science*, 440, 102-108.

Park, Y.K., Yoo, M.L., Lee, H.W., Park, S.H., Jung, S.C., Park, S.S., Kim, SC. 2012. Effects of operation conditiones on pyrolysis characteristics of agricultural residues. *Renewable Energy*, 42, 125-130.

Puziy, A.M., Poddubnaya, O.I., Martínez-Alonso, A., Suarez-García, F., Tascón, J.M.D. 2002. Synthetic carbons activated with phosphoric acid. I. Surface chemistry and ion binding properties. *Carbon* 40, 493-505.

Puziy, A.M., Poddubnaya, O.I., Martínez-Alonso, A., Castro Muñiz, A., Suarez-García, F., Tascón, J.M.D. 2007. Oxygen and phosphorus enriched carbons from lignocellulosic material. *Carbon* 45, 1941-1950.

Qi, Z., Jie, C., Tiejun, W., Ying, X. 2007. Review of biomass pyrolysis oil properties and upgrading research. *Energy Conversion and Management* 48, 87-92.

Qu, T., Guo, W., Shen, L., Xiao, J., Zhao, K. 2011. Experimental study of biomass pyrolysis based on three major components: hemicellulose, cellulose, and lignin. *Industrial and Engineering Chemistry Research,* 50, 10424-10433.

Ramos, M.E., Gonzalez, J.D., Bonelli, P.R., Cukierman, A.L. 2007. Effect of process conditions on physicochemical and electrical characteristics of denim-based activated carbon cloths. *Industrial and Engineering Chemistry Research,* 46, 1167-1173.

Ramos, M.E., Bonelli, P.R., Cukierman, A.L. 2011. Strategies for optimizing the development of cellulose-based activated carbon cloths by the chemical activation process, Chapter 17, in: Activated Carbon: Classifications, Properties and Applications. J. F. Kwiatkowski (Ed.). *Nova Science Publishers Inc.* N.Y., USA, 475-508.

Scipioni, G.P., Ferreyra, D.J., Acuña, M., Schmalko, M.E. 2010. Rebaudioside A release from matrices used in a yerba mate infusion. *Journal of Food Engineering,* 100, 627-633.

Schimmelpfennig, S., Müller, C., Grünhage, L., Koch, C., Kammann, C. 2014. Biochar, hydrochar and uncarbonized feedstock application to permanent grassland. Effects on greenhouse gas emissions and plant growth. *Agriculture, Ecosystems and Environment,* 191, 39-52.

Shi, L., Yu, S., Wang, F.C., Wang J. 2012. Pyrolytic characteristics of rice straw and its constituents catalyzed by internal alkali and alkali earth metals. *Fuel,* 96, 586-594.

Spokas, K.A. 2010. Review of the stability of biochar in soils: Predictability of O:C molar ratios. *Carbon Management,* 1, 289-303.

Strezov, V., Evans, T.J., Hayman, C. 2008. Thermal conversion of elephant grass (*Pennisetum Purpureum Schum*) to bio-gas, bio-oil and charcoal. *Bioresource Technology,* 99, 8394-8399.

Valente Nabais, J., Carrott, P., Ribeiro Carrott, M.M.L., Luz, V., Ortiz, A.L. 2008. Influence of preparation conditions in the textural and chemical properties of activated carbons from a novel biomass precursor: The coffee endocarp. *Bioresource Technology*, 99, 7224-7231.

Valente Nabais, J., Laginhas, C., Ribeiro Carrott, M.M.L., Carrott, P.J.M., Crespo Amorós, J.E., Nadal Gisbert,. A.V. 2013. Surface and porous characterisation of activated carbons made from a novelbiomass precursor, the esparto grass. *Applied Surface Science* 265, 919- 924.

Vázquez-Santos, M.B., Suárez-García, F., Martínez-Alonso, A., Tascón, J.M.D. 2012. Activated carbon fibers with a high heteroatom content by chemical activation of PBO with phosphoric acid. *Langmuir,* 28, 5850-5860.

Vernersson, S.T., Bonelli, P.R., Cerrella, E.G., Cukierman, A.L. 2002. *Arundo donax* cane as precursor for activated carbons preparation by phosphoric acid activation. *Bioresource Technology,* 83, 95-104.

Windeatt, J.H., Ross, A.B., Williams, P.T., Forster, P.M., Nahil, M.A., Singh, S. 2014. Characteristics of biochars from crop residues: Potential for carbon sequestration and soil amendment. *Journal of Environmental Management* 146, 189-197.

White, J.E., Catallo, W.J., Legendre, B.L. 2011. Biomass pyrolysis kinetics: A comparative critical review with relevant agricultural residue case studies. *Journal of Analytical and Applied Pyrolysis,* 91, 1-33.

Woolf, D., Amonette, J.E., Street-Perrott, F.A., Lehmann, J., Joseph, S. 2010. Sustainable biochar to mitigate global climate change. Nature Communications. DOI: 10.1038/ncomms1053 |www.nature.com/nature communications.

Zhang, J., Liu, J., Liu, R. 2015. Effects of pyrolysis temperature and heating time on biochar obtained from the pyrolysis of straw and lignosulfonate. *Bioresource Technology,* 176, 288-291.

Zimmerman, A. 2010. Abiotic and microbial oxidation of laboratory-produced black carbon (Biochar). *Environmental Science and Technology,* 44, 1295-1301.

In: Agricultural Wastes
Editor: Camille N. Foster
ISBN: 978-1-63482-359-3

Chapter 8

AGRICULTURAL SOLID WASTES IN AQUEOUS PHASE DYE ADSORPTION: A REVIEW

Sharmeen Afroze, Tushar Kanti Sen ***and Ming Ang***
Department of Chemical Engineering, Curtin University, Australia

ABSTRACT

Wastewater from many industries such as textile, leather, paper, printing, food, etc. contains large amount of hazardous dyes. Dyes are not biodegradable and photodegradable due to its synthetic origin and complex aromatic nature. Among various physiochemical processes, adsorption techniques are usually widely used to treat dyes laden wastewater. Although commercial activated carbon is the most widely used adsorbent with large success, its use is limited due to high cost and difficulties in regeneration. Therefore there have been explosive growths in research concerning the use of alternative cost effective non-conventional effective adsorbents in the removal of dyes from aqueous solution. In this research direction, agricultural by-product solid wastes which are available in large quantities worldwide with almost through away price are utilized as effective adsorbents in the removal of inorganics and organics from wastewater. The focus of this book chapter is to review extensive literature information about dyes, its classification and toxicity, various treatment methods and finally dye adsorption characteristics by various agricultural by-products solid wastes as adsorbents. The major objective of this chapter is to organize the scattered available information on the adsorptive removal of dyes from its aqueous solution by raw and treated agricultural by-products. Selectively widely used agricultural solid waste adsorbents in the removal of dyes have also been discussed in details here. Finally mechanism, kinetics and adsorptive behaviour of adsorbents under various physicochemical process parameters have been critically analysed and compared. Conclusions have been drawn from the literature reviewed and few suggestions for future research are proposed.

Keywords: adsorption characteristics; dye removal; low cost adsorbents; activated carbon

---

* GPO Box U1987, Bentley, WA 6845, Australia; E-mail: t.sen@curtin.edu.au; Tel. No: +61 8 92669052.

1. INTRODUCTION

The emission of dye bearing effluents from industries into water bodies and its effective treatments has become a major challenge for environmentalists, academics and engineers. Dyes are coloured, ionising and aromatic organic compound that can attach itself to surfaces or fabrics to impart colour. The residual dyes in effluents from different sources of industries such as textile [1, 2], paper and pulp manufacturing [2], agricultural sector [3, 4], food technology [5], light-harvesting arrays [6], photo electrochemical cells [7], hair colourings [8], dye and dye intermediates [2], pharmaceutical [2], tannery [9, 10], and kraft bleaching industries [2], etc. are considered a wide variety of organic pollutants introduced into the natural water resources or wastewater treatment systems [11]. Among all industrial sectors, textile industries are rated as high polluters, taking into consideration the volume of discharge and effluent composition [12]. The total dye consumption in textile industry worldwide is more than 10,000 tonnes/year and approximately 100 tonnes/year of dyes are discharged into water streams [13]. Recent studies indicate that approximately 12% of synthetic dyes are lost during manufacturing and processing operations. Approximately 20% of these lost dyes enter the industrial wastewaters [14]. The discharge of untreated dyes in the environment are considered to be worrying for both toxicological and esthetical reasons as dyes impede light penetration, damage the quality of the receiving streams, toxic to food chain organisms [15] and sometimes even carcinogenic [16, 17]. Furthermore, the dyes have a tendency to sequester metal and may cause micro-toxicity to fish and other organisms [18]. However, wastewater containing dyes are very difficult to treat as dyes are recalcitrant organic molecules, which are resistant to aerobic digestion, and are made to resist degradation by light, heat, and oxidizing agents [19, 20]. Furthermore, dyeing wastewater compositions are not simple solutions of dye in water, but include many other materials such as particulates, processing assistants, salts, surfactants, acids and alkalis [21]. Currently, several physical or chemical processes are used to treat dye laden wastewaters [22]. Dyes can be removed, physically by processes such as foam flotation, precipitation, flocculation, adsorption, activated coal, wood chips, silica gel, filtration by membranes, ion exchange, UV radiation, electro-kinetic coagulation and filtration, or chemically by processes of oxidation, peroxidation of salts of iron II (Fenton reaction), ozonization, photocatalytic degradation, UV-peroxide system, cucurbituril and by sodium hypochlorite [23] and also by sonochemical, electrochemical and by integrated chemical-biological degradation [24]. However, most of these methods, which simply accumulate or concentrate the dyes [23], present high cost and trigger secondary pollution, caused by the excessive use of chemical substances [25]. The advantages and disadvantages of some methods of dye removal from wastewaters are presented by Garg et al., [22]. Among all these methods, adsorption is considered as one of the most effective processes of advanced wastewater treatment which industries employ to reduce hazardous inorganic/organic presence in the effluent [26]. Many textile industries use commercial activated carbon for the treatment of dye waste. Activated carbons are expensive due to their regeneration and reactivation procedures. The current research is focused on the need to alternative to commercial activated carbon as the cost effective, but potential adsorbent. Many researchers have reported the feasibility of using various low cost adsorbents derived from natural materials, industrial solid wastes, agricultural by-products and biosorbents as viable alternatives to activated carbon for the treatment of contaminated wastewater containing different classes of dyes [23, 24, 27, 28].

This book chapter deals with the effectiveness of various agricultural solid wastes as low-cost adsorbents for dye removal from water and wastewater. It is well-known that natural materials, solid waste materials from industry and agricultural by-products can be employed as an inexpensive sorbents.The main objective of this review chapter is to provide up-to-date information on the application of various sustainable low cost alternative adsorbents such as agricultural solid waste, industrial solid waste, agricultural by-products, and biomass based cost effective activated carbon, and various other natural materials in the removal of dyes from aqueous phase. This review chapter also critically analyse the effectiveness of various adsorbents under different physiochemical process parameters and their comparative adsorption capacity is also presented. A compilation of relevant published data with respect to adsorption kinetics, isotherm models and adsorption capacity under various process conditions along with important findings are presented here. There are couple of review articles on dyes removal such as review article by Slokar and Majcen Le Marechal [29]; remedation of dyes in textile effluents by Robinson et al. [23]; pulp and papermill wastewater treatment by Pokhrel and Viraraghavan [30]; non-conventional low-cost adsorbents for dye removal by Crini [31] and Gupta and Suhas [32]; decolourization of dye wastewaters by biosorbents by Srinivasan & Viraraghavan [33]; biodegradation of synthetic dyes by Ali [34]; and adsorption of methylene blue on low-cost adsorbents by Rafatullah et al., [24], cationic and anionic dye adsorption by agricultural solid wastes by Salleh et al., [28]; removal of organic pollutants from wastewater by low cost adsorbents by Ali et al., [35], but most of the existing review articles are dealing either with specific system or lack of up-to-date information. However all these existing review articles helped us to develop this current book chapter review which provides more comprehensive up-to-date and critical review information on the adsorption of various dyes from aqueous solutions using wide range of adsorbents. The new aspect of this chapter write up is to cover up-to-date research results presentation on various dyes adsorption and its effectiveness on these wide ranges of adsorbents and also identify and critically analyse various operation conditions and their maximum adsorption capacity in the removal of several dyes. Also authors here tried to organize the scattered available information on the adsorptive removal of dyes from its aqueous solutions by raw and treated agricultural by-products and agricultural solid wastes.

1.1. Classification of Dyes

Commercial dyes are usually a mixture of large complex and have often unreported molecular structure and diverse properties and also vary widely in chemical composition [12]. There are several ways for classification of commercial dyes. It should be noted that each class of dye has a very unique chemistry, structure, and particular way of bonding. Agreeing with Agustina [36], dyestuff can be classified according to their origin (natural/synthetic), chemical and/or physical properties or characteristics related to the application process. Another categorization is based on the application. Dyes are also classified based on their particle charge upon dissolution in aqueous medium [37] such as cationic (all basic dyes), anionic (direct, acid, and reactive dyes), and non-ionic (dispersed dyes). A systematic classification of dyes according to chemical structure is the colour index [36]. However, due to the complexities of the colour nomenclature from the chemical structural system, the classification based on application is often favourable [32]. The classification based on chemical structure for the common class of the dyes is presented in Table 1, whereas Table 2 represents the classification based on dye application.

Table 1. Functional structure based dyes classification

Class	Functional Groups	Example & Molecular Formula
Azo dyes	–N=N–	Reactive Black 5, $C_{26}H_{21}N_5Na_4O_{19}S_6$
Anthraquinone dyes	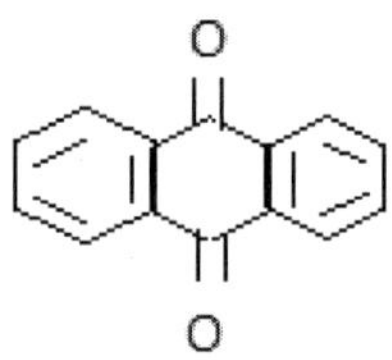	Reactive Blue 4, $C_{23}H_{14}Cl_2N_6O_8S_2$
Indigoid dyes	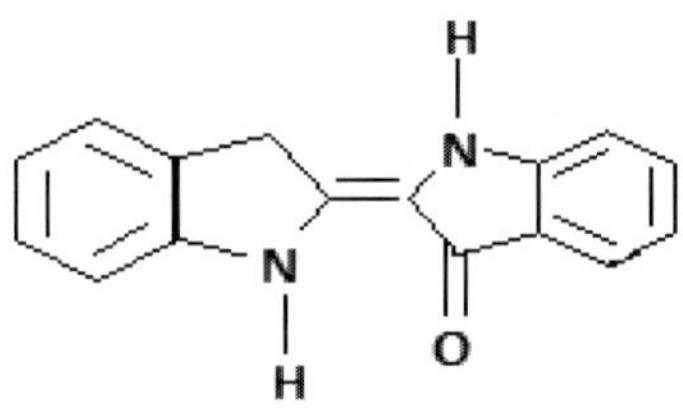	Vat Blue 5, $C_{16}H_6Br_4N_2O_2$
Nitroso dyes	–N=O	Acid green 1, $C_{30}H_{15}FeN_3Na_3O_{15}S_3$
Nitro dyes	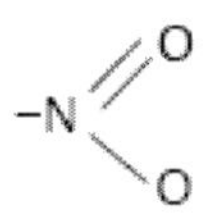	Acid Yellow 24, $C_{10}H_5N_2NaO_5$
Triarylmethane dyes	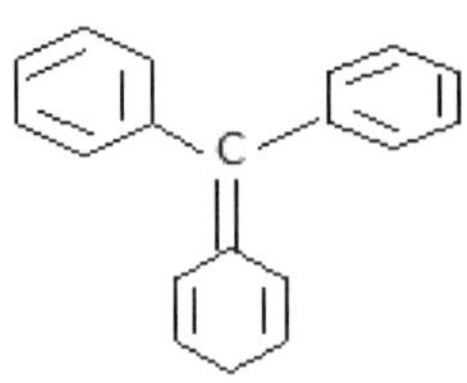	Basic red 9, $C_{19}H_{17}N_3.ClH$

Table 2. Application based dyes classification updated and modified from [38]

Class	Sources	Application Process	Functional Chemicals
Acid	Nylon, wool, silk, paper, inks and leather	Usually from neutral to acidic dye baths	Azo (including premetallized), anthraquinone, triphenylmethane, azine, xanthenes, nitro, nitroso
Basic	Paper, polyacrylonitrile, modified nylon, inks, polyester	Applied from acidic dye baths	Cyanine, Diaz cyanine, hemidiazacyanine, diphenymethane, azo, xanthenes
Direct	Cotton, rayon, paper, leather and nylon	Applied from neutral or slightly alkaline baths containing additional electrolytes	Azo, Phthalocyanine, stilbene and oxazine
Disperse	Polyester, polyamide, acetate, acrylic and plastics	Fine aqueous dispersion often applied by high temperature/pressure or low temperature carrier methods	Azo, anthraquinone, styryl, nitro and benzodifuranone
Fluorescent brighteners	Soaps, detergents, all fibres,paints,oil and plastic	From solution, dispersion or suspension in mass	Naphtha, stilbene, pyrazoles, coumarin
Food, cosmetics and drugs	Food, drug and cosmetics		Azo, anthraquinone, carotenoid and triarylmethane
Reactive	Cotton, wool, silk and nylon	Reactive site on dye reacts with functional group on fiber to bind dye covalently under influence of heat and pH(alkaline)	Azo, anthraquinone, phthalocyanine, oxazine and basic
Sulphur	Cotton and rayon	Aromatic substrate vatted with sodium sulphide and reoxidized to insoluble sulphur-containing products on fiber	Indeterminate structures
Vat	Cotton, rayon, wool	Water insoluble dyes solubilized by reducing with sodium hydrogen sulphite, then exhausted on fiber and reoxidized	Anthraquinone(including polycyclic quinones) and indigoids

1.2. Potential Dye Pollutants and Its Toxicity Effects

The evaluation of the toxicity of dyes is very important, mainly due to the different effects that they cause in the environment and the organisms exposed to them. Dyes are considered as potential pollutants because of their toxicity effects [32]. Generally dyes cause oral ingestion and inhalation, skin and eye irritation, and also carcinogenicity [39-41]. They impart colour to water which is visible to human eye and therefore, highly objectionable on aesthetic grounds. Not only this, dyes may cause micro-toxicity to fish and other organisms, also interfere with the transmission of light and upset the biological metabolism processes which cause the destruction of aquatic communities present in ecosystem [42]. Additionally dyes also can cause severe damage to human beings such as dysfunction of kidney, reproductive system, liver brain and central nervous system [43]. Generally basic dyes have been classified as toxic colorants [12]. Basic dyes have high intensity of colours and the tinctorial value of basic dyes is very high, less than 1 ppm of dye is greatly visible in solution [44]. Anthroquinone based dyes made from known carcinogens such as benzidine and other aromatic compounds are resistant to degradation due to their fused aromatic ring structure [2]. Disperse dyes have good ability to bioaccumulation [2]. Azo dyes are one of the more detrimental toxic classes because it is highly persistent in the aquatic environment, due to its chemical compositions, involving aromatic rings, azoic linkages and amino groups [45]. Many studies have shown that reactive dyes can cause allergic dermatoses and respiratory diseases [46-49].

In addition to being toxic, dye effluents also contain chemicals that has been informed to cause carcinogenesis, mutagenesis or teratogenesis and respiratory toxicity to various organisms [50, 51]. As such it is important to treat coloured effluents for the removal of dyes. Therefore, focuses on specific methods and technologies for the removal of dye bearing wastewater is desired and is under constant development, with increasing attention being paid to the ecological effects of dyestuffs regarding its use.

2. Dye Removal Techniques

Environmental contamination resulted from the emission of effluents of dyeing industries is a global problem, therefore different methods of effluents treatment have been used in an attempt to minimize the problems resulted from this contamination. Coagulation, foam flotation, precipitation, ozonation, ion exchange, filtration, solvent extraction, electrolysis, chemical oxidation, membrane separation technology, liquid-liquid extraction and adsorption on activated carbon have been used for the removal of dye contaminants from wastewater [32, 52]. The technologies can be divided into three broad categories: physical, chemical and biological [53]. The advantages and disadvantages for dyes and colour removal of industrial wastewater applied over the time into different industrial units are summarized in Table 3.

Table 3. Advantages and disadvantages of dyes removal techniques updated and modified from [2]

Separation Techniques	Advantages	Limitations
Physical-Chemical processes		
Precipitation, Coagulation-flocculation	Short detention time and low capital costs. Relatively good removal efficiencies.	Sometimes expensive, Agglomerates separation and treatment. Specific to operating condition.
Electrokinetic coagulation	Economically feasible	High sludge production.
Solid-Liquid physical adsorption such as:		
Activated carbon	Economically attractive. Good removal efficiency of wide variety of dyes.	Very expensive; cost intensive regeneration process
Peat	Effective adsorbent due to cellular structure. No activation required.	Surface area is lower than activated carbon.
Industrial solid wastes such as ash, red mud etc	Economically attractive. Good removal efficiency.	Slow kinetics. Specific surface area for adsorption are lower than activated carbon.
Raw wood chips/ Wood sawdust	Effective adsorbent due to cellular structure. Economically attractive. Good adsorption capacity for acid dyes.	Slow kinetics and activation may be required
Other agricultural solid wastes such as pine cone, pine leaves, eucalyptus bark etc.	Sustainable cost-effective good adsorbents. Moderate to high removal efficiency	Sometime slow kinetics and activation may be required
Silica gels	Effective for basic dyes	Side reactions prevent commercial application
Ion Exchange	Effective with ionic compounds removal and easy regeneration	Specific and effective for dyes removal and softening of water
Biological separation processes		
Aerobic process	Partial or complete decolourization for all classes of dyes.	Not cost-effective
Anaerobic process	Resistant to wide variety of complex coloured compounds. Bio gas produced is used for stream generation.	Longer acclimatization Phase, odour
Microbial processes	Good removal efficiency for low volumes and concentrations. Very effective for specific colour removal.	Culture maintenance is cost intensive. Cannot cope up with large volumes of wastewater.

Table 3. (Continued)

Separation Techniques	Advantages	Limitations
Advanced separation processes		
Electrochemical oxidation	No additional chemicals required and the end products are non-dangerous/hazardous.	High cost
Other advanced oxidation process	Complete mineralization ensured. Growing number of commercial applications. Effective pre-treatment methodology in integrated systems and enhances biodegradability.	Cost intensive process
Fenton process	High efficiency for colour removal	Generation of sludge and expensive
Membrane filtration	Removes all dye types; recovery and reuse of chemicals and water.	High running cost. Concentrated sludge production. Dissolved solids are not separated in this process.
Photo catalysis	Process carried out at ambient conditions. Inputs are no toxic and inexpensive. Complete mineralization with shorter detention times. No sludge production	Effective for small amount of coloured compounds. Expensive process.
Photochemical process	Effective oxidation at lab scale	Formation of by-products
Irradiation		Requires a lot of dissolved oxygen (O_2).
Sonication	Simplicity in use. Very effective in integrated systems.	Relatively new method and awaiting full scale application.
Redox mediators	Easily available and enhances the process by increasing. electron transfer efficiency	Concentration of redox mediator may give antagonistic effect. Also depends on biological activity of the system.
Engineered wetland systems	Cost effective technology and can be operated with huge volumes of wastewater	High initial installation cost. Requires expertise and managing during monsoon becomes difficult.

3.0. DYE ADSORPTION

Adsorption refers to a process wherein a species is concentrated on a solid surface from its liquid or gaseous surroundings. More precisely, adsorption is a physical/chemical separation process by which certain solute components of a fluid phase are attracted to the surface of a solid adsorbent and forms surface complex via physical or chemical bonds, and hence removing the said solute component from the fluid phase [54, 55]. Adsorption process usually occurs at interfacial layers which are regarded as two regions: the surface layer of the adsorbent (often simply called the adsorbent surface) and the adsorption space, in which the enrichment of the adsorptive can occur [56]. In general, adsorption processes may be classified into two groups: physical sorption and chemical sorption; depending on the nature of forces and bond formation are involved. Depending on the nature of the interactions ionic species and molecular species carrying different functional groups may be held to the surface through electrostatic attraction to sites of opposite charge at the surface or physiosorbed due to action of Van der Waals forces or chemisorbed involving strong adsorbate–adsorbent bonding which may lead to attachment of adsorbate molecules at specific functional group on adsorbent surface [57]. The main physical forces controlling adsorption are Van der Waals forces, hydrogen bonds, polarity, dipole-dipole ∏-∏ interaction, etc. [34].

Adsorption is one of the effective separation techniques for the removal of inorganic/organics from its aqueous phase. Adsorption has been found to be superior to other techniques for water re-use in terms of initial cost, flexibility and simplicity of design, ease of operation and insensitivity to toxic pollutants [24]. Adsorption also does not result in the formation of harmful substances [24]. Therefore considerable attention has been paid to adsorption technologies as efficient and versatile methods of removing dyes from wastewater effluents.

3.1. Factors Affecting Adsorption of Dye

Many physico-chemical factors influence the adsorption process and these include; adsorbate/adsorbent interaction, adsorbent surface area and pore structure, chemistry of the adsorbent surface, nature of the adsorbate, presence of other ions, particle size, pH, temperature, contact time, etc. [21, 31, 58]. Thus, the effects of these parameters are to be taken into account. Optimization of such conditions will greatly help in the development of industrial-scale dye removal treatment technology. In this section, some of the leading factors affecting adsorption of dyes are therefore discussed as follows:

3.1.1. Influence of Solution pH on Dye Adsorption

In wastewater treatment, the initial pH of dye solution is one of the most important factors that plays a vital role particularly on the adsorption capacity by influencing the chemistry of dye molecule and adsorbent in aqueous solution [59]. Initial pH of aqueous solution influences not only site dissociation, but also the solution chemistry of the dyes: hydrolysis, complexation by organic and/or inorganic ligands, redox reactions and precipitation are strongly influenced by pH [60]. The effect of solution pH on the adsorption can be determined by preparing adsorbent–adsorbate solution with fixed dye concentration and adsorbent dose but with different pH by adding NaOH (1 M) or HCl (1 M) solutions [27, 28].

Pavan et al. [61] studied the effect of solution pH on the adsorption of Crystal Violet (CV) by Formosa papaya seed powder and they noticed that at a pH range from 2 to 8, the dye removal percentage was maximum at pH 8. Argun et al. [62] studied the effect of solution pH for the removal of Reactive Blue 114 dye from aqueous solutions by using Pomelo (*Citrus grandis*) peel and noticed that the maximum adsorption was found at pH of 2.0. Tilaki et al. [63] studied the adsorption of Acid Orange 7 dyes by rice stem biomass and they found that Acid Orange 7 gives 98% removal efficiency at pH of 3 and removal percentage started to decrease below 50% as the pH increased. Zhao et al. [64] reported that adsorption of anionic dye Light Green (LG) onto surfactant modified peanut husk biomass was decreased with increase in solution pH. The compilation of different studies on the effect of solution pH on dye adsorption by agricultural by-product solid adsorbents is presented in Table 4.

The pH at which the surface charge is zero is called the point of zero charge (pHpzc). The point of zero charge (pzc) is relating to the adsorption phenomenon and it describes the condition when the electrical charge density on a surface is zero. Therefore the adsorption ability of the surface and the type of surface active centres are indicated by the point of zero charge (pH_{pzc}) [65] for systems where H^+/OH^- are the potential-determining ions. Due to presence of functional groups such as OH^- group on adsorbent surface, cationic dye adsorption is favoured at $pH>pH_{pzc}$, whereas, anionic dye adsorption is favoured at $pH< pH_{pzc}$ [66]. In order to understand the adsorption mechanism, the point of zero charge (pH_{pzc}) of various adsorbents prepared from agricultural solid wastes was investigated by many researchers [61, 67, 68]. Han et al. [69] studied the adsorption of Methylene blue (MB) onto poplar leaf and they found that the zero point of charge (pH_{pzc}) for the poplar leaf was around 5.6, while the maximum adsorption capacity of MB was at pH 9, in other word $pH>pH_{pzc}$. The adsorption of Blue Remazol onto babassu coconut mesocarp has been studied by Vieira et al. [70] and they found that the maximum adsorption capacity of Blue Remazol was at pH 1, while the point of zero charge (pH_{pzc}) for the babassu coconut mesocarp was 6.7.

Table 4. Influence of solution pH on dye adsorptive behaviour of different agricultural by-product adsorbents

Adsorbents	Dye	pH range	Percentage (%) of removal range	References
Coffee waste	Toluidine blue	2.0-10.5	50-95	[71]
Formosa papaya seeds	Crystal violet	2.0-8.0	41.3-97.3	[61]
Coffee husk	Fast green	2-5	Decrease	[60]
Parsley stalks	Methylene blue	2-10	20.72-82.84	[72]
Cucumber peels	Methylene blue	2-10	13.44-77.70	[72]
Cinnamomum camphora sawdust	Malachite green	2.46-7.18	52.0-90.8	[73]
Potato peel waste	Acid blue 1	2-11	Decrease	[74]
Garlic peel	Methylene blue	4-10	Increase	[75]
Activated- Rice husk	Acid yellow 36	2-9	80-45	[76]
Bagasse fly ash	Orange G	3-12	Decrease	[77]
Pine cone	Congo red	3.55-10.95	60.5-5.75	[78]
Spent coffee grounds	Methylene blue	3-11	81-95	[79]
Modified sawdust	Methylene blue	2-11	Increase	[80]
Pine leaves	Methylene blue	2-11	20-80	[13]

3.1.2. Influence of Initial Dye Concentration on Dye Adsorption

It is well known that dye solute concentration plays an important role in the adsorption process. The amount of adsorption for dye removal is highly dependent on the initial dye concentration. The effect of initial dye concentration depends on the immediate relation between the concentration of the dye and the available sites on an adsorbent surface. The effect of initial dye concentration can be studied by preparing adsorbent–adsorbate solution with fixed adsorbent dose and different initial dye concentration for different time intervals and shaken until equilibrium [27, 28].

In adsorbent-adsorbate solution, the initial dye concentration provides the driving force to overcome the resistance to the mass transfer of dye between the aqueous and the solid phase. At low concentrations, adsorption sites take up the available dye more quickly. However, at higher concentrations, dye needs to diffuse to the sorbent surface by intra-particle diffusion. Also, the steric repulsion between the solute molecules may slow down the adsorption process [81]. However, after the initial adsorption stages, a competition occurs between dye molecules for the adsorption sites on the surface of adsorbent, resulting in a decrease of adsorption rate [61]. Therefore, in general, it is evident that the higher percentage of dye removal decreases with an increase in the initial dye concentration. However, the amount of dye adsorbed per unit adsorbent mass increases with increase in initial dye concentration due to the decrease in resistance to the uptake of solute from solution of dye [27, 57].

Table 5. The results of various reported studies on the initial dye concentration effect on dye adsorption by various agricultural solid waste adsorbents

Adsorbents	Dye	Initial dye concentration range (mg/L)	Percentage (%) of removal range	References
De-oiled algal biomass	Methylene blue	25-125	Decrease	[85]
Neem leaf	Malachite green	10-20	Decrease	[86]
Coconut coir (acid treated)	Acid Orange 7	5-20	Decrease	[87]
Saw dust	Acid Yellow 23	1-15	97-71	[88]
Oil palm tree sawdust (tartaric acid modified)	Malachite green	100-300	99.97-83.46	[83]
Pine leaves	Methylene blue	10-90	96-41	[13]
Sugarcane bagasse	Rhodamine B	100-500	99.1-87.1	[89]
Sugarcane bagasse	Basic Blue 9	250-500	94-55.5	[89]
Modified sawdust	Methylene blue	25-500	91.2-66.3	[80]
Rice husk	malachite green	10-30	82.5-71	[90]
Raw mango Seed	Methylene blue	50-250	99.1-92.5	[82]
Modified mango seed	Methylene blue	50-250	99.9-96.9	[82]
Apricot seed	Astrazone Black	50-500	91-62	[91]
Coconut bunch waste	Methylene blue	50-500	57-13	[92]
Sunflower seed hull	Methyl violet	25-300	99.9-45	[81]

Senthil Kumar et al. [82] studied the effect of initial dye concentration on the adsorption of Methylene Blue (MB) dye by Raw mango seeds and they reported that at an initial dye concentration range from 50 mg/L to 250 mg/L, the percentage of dye removal was decreased from 99.136% to 92.523%. Low et al. [83] studied the effect of initial dye concentration on the adsorption of methylene blue (MB) dye by modified bagasse and they reported that at an initial dye concentration range from 100 mg/L to 300 mg/L, the percentage of dye removal was decreased from 99.94% to 86.59%. Zhang et al. [84] studied the adsorption of Methyl Orange by Chitosan/Alumina interface and it was found that the percentage of dye removal decreased from 99.53% to 83.55% with the increase of MB dye concentration. Table 5 presented the results of various reported studies on the of initial dye concentration effect on dye adsorption.

3.1.3. Influence of Solution Temperature on Dye Adsorption

Temperature is an important parameter for adsorption processes. Changing in temperature changes the equilibrium capacity of the adsorbent for a particular adsorbate. To study the effect of temperature on the adsorption of dye process, it can be carried out by preparing adsorbent–adsorbate solution with fixed dye concentration and adsorbent dose but with different temperature [27].

Table 6. Influence of temperature on the adsorption of dyes using various agricultural solid waste adsorbents

Adsorbents	Dye	Temperature range (°K)	Type of process	References
Cone of pinus bruitia	Congo red	295-333	Endothermic	[93]
Peanut hull	Novacorn golden yellow	303-333	Exothermic	[94]
Modified magnetic corn straw	Methylene blue	303-323	Exothermic	[95]
Rice husk (nitric acid treated)	Malachite green	303-333	Endothermic	[90]
Saw dust	Acid Yellow 23	298-318	Endothermic	[88]
Sugarcane bagasse	Rhodamine B	303-323	Endothermic	[89]
Sugarcane bagasse	Basic Blue 9	303-323	Endothermic	[89]
Treated Rice husks	Methylene blue	293-313	Endothermic	[96]
Peanut husk	Indosol Black	303-333	Exothermic	[97]
Activated Bamboo waste	Reactive Black 5	303-323	Endothermic	[98]
Pine leaves	Methylene blue	313-333	Endothermic	[13]
Pine Cone biomass	Methylene blue	303-333	Endothermic	[99]

Increasing the temperature is known to increase the rate of diffusion of the adsorbate molecules across the external boundary layer and in the internal pores of the adsorbent particle, owing to the decrease in the viscosity of the solution [75]. If the amount of adsorption increases with increasing temperature then the adsorption is an endothermic process. This may be a result of increase in the mobility of the dye molecule with an increase in their kinetic energy, as well as an enhanced rate of intraparticle diffusion of sorbate with the rise in temperature [93]. An increasing number of molecules may also acquire sufficient energy to undergo an interaction with active sites at the surface [75, 93]. Whereas the decrease of adsorption capacity with increasing temperature indicates that the adsorption is an exothermic process. This may be due to increasing temperature decrease the adsorptive forces between the dye species and the active sites on the adsorbent surface as a result of decreasing the amount of adsorption [28]. Argun et al. [62] reported that the adsorption capacity of Pomelo peel for the removal of Reactive Blue 114 dye increased with increase in solution temperature. Table 6 showed compilation results of various studies on the effect of temperature on dye adsorption by various agricultural solid waste adsorbents.

3.1.4. Influence of Adsorbent Dosage on Dye Adsorption

Adsorbent dosage is also an important parameter for the optimisation of biosorption system. The effect of the amount of adsorbent on the adsorption process can be determined by preparing adsorbent–adsorbate solution with different amount of adsorbents added to fixed initial dye concentration and fixed initial solution pH and then shaken together until equilibrium time [27]. Generally the percentage of dye removal increases with increasing adsorbent dosage. This higher percentage removal at higher adsorbent dosages is due to the increase in surface area resulting from the increase in adsorbent mass, thus increasing the number of active adsorption sites for adsorption.

Table 7. Influence of adsorbent dosage on the percentage removal of dye using various agricultural solid waste adsorbents

Adsorbents	Dye	Adsorbent dosage	Percentage (%) of removal range	References
Palm tree flower	Methylene blue	0.05-0.3 g	40.75-91.65	[102]
Watermelon seed hull	Methylene Blue	0.1-1.0 g/50 mL	27.71-97.65	[72]
Modified sawdust	Methylene blue	1.5-5 g	34.4-96.6	[80]
Raw Mango Seed	Methylene blue	0.1-1.2 g	99.5-68	[82]
Modified mango seed	Methylene blue	0.1-1.2 g	99.8-79	[82]
Tea waste	Basic yellow 2	2-20 g.L-1	19-60	[103]
Tea waste	Acid orange 7	2-20 g.L-1	90-99	[103]
Pine cone	Congo red	0.01-0.03 mg	13.45-18.96	[78]
Loquat seed	Reactive black 5	5-30 g.L-1	83.52-90.36	[104]
Rice husk	Direct Red 31	0.05-0.5 g	50.11-90.96	[105]
Rice husk	Direct Orange 26	0.04-0.1 g	35.57-68.63	[105]
Spent tea bags	Methylene blue	0.05-0.6 g	43.8-95.5	[106]
Papaya Seeds	Methylene blue	0.05-1.0 g	43.6-80	[107]
Cashew nut shells	Congo red	5-30 g.L-1	56.3-99.3	[108]
Treated saw dust	Brilliant Green	1-30 g.L-1	61-99.9	[109]
Modified sugarcane Bagasse	Methyl red	0.2-1.0 g/100 mL	67.8-83.2	[110]
Indian rosewood sawdust	Methylene blue	0.2-1.0 g/100 mL	45.1-97.1	[22]
Rice hull	Reactive orange 16	0.02-0.08 g	21.7-56.2	[111]

Feng et al. [100] studied the effect of adsorbent dose on the removal of Methylene blue dye by Swede rape straw and they observed dye removal percentage increased with decreased adsorbent dose. Ata et al. [101] studied the adsorption of Coomassie Brilliant Blue on wheat bran at different amount of adsorbent and they concluded that the percentage of adsorption increased by increasing the amount of adsorption using various agricultural adsorbents. Table 8 shows reported studies on the effect of adsorbent dosage on the percentage of dye removal onto various agricultural adsorbents. Franca et al. [79] reported the effects of doses on the removal of Methylene Blue (MB) dye using spent coffee grounds. The compilation results evaluated from many research studies on the effect of doses on dye adsorption by various agricultural solid waste adsorbents are presented in Table 7.

3.1.5. Influence of Presence of Salts on Dye Adsorption

Presence of electrolytes in solution influences strongly interfacial potential, thus it influences also adsorption process. The wastewater containing dye has commonly higher salt concentration; therefore ionic strength effect is of some importance in dye adsorption onto adsorbents. The effect of ionic strength on adsorption can be carried out by adding different doses of salts (normally NaCl) to the adsorbent–adsorbate solution and shaking up to equilibrium time [112]. Generally the removal efficiency of adsorbent is decreased with the increase in salt (NaCl) concentration. This behaviour is attributed to the competitive effect between adsorbent ions and Na^+ for the limited active sites available during the adsorption process [113]. This implied that the mechanism is mainly ion exchange and the activity of adsorption sites reduce as the salt concentration increases [114]. Additionally, the electrostatic interaction between sorbent and sorbate may be screened by Na^+, and the great ionic strength affected on the activity coefficient of adsorbents [73]. The increase in salt concentration changes the equilibrium constant between the interface and the bulk of the liquid ultimately affects adsorption. Further percentage removal efficiency of dye adsorption with various salt concentrations is due to the change in diffuse double layer thickness of Gouy-Chapman adsorption model and hence change in electrostatic force of attraction. Theoretically, when the electrostatic forces between the adsorbent surface and adsorbate ions are attractive, an increase in ionic strength will decrease the adsorption capacity. Conversely, when the electrostatic attraction is repulsive, an increase in ionic strength will increase adsorption [112]. This behaviour may be due to the dye dimerization in solution or formation of dye aggregation induced by the effect of salt ions which is known as “salting-out-effect” [115, 116]. This phenomenon comprises of two facts: (i) reduction of dye solubility due to bulk phase addition of salts and (ii) stimulate the MB adsorption at the interface by changing the equilibrium constant between adsorbed dye molecules at interface and those being dissolved in bulk phase [116]. Basically, various types of short range forces such as Van der Waals force, dipole-dipole and ion-dipole force increased with salt concentration during dye dimerization [116]. The reported studies on the effect of ionic strength on the percentage of dye removal using various agricultural solid waste materials as adsorbents are shown in Table 8.

Table 8. The results of various reported studies on the effect of ionic strength on dye adsorption by various agricultural solid waste adsorbents

Adsorbents	Dye	Initial salt (NaCl) concentration range (mg/L)	Percentage (%) of removal range	References
Cinnamomum camphora sawdust	Malachite green	0-0.05	89.21-64.75	[73]
Modified peanut husk	Light green	0-0.1	decrease	[64]
Sugarcane bagasse	Foron blue	0.2-1%	increase	[117]
Pirina (treated with nitric acid)	Remazon brilliant blue	0.05-0.3	96.45-86.15	[118]
Pleurotus ostreatus	Malachite green	0.05-0.2 M	87.22-49.01	[119]
Modified breadnut peel	Malachite green	0.0-1.0 M (KNO_3)	decrease	[120]
Modified phoenix tree leaves	Crystal violet	0.0-0.1 M	decrease	[114]
Fallen leaves	Methylene blue	0.0-0.1 M	increase	[121]
Lotus leaf	Malachite Green	0-0.2 M	81.7-62.5	[122]
Melon peel	Methylene blue	0-500 (K_2SO_4)	increase	[123]
Pine cone	Methylene blue	0-200	decrease	[99]
Potato peel	Malachite green	0-2.0 g	decrease	[113]

3.2. Kinetics of Dye Adsorption

The kinetics of dyes adsorption onto adsorbent materials is prerequisite for choosing the best operating conditions for the full-scale batch process [27]. The study of adsorption dynamics describes the solute uptake rate and evidently this rate controls the residence time of the adsorbate uptake at the solid-solution interface. To design the adsorption system, this solute uptake rate plays the most significant role by determining the residence time required for completing the adsorption reaction and can be enumerated from kinetic analysis. Therefore the adsorption rate is an important factor for a better choice of material to be used as an adsorbent; where the adsorbent should have a large adsorption capacity and a fast adsorption rate [112]. In order to study the mechanism of sorption and potential rate determining steps, most of the adsorption studies used pseudo-first-order and pseudo-second-order models which are presented below:

3.2.1. Lagergren Pseudo-First-Order and Pseudo-Second Order Kinetic Model

The linearized integral form of the pseudo-first-order model is generally expressed as [124] :

$$\log (q_e - q_t) = \log q_e - \frac{k_1}{2.303} t \qquad (1)$$

where q_t and q_e are the adsorption capacity at time t and at equilibrium, respectively (mg g^{-1}), k_1 is the rate constant of pseudo-first-order adsorption (min^{-1}) and t is the contact time (min).

the slope and intercept of a linear plot of log ($q_e - q_t$) versus t will give the value of k_1 and q_e respectively.

Similarly linearized form of Lagergren pseudo-second-order adsorption kinetic is as follows [125]:

$$\frac{t}{q_t} = \frac{1}{k_2 q_e^2} + \frac{1}{q_e} t \qquad (2)$$

where k_2 (g mg^{-1} min^{-1}) is pseudo-second-order rate constant of adsorption. A plot between t/q_t vs. t gives the value of rate constant k_2 (g/mg min), initial sorption rate h (mg/g-min) and also q_e (mg/g) can be calculated.

The constant k_2 is used to calculate the initial sorption rate h, at t→0, as follows:

$$h = k_2 q_e^2 \qquad (3)$$

Thus, the rate constant k_2, initial adsorption rate h and predicted q_e can be calculated from the plot of t/q_t versus time t using Eq. (2).

3.2.2. Intra-Particle Diffusion Kinetic Model and Mechanism of Solute Adsorption

The mechanism of adsorption techniques involve four steps: migration of dye molecules from bulk solution to the surface of the sorbent, diffusion through the boundary layer to the surface of the sorbent, adsorption at a site, and intra-particle diffusion into the interior of the sorbent [126]. According to Weber and Morris [127], the amount of adsorption varies proportional with $t^{1/2}$ for most adsorption processes and can be expressed as:

$$q_t = K_{id} t^{0.5} \qquad (4)$$

where q_t is the adsorption capacity at time t, $t^{0.5}$ is the half-life time in second, K_{id} (mg/g min$^{0.5}$) is the rate constant of intraparticle diffusion. A plot q_t versus $t^{0.5}$ will give a linear relationship where K_{id} can be determined from the slope of the plot.

Usually the kinetic adsorption data is better represented by pseudo-second-order model for most of dye adsorption systems. Zolgharnein et al. [128] studied the adsorption of Methyl Violet by *Platanus Carpinifolia* tree leaves and they found that the linear correlation coefficient R^2 values of the Pseudo-second-order model was higher (0.99) fitted than the Pseudo-first-order kinetic (0.79). Dursun et al. [129] evaluated the adsorption of Remazol Black B dye using sugar beet pulp and they found that the applicability of Pseudo-second over Pseudo-first order models. Rebitanim et al. [130] studied the use of raw empty fruit bunch biomass for adsorption of Methylene blue and they found that the values of the regression correlation coefficient R^2 for Pseudo-first-order model were between 0.83 and 0.86, while the values of R^2 for the second order model were between 0.992 and 0.999 indicating applicability of second order model was more accurate. Table 9 presented compilation results of the applicability of Pseudo-second-order Kinetic models for dye adsorption using several agricultural solid waste adsorbents.

Table 9. Various Kinetic studies for dye adsorption by various agricultural solid waste adsorbents

Adsorbents	Dye	Fitted Kinetic Model	References
Acer tree leaves	Methylene blue	Pseudo-second-order	[131]
Reed	Basic yellow 28	Pseudo-second-order	[132]
Peach gum	Methyl violet	Pseudo-second-order	[133]
Indian jujube seed powder	Acid blue 25	Pseudo-second-order	[134]
Modified Oil palm tree sawdust	Malachite green	Pseudo-second-order	[83]
Modified phoenix tree leaves	Crystal violet	Pseudo-second-order	[114]
Raw and modified pine cone	Methylene blue	Pseudo-second-order	[135]
Raw and modified mango seed	Methylene blue	Pseudo-second-order	[82]
Treated Onion Skins	Methylene blue	Pseudo-second-order	[136]
Almond shell	Methyl violet	Pseudo-second-order	[137]
Treated saw dust	Brilliant green	Pseudo-second-order	[109]
Rice husk	Direct orange 26	Pseudo-second-order	[105]
Wood apple shell	Crystal violet	Pseudo-second-order	[67]
Beech wood sawdust	Basic blue 86	Pseudo-second-order	[138]

3.3. Equilibrium Dye Adsorption Isotherms

Adsorption isotherms are crucial in understanding the mechanism of adsorption and the analysis of the isotherm data by fitting them to different isotherm models is an important step to understand the suitable model that can be used for design purposes. Among several isotherm models presented in the literature [139], Langmuir and Freundlich models are the earliest and simplest known relationships describing the adsorption equation [140] and therefore most widely used to describe the adsorption isotherm.

3.3.1. Langmuir Adsorption Isotherm Model

The Langmuir model of adsorption is obtained on the basis of ideal assumption that the intermolecular forces decrease rapidly with distance, and consequently, predicts the existence of monolayer coverage of adsorbate at the outer surface of the adsorbent, which is assumed to be structurally homogeneous [141]. The activities of the surface sites are proportional to their concentration and the number of sorption sites is fixed; no further adsorption can take place after the saturation point. For solid/liquid systems, the linear form of the Langmuir equation is:

$$q_e = \frac{q_m K_a C_e}{1 + K_a C_e} \quad (5)$$

where q_e is the amount of dye adsorbed at equilibrium time (mg/g), C_e is equilibrium concentration of dye in solution (mg L^{-1}), q_m is maximum adsorption capacity (mg/g) and K_a is isotherm constants for Langmuir (L mg^{-1}).

The linearized form of Langmuir isotherm can be written in two different forms:

A) Langmuir-I Isotherm Model

$$\frac{C_e}{q_e} = \frac{C_e}{q_m} + \frac{1}{K_a q_m} \tag{6}$$

The slop and intercept of plot between C_e/q_e vs. C_e will give q_m and K_a respectively.

B) Langmuir-II Isotherm Model

$$\frac{1}{q_e} = \frac{1}{K_a q_m}\frac{1}{C_e} + \frac{1}{q_m} \tag{7}$$

In this form q_m and K_a are determined from plotted between $1/q_e$ vs. $1/C_e$. The separation factor R_L can be found from Langmuir isotherm which is:

$$R_L = \frac{1}{1+K_a C_0} \tag{8}$$

where K_a is the Langmuir constant and C_0 is the initial MB dye concentration (ppm).

R_L indicates the nature of adsorption [142] as indicated below:

- unfavorable $R_L > 1$;
- linear $R_L = 1$;
- favorable $0 < R_L < 1$;
- irreversible $R_L = 0$.

3.3.2. Freundlich Adsorption Isotherm Model

The Freundlich adsorption isotherm model considers a heterogeneous adsorption surface that has unequal available sites with different energies of adsorption. The Freundlich adsorption isotherm model can be represented as follows [143]:

$$q_e = K_f\, C_e^{\frac{1}{n}} \tag{9}$$

The linearized form of Freundlich can be expressed as:

$$\ln q_e = \ln K_f + \frac{1}{n}(\ln C_e) \tag{10}$$

where q_e is the amount of metal ion adsorbed at equilibrium time (mg/g), C_e is equilibrium concentration of dye in solution (mg L^{-1}). K_f is the capacity of the adsorbent and n is the intensity of adsorption constant for Freundlich. The plot of $\ln q_e$ versus $\ln C_e$ is employed to determine the K_f and n from intercept and slope respectively. Generally the value of the linear correlation coefficient R^2 gives the indication which model to be chosen to give best-fit. Nausheen et al. [144] used peanut husk for the removal of Drimarine Red HF-3D dye from its aqueous solution and they have reported that the maximum monolayer adsorption capacity was 142.857 mg/g. Table 10 shows various fitted isotherm models for dye adsorption by various agricultural adsorbents.

Table 10. Various fitted isotherm models for dyes adsorption by various agricultural solid waste adsorbents

Adsorbents	Dye	Fitted Isotherm Model	References
Cucumber peel	Methylene blue	Langmuir	[72]
Pistachio shell	Remazol red	Freundlich	[145]
Pumpkin husk	Reactive red 120	Freundlich	[146]
Bagasse	Indosol Turquoise FBL	Langmuir	[147]
Sugar beet pulp	Remazol black B	Langmuir	[129]
Coconut coir (acid treated)	Acid orange 7	Langmuir and Freundlich	[87]
Poplar leaf	Methylene blue	Langmuir	[69]
Pine cone	Methylene blue	Langmuir	[148]
Green pea peels	Methylene blue	Langmuir	[149]
Pine leaves	Basic red 46	Langmuir	[150]
Modified barley straw	Acid Blue	Langmuir	[151]
Gulmohar plant leaf	Methylene blue	Langmuir	[152]
Pine cone (acid treated)	Congo red	Freundlich	[78]

3.4. Dye Adsorption Capacities of Various Agricultural Solid Waste Adsorbents

At present, the interests in utilization of cheap alternatives have been significantly increased and many attempts have been made by researchers on feasibility of biosorption potential of agricultural solid waste materials (either in its raw or modified form) as an economic and eco-friendly options. Waste treatment by adsorption using agricultural solid wastes as adsorbents has double benefits i.e., water treatment and waste management. The basic components of the agricultural waste materials include hemicellulose, lignin, lipids, proteins, simple sugars, water, hydrocarbons, and starch, containing variety of functional groups. Agricultural materials particularly those containing cellulose shows potential sorption capacity for various pollutants. In the review article, Bhatnagar et al. [153] mentioned that agricultural waste materials being economic and eco-friendly due to their unique chemical composition, availability in abundance, renewable nature and low cost are viable option for water and wastewater remediation. Agricultural waste is a rich source for activated carbon production due to its low ash content and reasonable hardness [154], therefore, conversion of agricultural wastes into low-cost adsorbents is a promising alternative to solve environmental problems and also to reduce the preparation costs. In the last several decades, various agricultural wastes have been explored as low-cost adsorbent. These agricultural waste materials can be used in their natural form or after some physical or chemical modification. Different physical or chemical treatment was applied to the raw biomass adsorbents for improving their adsorption capacities which also reported [155]. Numerous studies have been conducted to develop cheaper and effective adsorbents from a variety of agricultural solid waste materials for the removal of dyes from wastewater. The various effective reported agricultural biomass adsorbents such as the shells and/or stones of fruits like olive wastes [156], pineapple stem [157], ground hazelnut shells [158, 159], banana waste [160, 161], raw date pits [162], coconut bunch waste [92], mango seed kernel [163], lemon peel [164], sawdust [22, 165], sugarcane [166], bagasse pith [55], waste banana pith [167], grass waste [168, 169], guava leaves [170, 171], gulmohar (*Delonix regia*) plant leaf powder [152], pine leaves [13], cotton waste [172, 173], green pea peels [149], broad bean peels [174], loquat seed [104], castor seed shell [126], pumpkin seed hull [175], dehydrated peanut hull [176], coconut husk [177], coffee husks [178], peanut hull [179], pine cone biomass [78, 99, 180], and waste resulting from the production of cereals such as rice barn [181], rice straw [182], wheat barn [181], wheat shells [183], wheat straw [184]; have been successfully employed for the removal of dyes from aqueous solutions for wastewater treatment. Table 11-13 summarized the up-to-date information on the dye adsorption capacity of various agricultural solid wastes. Readers are encouraged to go through many review articles on the removal of dyes by agricultural wastes adsorbents by Sharma et al. [57], Salleh et al. [28], Bharathi and Ramesh [112] and Yagub et al. [27]. However selectively widely used agricultural biomasses and their effective use in the removal of dyes are presented below:

3.4.1. Rice Husk

Rice husk is an agricultural waste and a by-product of the rice milling industry. Rice husk is insoluble in water, has good chemical stability, high mechanical strength, and possesses a granular structure, making it a good adsorbent material [57]. The utilization of this source of agricultural waste would solve both a disposal problem as well as access to a cheaper material for adsorption in water pollutants control system [185]. Safa et al. [105] studied rice husk for the removal of Direct Red-31 and Direct Orange-26 from aqueous solutions in a batch mode. Experiments were carried out as function of pH, biosorbent dose, particle size of biosorbent, initial dyes concentration, contact time and temperature. The equilibrium biosorption data were analyzed by Langmuir, Freundlich, Temkin, Doubinin–Radushevich (D–R) and Harkins–Jura isotherm models. The results indicated that the Langmuir model provided the best correlation ensuring adsorption capacities of 129.87 mg g^{-1} for Direct Red-31 and 66.67 mg g^{-1} for Direct Orange-26 dyes and pseudo-second-order kinetic equation explained the biosorption kinetics of dyes on rice husk well. Lakshmi et al. [186] carried out study of the adsorptive characteristics of Indigo Carmine (IC) dye from aqueous solution onto Rice Husk Ash (RHA). Batch experiments were carried out to determine the influence of parameters like initial pH, contact time, adsorbent dose, and initial dye concentration on the removal of IC. The pseudo-second order kinetic model represented the adsorption kinetics of IC on to RHA. Equilibrium isotherms were analysed by Freundlich, Langmuir, Temkin, and Redliche Peterson models using a nonlinear regression technique. Adsorption of IC on RHA was favourably influenced by an increase in the temperature of the operation. The positive values of the change in entropy (ΔS^o) and heat of adsorption (ΔH^o); and the negative value of change in Gibbs free energy (ΔG^o) indicate feasible and spontaneous adsorption of IC on to RHA. An adsorption capacity in the range of 29.3–65.9 mg g^{-1} was reported at different temperatures

(293–323 K). The possible utilization of rice husk ash as an adsorbent for methylene blue dye from aqueous solutions has been investigated by Chandrasekhar and Pramada [187]. The highest adsorption capacity was found to be approximately 690 mg g^{-1}. Kaur and Prasad [188] studied the removal of acidic dyes, viz. acid violet 54, acid violet 17, acid blue 15, acid violet 49 and acid red 119 from aqueous solutions using rice husk ash. The adsorption capacity was found to vary from 99.4 to 155 mg g^{-1}, suggesting rice husk ash as a good adsorbent for dyes removal. McKay et al. [173] examined the potential of rice husk for the removal of two basic dyes, safranine and methylene blue, and adsorption capacity of 838 and 312 mg g^{-1}, respectively was reported.

3.4.2. Sawdust

Sawdust, an abundant by-product of the wood industry is either used as solid fuel for cooking or as packing material which contains various organic compounds (lignin, cellulose and hemicellulose) with polyphenolic groups that might be useful for binding dyes through different mechanisms [24]. Sawdust is easily available at the countryside at zero or negligible price [22]. The role of sawdust materials in the removal of pollutants from aqueous solutions has been reviewed recently [165]. The efficiency of chir pine sawdust (CPS) for adsorptive removal of the dyes, congo red (CR) and basic violet 1 (BV), from aqueous solution was evaluated by Khan et al. [189]. The adsorption of BV obeyed the Langmuir isotherm model while the Freundlich isotherm fitted the equilibrium data of CR adsorption. The Langmuir monolayer adsorption capacities of CPS for BV and CR were 11.3 and 5.8 g kg^{-1}, respectively. The kinetic data for CR were best fitted to the Lagergren pseudo-first order model and for BV to the pseudo-second-order model. Dulman and Cucu-Man [138] investigated the effect of Beech wood sawdust on the adsorption of six reactive dyes in aqueous solution, namely C.I. Direct Blue 6, C.I. Direct Brown 2, C. I. Direct Green 26, C.I. Direct Brown, C.I. Reactive Red 3. C.I. Basic Blue 86. The percentage removal of the reactive dyes Direct Brown 2 and Direct Brown decreased with increase in pH (above pH 10) and the maximum removal rate was occurred at pH 3. Consequently, the percentage color removal of Direct Brown 2 and Direct Brown dropped from 98.6 to 34.7 % and 94.4 to 28.5 %, respectively. For Basic Blue the sorption has a maximum value of 97 % at pH 4.43–7.06. Garg et al. [22] investigates the potential use of Indian Rosewood (*Dalbergia sissoo*) sawdust, pretreated with formaldehyde and sulfuric acid, for the removal of methylene blue dye from simulated wastewater. Higher adsorption percentages were observed at lower concentrations of methylene blue. Optimum pH value for dye adsorption was determined as 7.0 for both the adsorbents. The adsorption of methylene blue followed a first order rate equation and fit the Lagergren equation well. Khattri and Singh [190] noted that the adsorption capacity of Neem sawdust for the removal of dyes (Crystal Violet, Methylene Blue, Malachite Green and Rhodamine B) was highly concentration dependent. Ho and McKay [191] also showed that the sorption capacity of basic dye is much higher than that of acid dye because of the ionic charges on the dyes and the ionic character of sawdust. Most of the studies showed that sawdust in natural form or modified form is highly efficient for the removal of Methylene Blue (MB) [24].

3.4.3. Sugarcane Bagasse

Sugarcane bagasse is the fibrous matter that remains after sugarcane or sorghum stalks are crushed to extract their juice. Since sugarcane is a renewable plant, its bi product –

bagasse is a great raw material to use as an adsorbent product. The study conducted by Saad et al. [110] to find out the possibility of using phosphoric acid (H_3PO_4) treated sugarcane bagasse to remove methyl red dye. The efficiency of dye removal by H_3PO_4 treated bagasse was less than activated carbon and followed by untreated bagasse. Study on the effect of pH indicated that activated carbon decolourization remained 100% for all pH values, whereas for both treated and untreated bagasse, lowest percentage of dye removal was recorded at pH of 2. Dye adsorption was significantly increased between the pH value of 3 and 6 and then gradually decreased in the pH range of 7–10. This trend illustrated that ion exchange mode and electrostatic attraction between dye anions and negatively charged surface of biosorbent. The kinetics of methyl red adsorption followed the pseudo-first-order kinetic expression and Langmuir isotherms model fit well the experimental data. Azhar et al. [192] in one of his papers has reported that adsorbents prepared from sugarcane bagasse, an agro industries waste was successfully used to remove the methyl red from aqueous solution in a batch reactor. This study investigated the potential use of sugarcane bagasse, pre-treated with formaldehyde (PCSB) and sulphuric acid (PCSBC), for the removal of methyl red from simulated wastewater. Rachakornkji et al. [193] explored bagasse pith from sugarcane industry without any pre-treatment for the removal of three reactive dyes, Remazol Black B, Remazol Brilliant Blue and Remazol Brilliant Red from aqueous solutions. High percentage removal was observed for the adsorption of reactive dyes in the order of 58.48–98.03 % for RR Black, 46.15–93.47 % for RB Blue and 46.30–94.60 % for RB Red, respectively.

3.4.4. Fruit Waste

Peel, also known as skin, is the outer protective layer of a fruit or vegetable, currently gaining wide attention as adsorbent in water treatment. Peels of different fruits such as, orange, banana, watermelon, cassava, mango, etc. can be effectively used as adsorbents for wastewater treatment [57].

Formosa papaya (*Carica papaya* L.) seed powder (FPSP), a solid by-product from agricultural activities, was investigated by Pavan et al. [61] as an alternative adsorbent for the removal of crystal violet (CV) from aqueous solutions. The FPSP was characterized by specific surface area (BET), scanning electron microscopy (SEM), infrared spectroscopy (FT-IR), thermal analysis (TGA) and Boehm titration techniques. The effects of initial pH of solution, adsorbent dosage, contact time and initial dye concentration on CV adsorption were studied using batch contact mode at 25°C. The pseudo-second order model agreed very well with the kinetic data. The equilibrium adsorption was analyzed by Langmuir, Freundlich and Redlich–Peterson isotherms. The adsorption of CV onto FPSP was well fitted using Langmuir isotherm with a maximum adsorption capacity of 85.99 mg g^{-1}.

Asgher and Bhatti [194] compared raw, immobilized and acetic acid-treated citrus waste to remove reactive blue 19 and reactive blue 49 from aqueous solution. Very excellent performance was achieved for both dyes. The physio-sorption was the predominant mechanism of removing reactive blue 19 and reactive blue 49 biosorption using citrus waste. The biosorption process of both dyes adequately followed all Langmuir, Freundlich and Temkin isotherms.

Acetone-treated capsicum seeds were used for reactive blue 49 uptakes in batch and continuous column systems. This process well described by Langmuir model in both batch and continuous modes. In batch biosorption experiments, ionic strength increase made a slight decrease in dye removal efficiency, while there were no significant changes obtained by increasing initial dye concentration [195].

Akar et al. [196] studied RR198 biosorption onto olive pomace in synthetic and real wastewater treatment. They found this process was spontaneous and endothermic in nature by calculating the thermodynamic parameters and well fitted by Langmuir isotherms models. RR198 biosorption onto olive pomace was independent of ionic strength in the concentration range of 0.01–0.15 M, whereas, it decreased in the ionic strength over 0.15 M. Besides, there was no tangible decrease in biosorption capacity of olive pomace when it was utilized for treating real wastewater containing several interfering species.

Garlic peel was investigated by Hameed and Ahmad [75] for the removal of Methylene blue from aqueous solution. Equilibrium isotherms were determined and analyzed using the Langmuir, Freundlich and Temkin isotherms. The maximum monolayer adsorption capacities were found to be 82.64, 123.45 and 142.86 mg g^{-1} at 303, 313 and 323 K, respectively. For Methylene blue, author observed that the adsorption capacity was higher due to the presence of polar functional group.

Dogan et al.[159] studied hazelnut shell, an agricultural waste, without any pretreatment for the removal of Methylene blue with respect to the initial dye concentration, pH, ionic strength, particle size and temperature. Kinetic studies showed that the kinetic data were well described by the pseudo-second-order kinetic model. Significant increases in initial adsorption rate were observed with the increase in temperature followed by pH and initial MB concentration.

Jack fruit peel has been investigated as adsorbents for removal of Methylene blue by Hameed [197]. The effect of different system variables like adsorbent dose, initial dye concentration, contact time and pH were evaluated and found that as the amount of adsorbent increased, the percentage of dye removal increased accordingly. Low concentrations of Methylene blue favored high adsorption percentages and the optimum pH value for dye adsorption was found to be 4.0. Among four different types of linearized Langmuir isotherm, the Freundlich isotherm and the Temkin isotherm, the equilibrium biosorption data were best fitted with the type 2 Langmuir model. The sorption capacity of Methylene blue on jack fruit peel was found to be 285.713 mg g^{-1}.

Pavan et al. [198] reported that the use of yellow passion fruit (YPFW), a powdered solid waste, was tested as biosorbent for the removal of a cationic dye MB from aqueous solution. Adsorption of MB onto this low-cost natural adsorbent was studied by batch adsorption at 25°C. The effects of shaking time, biosorbent dosage, and pH on adsorption capacities were studied. In alkaline pH region, the adsorption of MB is favourable. The adsorption kinetics was better fitted to pseudo-first-order and ion exchange kinetic models. The equilibrium data was fitted to Langmuir, Freundlich, Sips, and Redlich–Peterson isotherm models. The maximum amount of MB is absorbed by YPFW biosorbent was 44.70 mg g^{-1}.

Banana stalks were studied as adsorbents by Hameed et al. [81] for basic dyes in aqueous solutions with equilibrium isotherms and kinetic adsorptions. High adsorption capacity of 243.90 mg g^{-1} was observed and authors suggested that banana stalks consists of cellulose and lignin; it is the polyol structure of cellulose-based materials that has relatively strong chemical adsorption of cations such as metal ions and organic bases as well as physical adsorption of other material such as acidic and anionic compounds.

Orange peels have been investigated as an adsorbent by Sivaraj et al. [199] for the removal of an acid dye: acid violet17. The adsorption capacity was 19.88 mg g^{-1} at initial pH of 6.3. The equilibrium time was found to be 80 min for 10, 20, 30, and 40 mg L^{-1} dye concentration, respectively. A maximum removal of 87% was obtained at pH 2.0 for an adsorbent dose of 600 mg 50 mL^{-1} of 10 mg L^{-1} dye concentration.

3.4.5. Plant Waste

Leaf biomass, an agricultural waste, widely available was studied as an alternative adsorbent for different pollutants as well as dyes by various investigators (Zolgharnein et al. [128]; Belay and Abebe [86]; Ponnusami et al. [152]; Weng et al. [200]; Han et al. [201]). The plant leaves have porous structure which can effectively adsorb dye molecules. The functional groups on the leaf surface can attract ionic dye molecules of opposite charge which lead to increase in dye removal efficiency.

Zolgharnein and Bagtash [131] studied Acer tree leaves as adsorbent for the removal of Methylene blue from aqueous solution. The materials were studied without chemical treatment. Authors found that the adsorption was favourable at lower pH and the biosorption is dominated by chemical adsorption. It was also found that both the Langmuir and Freundlich models were applicable for describing the adsorption equilibrium of MB by the adsorbent.

Phoenix tree leaf, modified with NaOH solution was also been studied as an adsorbent by Ren et al. [114], for the removal of Crystal violet from aqueous solution. The Phoenix tree leaves contain abundant floristic fiber, protein and some functional groups such as carboxyl, hydroxyl and amidogen, etc., which make favourable biosorption process. The leaf is a waste product with practically no cost but its adsorption capacity of 510.3 mg g^{-1} at 293 K was obtained.

The use of Palm kernel fiber, a readily available agricultural waste product, for the sorption of Methylene Blue from aqueous solution and the possible mechanism of sorption has been investigated by Ofomaja [202]. Analysis of the kinetic data at different sorbent dose revealed that the pseudo-second-order kinetic model was to be better fit the experimental data with high correlation coefficients at the various fiber doses used. It was found that the change in hydrogen ion concentration and increase in sorption temperature were directly related to the amount of dye sorbed, and activation energy was calculated as −39.57 kJ mol^{-1}, indicating that the dye uptake is chemisorption, involving valence forces through sharing or exchange of electrons between sorbent and sorbate as covalent forces.

The development, characterization, and application of adsorbents prepared from avocado kernel seeds were discussed by Elizalde-González et al. [203], and they come to the conclusion that a mayor adsorption capacity of the non-carbonized adsorbent in comparison with carbonized samples is due to the greater amount of surface acidic groups.

Hamdaoui et al. [204] demonstrated that the dead leaves of plane tree can be successfully used for the removal of hazardous dye, Malachite Green (MG), from aqueous solutions in their study. The batch sorption process was found to be dependent upon contact time, sorbent dose, initial dye concentration, ionic strength and temperature. The obtained value of the activation energy was very low (7.13 kJ/mol), which indicated not only an activated process but a physical sorption.

Tree fern; dark brown coloured complex material (containing lignin and cellulose as major constituents) is naturally and commercially available all over the world. This variety of tree fern is generally marketed for horticultural purposes because of its ability of adsorption to retain water and manure for plants [57]. Tree fern, an agricultural by-product, has been currently investigated to remove dyes from aqueous solutions (McKay et al. [205]; Han et al. [206]; Ofomaja [207]; Ofomaja and Ho [208]).

3.4.6. Bark Material

Another abundant forest residual waste product from the timber industry is bark material, a polyphenol-rich material. Bark which is high in tanning content, freely available and literally carries no cost of processing has made itself an effective and attractive adsorbent material in removing dyes from wastewater solutions.

Adsorption of Methyl red (MR) and Methyl Orange (MO) was conducted by DIM [209] using Neem tree bark powder (NBP), Mango tree bark powder (MBP) and Locust bean tree bark powder (LBP). The adsorption isotherm revealed that the adsorption of dyes using neem tree bark powder, mango tree bark powder and locust-bean tree bark powder fitted well into the Langmuir isotherm which signifying the monolayer coverage of dyes onto adsorbent particles and also homogenous distribution of active sites on the materials.

Dave et al. [210] studied adsorption of Eriochrome Black T (EBT), an azo dye onto eucalyptus bark in batch mode. The investigations were carried out to study the effects of pH, adsorbate concentration, adsorbent dosage, contact time and temperature after treating the bark with 1% NaOH solution. The total organic content of bark was 92.7%. The kinetic studies showed that the adsorption reaction is of first order and almost 77.33% removal efficiency of EBT over eucalyptus bark could be achieved at low concentrations at all temperatures.

Srivastava and Rupainwar [211] investigated the use of wood wastes; neem bark and mange bark as the low cost adsorbent materials for the removal of Malachite Green (MG) from its aqueous solution. For the both adsorbents, the adsorption rate of MG increased with temperature which indicting the reaction to be spontaneous and endothermic in nature. The experimental data produced perfect fit with the Langmuir isotherm for both the adsorbents suggesting monolayer coverage of MG.

In their other work, Srivastava and Rupainwar [212] conducted batch experiments to investigate the potential use of Eucalyptus bark powder (EBP) as an adsorbent for adsorption of industrially important dyes namely malachite green, indigo carmine and methylene blue from wastewater. The operating variables studied were initial dye concentration, pH, temperature and contact time. It was noted that adsorption of all the dyes on eucalyptus bark powder increases with an increase in pH and temperature which was related with the zeta potential of EBP. The equilibrium data are fitted to Langmuir in comparison to Freundlich isotherm equations. Adsorption isotherm modelling shows that the interaction of the dyes with Eucalyptus bark powder surface is localized monolayer adsorption with adsorption capacities as 4.58×10^4 mol.g^{-1}, 5.87×10^4 mol.g^{-1} and 4.87×10^4 mol.g^{-1} for malachite green, indigo carmine and methylene blue respectively. Mckay et al. [173] used the teak wood bark as an adsorbent to remove the MB from aqueous solutions. The monolayer saturation capacity for MB onto teak wood bark is 914.59 mg g^{-1}. Table 11 presented the compilation results of various agricultural by-products in the removal of dyes and their adsorption capacity.

Table 11. Acid dyes adsorption capacities q_m (mg/g) of agricultural solid waste adsorbents

Material	Acid Dye	Maximum adsorption capacity, q_m (mg/g)	References
Indian jujuba seed powder	Acid blue 25	54.95	[134]
Rice stem	Acid orange 7	10.2	[63]
Potato peel	Acid blue 113, acid black 1	11.71, 1.79	[74]
Barley straw	Acid blue 40	$1.02\times10^{-4*}$	[213]
Pine sawdust	Acid yellow 132, acid blue 256	398.8, 280.3	[214]
Coir pith	Acid brilliant blue	16.67	[215]
Banana peel	Acid orange 52	21	[161]
Orange peel	Acid orange 52	20.5	[161]
Orange peel	Acid violet 17	19.88	[199]
Egyptian bagasse pith	Acid red 114, acid blue 25	20, 17.5	[216]
Egyptian bagasse pith	Acid blue 25	14.4	[217]
Banana pith	Methyl orange	21	[161]
Rice husk ash	Acid violet 54, acid violet 17, acid blue 15, acid violet 49	99.4-155	[188]
De-oiled soya	Methyl Orange	16.7	[218]
Maize cob	Acid dyes	47.7, 41.4	[219]
Hazelnut shell	Acid blue 25	60.2	[158]
Wood sawdust	Acid blue 25	5.92	[220]
Sawdust-walnut	Acid blue 25	36.98	[158]
Sawdust-oak	Acid blue 25	27.85	[158]
Sawdust-pitch pine	Acid blue 25	26.19	[158]
Bagasse pith	Acid blue 25	17.5	[216]

* mol g^{-1}

Table 12. Basic dyes adsorption capacities q_m (mg/g) of agricultural solid waste adsorbents

Material	Basic Dye	Maximum adsorption capacity q_m (mg/g)	References
Formosa papaya seed	Crystal violet	85.99	[61]
Coffee waste	Toluidine blue	142.5	[71]
Chir pine sawdust	Basic violet 1	11.3	[189]
Peach gum	Methylene blue, methyl violet	298, 277	[133]
Neem (Azadiracta indica) bark (acid treated)	Malachite green	500	[221]
Reed	Basic yellow 28	181	[132]
Modified Bagasse	Methylene blue	69.93 (at 60°C)	[83]
Modified Oil palm tree sawdust	Malachite green	65.36 (at 60°C)	[83]
Platanus fallen leaves	Methylene blue	101.01 at 50°C)	[121]
Bagasse fly ash	Orange G, methyl violet	18.79, 26.24	[77]
Tree fern	Basic red 13	408	[222]

Material	Basic Dye	Maximum adsorption capacity qm (mg/g)	References
Egyptian bagasse pith	Basic blue 69	168	[217]
Coffee residues	Basic blue 3G	179	[223]
Poplar leaf	Methylene blue	135.35	[69]
Oil palm fibre	Methylene blue	382.32	[224]
Swede rape straw	Methylene blue	246.4	[100]
Grapefruit peel	Crystal violet	254.16	[68]
Skin almond waste	Crystal violet	85.47	[225]
Wheat bran	Crystal violet	80.37	[226]
Tomato plant root	Crystal violet	94.7	[227]
Coniferous pinus bark	Crystal violet	32.78	[228]
Citrus sinensis bagasse	Methylene blue	96.4	[229]
Papaya seeds	Methylene blue	99.3	[107]
Peanut hull	Methylene blue	68.06	[179]
Banana peel	Methylene blue	20.8	[161]
Pineapple stem	Methylene blue	119.05	[157]
Garlic peel	Methylene blue	82.64	[75]
Coconut bunch waste	Methylene blue	70.92	[92]
Coffee husk	Methylene blue	90.1	[178]
Rubber seed shell	Methylene blue	82.64	[126]
Fallen phoenix tree leaves	Methylene blue	80.9	[201]
Ground hazelnut shells	Methylene blue	76.9	[158]
Walnut sawdust	Methylene blue	59.17	[158]
Yellow passion fruit waste	Methylene blue	44.7	[198]
Olive pomace	Methylene blue	42.3	[230]
Rice husk	Methylene blue	40.59	[231]
Cherry sawdust	Methylene blue	39.84	[158]
Coconut coir	Methylene blue	15.59	[232]
Indian rosewood sawdust	Methylene blue	11.8	[22]
Pineapple leaf powder	Crystal violet	78.22	[233]
Sawdust	Crystal violet	37.83	[234]
Rice husk	Crystal violet	44.87	[235]
Empty Fruit Bunch	Methylene blue	50.76	[130]
Orange Peel	Methylene blue	18.6	[161]
Mango seed kernel	Methylene blue	142.86	[163]
Jute processing waste	Methylene blue	16.6	[236]
Pine tree leaves	Basic red 46	71.94	[150]
Pine cone	Basic red 46	73.53	[237]
Canola hull	Basic red 46	49.00	[238]
Princess tree leaf	Basic red 46	43.1	[239]
Natural rice husk	Methylene blue	19.77	[240]
Spent rice biomass	Methylene blue	8.3	[155]
Pine cone	Methylene blue	109.89	[99]
Modified pine cone	Methylene blue	142.24	[135]
Modified sawdust	Methylene blue	111.46	[80]

Table 13. Adsorption capacities q_m (mg/g) of agricultural solid waste adsorbents for the removal of other dyes (apart from acid and basic)

Material	Dyes (other than acid and basic)	Maximum adsorption capacity q_m (mg/g)	References
Chir pine sawdust	Congo red	5.8	[189]
Modified peanut husk	Light green	146.2±2.4	[64]
Pomelo (Citrus grandis) peel	Reactive blue 114	16.3 (at 30°C)	[62]
Peanut hull	Novacorn golden yellow	7.28	[94]
Bagasse fly ash	Indole	85.66 (at 45°C)	[241]
Sugarcane bagasse (raw)	Foron Blue E-BL	81.3	[117]
Peanut husk (HNO_3 treated)	Drimarine red HF-3D	95.24	[144]
Pistachio shell	Remazol red	108 (at 20°C)	[145]
Pirina	Remazol brilliant blue R	23.63	[118]
Pumpkin husk	Reactive red 120	98.61	[146]
De-oiled soya	Eosin yellow	$12.16\times10^{-5*}$	[242]
Bagasse	Indosol Turquoise FBL	65.09	[243]
Loquat seed	Reactive black 5	13.75	[104]
Peanut hull	Reactive black 5	55.55	[244]
Orange peel	Direct red 23, direct red 80	10.72, 21.05	[245]
Coir pith	Direct red 28	6.72	[246]
Palm oil ash	Disperse blue, disperse red	49.5, 61.35	[247]
Soy meal hull	Direct red 80, direct red 81	178.57, 120.48	[248]
Sunflower stalk	Direct red 28, direct blue	37.78, 26.84	[249]
Banana peel	Direct red 28	18.2	[161]
Orange peel	Direct red 28	14.0	[161]
Eggshells	Congo Red	69.45	[250]
Rhizophora apiculata bark	Direct Red 23	21.55	[251]
Rice husk	Direct Red 23	4.35	[252]

* mol g^{-1}

CONCLUSION

This review article presented a wide range of agricultural solid waste materials as adsorbents for the treatment of industrial effluents that contain a large number of organic dyes using adsorption technique. These inexpensive, abundant, locally available and effective raw and treated natural waste materials from agriculture, plant waste, fruit waste and agricultural by-products are an alternative to the traditionally available aqueous waste processing methods such as coagulation, flocculation, chemical oxidation, biological treatment, etc. especially in alternative to expensive commercial activated carbon in the removal of dyes from dye bearing wastewater effluents. From the large number of published literature reviewed here, it is observed that the mechanism and kinetics of dyes adsorption on various agricultural solid waste adsorbents depends on the chemical nature of the materials and various physico-chemical experimental conditions such as solution pH, initial dye concentration, adsorbent dosage, temperature, ionic strength of the system, etc. Therefore the mechanism and dye adsorption behaviour of various adsorbents under these physico-chemical process parameters have been critically analysed in this review chapter. Therefore the widespread uses of low-cost agricultural solid waste adsorbents today are undoubtedly offer a lot of promising benefits for industrial wastewater treatment in commercial purposes due to their local availability, technical feasibility, engineering applicability, and cost effectiveness. This review article may help to propose optimum operating conditions for future studies on water and wastewater toxification, as well as to improve environmental issues of solid and aquatic industrial waste disposal. This comprehensive review analysis also revealed some important issues which need to be consider for future work such as (i) the research should not limit to only lab scale batch studies, but pilot-plant studies should also be conducted utilizing low-cost adsorbents to check their feasibility on commercial scale (ii) as the dye effluents contain several other pollutants, attention need to be given to adsorption of mixed dyes and the agricultural solid wastes should further be investigated for their efficiency using mixed dye effluents from industries (iii) only a few reported studies made some attempts to study the morphology of the adsorbent. The adsorption performance of different adsorbents not only depends on the experimental conditions and analytical methods (column, reactor, and batch techniques) but also depends on the surface morphology of the adsorbent, surface area, particle size and shape, micropore and mesopore volume, etc., therefore the characterization of adsorbents needs to be investigated properly (iv) most of the reported studies are performed in the batch process; therefore designing of the continuous flow systems are required to carry out the feasibility of various low cost alternative adsorbents at an industrial scale.

REFERENCES

[1] Sokolowska-Gajda, J., Freeman, H.S. and A. Reife, Synthetic dyes based on environmental considerations. Part 2: Iron complexes formazan dyes. *Dyes and pigments,* 1996. 30(1): p. 1-20.

[2] Carmen, Z. and Daniela, S., Textile organic dyes–characteristics, polluting effects and separation/elimination procedures from industrial effluents–a critical overview. *INTECH,* 2012: p. 978-953.

[3] Kross, B.C., Nicholson, H.F. and Ogilvie, L.K., Methods development study for measuring pesticide exposure to golf course workers using video imaging techniques. *Applied Occupational and Environmental Hygiene,* 1996. 11(11): p. 1346-1350.

[4] Cook, S.M.F. and Linden, D.R., Use of rhodamine WT to facilitate dilution and analysis of atrazine samples in short-term transport studies. *Journal of Environmental Quality,* 1997. 26(5): p. 1438-1440.

[5] Bhat, R. and Mathur, P., Changing scenario of food colours in India. *Current Science,* 1998. 74(3): p. 198-202.

[6] Bensalah, N., Alfaro, M. and Martínez-Huitle, C., Electrochemical treatment of synthetic wastewaters containing Alphazurine A dye. *Chemical Engineering Journal,* 2009. 149(1): p. 348-352.
[7] Wróbel, D., Boguta, A. and Ion, R.M., Mixtures of synthetic organic dyes in a photoelectrochemical cell. *Journal of Photochemistry and Photobiology A: Chemistry,* 2001. 138(1): p. 7-22.
[8] Scarpi, C., Ninci, F., Centini, M. and Anselmi, C., High-performance liquid chromatography determination of dir Schuler, CA, Anthony RG, and Ohlendorf. HM 1990. Selenium in Wetlands and Waterfowl Foods and Kesterson Reservoir, California, 1984. *Archives of Environmental Contamination and Toxicology*, 1998. 29: p. 845-853.
[9] Kabdaşli, I., Tünay, O. and Orhon, D., Wastewater control and management in a leather tanning district. *Water Science and Technology,* 1999. 40(1): p. 261-267.
[10] Tünay, O., Kabdaşli, I., Orhon, D. and Cansever, G., Use and minimization of water in leather tanning processes. *Water Science and Technology*, 1999. 40(1): p. 237-244.
[11] Ravikumar, K., Deebika, B. and Balu, K., Decolourization of aqueous dye solutions by a novel adsorbent: application of statistical designs and surface plots for the optimization and regression analysis. *Journal of Hazardous Materials,* 2005. 122(1): p. 75-83.
[12] El Qada, E.N., Allen, S.J. and Walker, G.M., Adsorption of basic dyes from aqueous solution onto activated carbons. *Chemical Engineering Journal*, 2008. 135(3): p. 174-184.
[13] Yagub, M.T., Sen, T.K. and Ang, H., Equilibrium, Kinetics, and Thermodynamics of Methylene Blue Adsorption by Pine Tree Leaves. *Water, Air, & Soil Pollution*, 2012. 223(8): p. 5267-5282.
[14] Demirbas, A., Agricultural based activated carbons for the removal of dyes from aqueous solutions: a review. *Journal of Hazardous Materials,* 2009. 167(1): p. 1-9.
[15] Padmesh, T.V.N., Vijayaraghanvan, K., Sekaran, G. and Velan, M., Biosorption of Acid Blue 15 using fresh water macroalga Azolla filiculoides: *Batch and column studies. Dyes and Pigments,* 2006. 71(2): p. 77-82.
[16] O'Neill, C., Hawkes, F.R., Hawkes, D.L., Lourenço, N.D., Pinheiro, H.M. and Delée, W., Colour in textile effluents–sources, measurement, discharge consents and simulation: a review. *Journal of Chemical Technology and Biotechnology*, 1999. 74(11): p. 1009-1018.
[17] Vandevivere, P.C., Bianchi, R. and Verstraete, W., Review: Treatment and reuse of wastewater from the textile wet-processing industry: Review of emerging technologies. *Journal of Chemical Technology and Biotechnology,* 1998. 72(4): p. 289-302.
[18] Walsh, G.E., Bahner, L.H. and Horning, W.B., Toxicity of textile mill effluents to freshwater and estuarine algae, crustaceans and fishes. *Environmental Pollution Series A, Ecological and Biological*, 1980. 21(3): p. 169-179.
[19] Sun, Q. and Yang, L., The adsorption of basic dyes from aqueous solution on modified peat–resin particle. *Water Research,* 2003. 37(7): p. 1535-1544.
[20] Ravi Kumar, M. N.V., Rajakala, S.T., Durga, B.K. and Dutta, P.K., Trends in color removal from textile mill effluents. *Colourage,* 1998. 45(8).
[21] Smith, B., Koonce, T. and Hudson, S., Decolorizing dye wastewater using chitosan. *American Dyestuff Reporter,* 1993. 82(10): p. 18-36.

[22] Garg, V., Amita, M., Kumar, R. and Gupta, R., Basic dye (methylene blue) removal from simulated wastewater by adsorption using Indian Rosewood sawdust: a timber industry waste. *Dyes and Pigments,* 2004. 63(3): p. 243-250.

[23] Robinson, T., McMullan, G., Marchant, R. and Nigam, P., Remediation of dyes in textile effluent: a critical review on current treatment technologies with a proposed alternative. *Bioresource Technology,* 2001. 77(3): p. 247-255.

[24] Rafatullah, M., Sulaiman, O., Hashim, R. and Ahmad, A., Adsorption of methylene blue on low-cost adsorbents: A review. *Journal of Hazardous Materials,* 2010. 177(1-3): p. 70-80.

[25] Jadhav, J.P., Parshetti, G.K., Kalme, S.D. and Govindwar, S.P., Decolourization of azo dye methyl red by Saccharomyces cerevisiae MTCC 463. *Chemosphere,* 2007. 68(2): p. 394-400.

[26] Kant, R., Adsorption of Dye Eosin from an Aqueous Solution on two Different Samples of Activated Carbon by Static Batch Method. *Journal of Water Resource and Protection,* 2012. 4(2): p. 93-98.

[27] Yagub, M.T., Sen, T.K., Afroze, S. and Ang, H.M., Dye and its removal from aqueous solution by adsorption: A review. *Advances in Colloid and Interface Science,* 2014. 209: p. 172-184.

[28] Salleh, M.A.M., Mahmoud, D.K., Karim, W.A.W.A. and Idris, A., Cationic and anionic dye adsorption by agricultural solid wastes: A comprehensive review. *Desalination,* 2011. 280(1): p. 1-13.

[29] Slokar, Y.M. and Majcen, A.M.L., Methods of decoloration of textile wastewaters. *Dyes and pigments,* 1998. 37(4): p. 335-356.

[30] Pokhrel, D. and Viraraghavan, T., Treatment of pulp and paper mill wastewater—a review. *Science of the Total Environment,* 2004. 333(1): p. 37-58.

[31] Crini, G., Non-conventional low-cost adsorbents for dye removal: a review. *Bioresource Technology,* 2006. 97(9): p. 1061-1085.

[32] Gupta, V., Application of low-cost adsorbents for dye removal–A review. *Journal of Environmental Management,* 2009. 90(8): p. 2313-2342.

[33] Srinivasan, A. and Viraraghavan, T., Decolorization of dye wastewaters by biosorbents: A review. *Journal of Environmental Management,* 2010. 91(10): p. 1915-1929.

[34] Ali, H., Biodegradation of synthetic dyes—a review. *Water, Air, & Soil Pollution,* 2010. 213(1): p. 251-273.

[35] Ali, I., Asim, M. and Khan, T.A., Low cost adsorbents for the removal of organic pollutants from wastewater. *Journal of Environmental Management,* 2012. 113: p. 170-183.

[36] Agustina, T.E., AOPs Application on Dyes Removal, in Wastewater Reuse and Management. 2013, *Springer*. p. 353-372.

[37] Hadi, M., Samarghandi, M.R. and McKay, G., Simplified fixed bed design models for the adsorption of acid dyes on novel pine cone derived activated carbon. *Water, Air, & Soil Pollution,* 2011. 218(1-4): p. 197-212.

[38] Kochher, M.S. and Kumar, S., Screening For Potential Textile Dye Decolorizing Bacteria. *International Journal for Science and Emerging Technologies with Latest Trends,* 2012. 2(1): p. 36-48.

[39] Christie, R., Environmental aspects of textile dyeing. *Woodhand Publishing*, 2007: Elsevier.

[40] Hatch, K.L. and Maibach, H., Dyes as contact allergens: a comprehensive record. *Textile Chemist and Colorist & American Dyestuff Reporter,* 1999. 1(2): p. 53-59.

[41] Rai, H.S., Bhattacharyya, M.S., Singh, J., Bansal, T.K., Vats, P. and Banerjee, U.C., Removal of dyes from the effluent of textile and dyestuff manufacturing industry: a review of emerging techniques with reference to biological treatment. Critical Reviews in Environmental Science and Technology, 2005. 35(3): p. 219-238.

[42] Kuo, W., Decolorizing dye wastewater with Fenton's reagent. *Water Research,* 1992. 26(7): p. 881-886.

[43] Kadirvelu, K., Kavipriya, M., Karthika, C., Radhika, M., Vennilamani, N. and Pattabhi, S., Utilization of various agricultural wastes for activated carbon preparation and application for the removal of dyes and metal ions from aqueous solutions. *Bioresource Technology,* 2003. 87(1): p. 129-132.

[44] McKay, G., Adsorption of dyestuffs from aqueous solutions with activated carbon I: Equilibrium and batch contact-time studies. *Journal of Chemical Technology and Biotechnology,* 1982. 32(7-12): p. 759-772.

[45] Ventura-Camargo, B.d.C. and Marin-Morales, M.A., Azo Dyes: Characterization and Toxicity– A Review. Textiles and Light Industrial Science and Technology (TLIST), 2013. 2(2).

[46] Hatch, K.L. and Maibach, H.I., Textile dye dermatitis. *Journal of the American Academy of Dermatology,* 1995. 32(4): p. 631-639.

[47] Manzini, B.M., Motolese, A., Conti, A., Ferdani, G. and Seidenari, S., Sensitization to reactive textile dyes in patients with contact dermatitis. *Contact Dermatitis,* 1996. 34(3): p. 172-175.

[48] Wilkinson, S. and McGechaen, K., Occupational allergic contact dermatitis from reactive dyes. *Contact Dermatitis*, 1996. 35(6): p. 376-376.

[49] Klemola, K., Pearson, J. and Lindstrom-Seppä, P., Evaluating the toxicity of reactive dyes and dyed fabrics with the HacaT cytotoxicity test. *Autex Res. J,* 2007. 7: p. 217-223.

[50] Mathur, N. and Bhatnagar, P., Mutagenicity assessment of textile dyes from Sanganer (Rajasthan). *Journal of Environmental Biology,* 2007. 28(1): p. 123-126.

[51] Novotný, Č., Dias, N., Kapanen, A., Malachová, K., Vándrovcová, M., Itävaara, M. and Lima, N., Comparative use of bacterial, algal and protozoan tests to study toxicity of azo-and anthraquinone dyes. *Chemosphere,* 2006. 63(9): p. 1436-1442.

[52] Bhatnagar, A. and Minocha, A., Conventional and non-conventional adsorbents for removal of pollutants from water-A review. *Indian Journal of Chemical Technology,* 2006. 13(3): p. 203-217.

[53] Ghoreishi, S. and Haghighi, R., Chemical catalytic reaction and biological oxidation for treatment of non-biodegradable textile effluent. Chemical Engineering Journal, 2003. 95(1): p. 163-169.

[54] McCabe, W.L., Smith, J.C. and Harriott, P., Unit operations of chemical engineering. Vol. 5. 1993: *McGraw-Hill* New York.

[55] McKay, G., Guendi, M.E. and Nassar, M., Adsorption model for the removal of acids dyes from effluent by bagasse pith using a simplified isotherm. *Adsorption Science and Technology,* 1997. 15(10): p. 737-752.

[56] Sing, K.S., Reporting physisorption data for gas/solid systems with special reference to the determination of surface area and porosity (Recommendations 1984). *Pure and Applied Chemistry,* 1985. 57(4): p. 603-619.

[57] Sharma, P., Kaur, H., Sharma, M., Sahore, V., A review on applicability of naturally available adsorbents for the removal of hazardous dyes from aqueous waste. *Environmental Monitoring and Assessment,* 2011. 183(1-4): p. 151-195.

[58] Allen, S. and Koumanova, B., Decolourisation of water/wastewater using adsorption. *Journal of the University of Chemical Technology and Metallurgy,* 2005. 40(3): p. 175-192.

[59] Kaur, S., Rani, S. and Mahajan, R.K., Adsorption kinetics for the removal of hazardous dye congo red by biowaste materials as adsorbents. *Journal of Chemistry,* 2012. 2013.

[60] Ahalya, N., Chandraprabha, M.N., Kanamadi, R.D., Ramachandra, T.V., Adsorption of Fast Green on to Coffee Husk. *Journal of Chemical Engineering and Research*, 2014. 2(1): p. 201-207.

[61] Pavan, F.A., Camacho, E.S., Lima, E.C., Dotto, G.L., Branco, V.T.A. and Dias, S.L.P., Formosa papaya seed powder (FPSP): Preparation, characterization and application as an alternative adsorbent for the removal of crystal violet from aqueous phase. *Journal of Environmental Chemical Engineering,* 2014. 2(1): p. 230-238.

[62] Argun, M.E., Güclü, D. and Karatas, M., Adsorption of Reactive Blue 114 dye by using a new adsorbent: Pomelo peel. *Journal of Industrial and Engineering Chemistry,* 2014. 20(3): p. 1079-1084.

[63] Tilaki, D., Ali, R., Balarak, D. and Ghasemi, M., Removal of Acid Orang 7 (AO7) Dye from Aqueous Solution by Using of Adsorption on to Rice Stem Waste: Kinetic and Equilibrium Study. *Iranian Journal of Health Sciences,* 2014. 2(1): p. 51-61.

[64] Zhao, B., Xiao, W., Shang, Y., Zhu, H. and Han, R., Adsorption of light green anionic dye using cationic surfactant-modified peanut husk in batch mode. *Arabian Journal of Chemistry,* 2014.

[65] Liu, W., Yao, C., Wang, M., Ji, J., Ying, L. and Fu, C., Kinetics and thermodynamics characteristics of cationic yellow X-GL adsorption on attapulgite/rice hull-based activated carbon nanocomposites. *Environmental Progress & Sustainable Energy,* 2012. 32(3): p. 655-662.

[66] Tan, J., Zhang, X., Wei, X. and Wang, L., Removal of Malachite Green from Aqueous Solution using Waste Newspaper Fiber. *Bioresources,* 2012. 7(3): p. 4307-4320.

[67] Jain, S. and Jayaram, R.V., Removal of basic dyes from aqueous solution by low-cost adsorbent: Wood apple shell (Feronia acidissima). *Desalination,* 2010. 250(3): p. 921-927.

[68] Saeed, A., M. Sharif, and M. Iqbal, Application potential of grapefruit peel as dye sorbent: Kinetics, equilibrium and mechanism of crystal violet adsorption. *Journal of Hazardous Materials,* 2010. 179(1): p. 564-572.

[69] Han, X., Niu, X. and Ma, X., Adsorption characteristics of methylene blue on poplar leaf in batch mode: Equilibrium, kinetics and thermodynamics. *Korean Journal of Chemical Engineering,* 2012: p. 1-9.

[70] Vieira, A.P., Santana, S.A.A., Bezerra, C.W.B., Silva, H.A.S., Chaves, J.A.P., de Melo, J.C.P., da Silva Filho, E.C. and Airoldi, C., Kinetics and thermodynamics of textile dye adsorption from aqueous solutions using babassu coconut mesocarp. *Journal of Hazardous Materials,* 2009. 166(2): p. 1272-1278.

[71] Lafi, R., Ben F.A., Hafiane, A., Hameed, B.H., Coffee waste as potential adsorbent for the removal of basic dyes from aqueous solution. *Korean Journal of Chemical Engineering,* 2014: p. 1-9.
[72] Akkaya, G. and Güzel, F., Application of some domestic wastes as new low-cost biosorbents for removal of methylene blue: kinetic and equilibrium studies. *Chemical Engineering Communications,* 2014. 201(4): p. 557-578.
[73] Wang, H., Yuan, X., Wu, Z., Wang, L., Peng, X., Leng, L. and Zeng, G., Removal of basic dye from aqueous solution using Cinnamomum camphora sawdust: kinetics, isotherms, thermodynamics and mass-transfer processes. *Separation Science and Technology,* 2014 (just-accepted).
[74] Hoseinzadeh, E., Samarghandi, M.R., McKay, G., Rahimi, N. and Jafari, J., Removal of acid dyes from aqueous solution using potato peel waste biomass: a kinetic and equilibrium study. *Desalination and Water Treatment,* 2013 (ahead-of-print): p. 1-8.
[75] Hameed, B. and Ahmad, A., Batch adsorption of methylene blue from aqueous solution by garlic peel, an agricultural waste biomass. *Journal of Hazardous Materials,* 2009. 164(2): p. 870-875.
[76] Maarof, I.H., H., Hameed, B.H. and Ahmad, A.L., Aqueous-Phase Adsorption of Phenolic Compounds on Activated Carbon. 2003.
[77] Mall, I.D., Srivastava, V.C. and Agarwal, N.K., Removal of Orange-G and Methyl Violet dyes by adsorption onto bagasse fly ash—kinetic study and equilibrium isotherm analyses. *Dyes and Pigments,* 2006. 69(3): p. 210-223.
[78] Dawood, S. and Sen, T.K., Removal of anionic dye Congo red from aqueous solution by raw pine and acid-treated pine cone powder as adsorbent: Equilibrium, thermodynamic, kinetics, mechanism and process design. *Water Research*, 2012. 46(6): p. 1933-1946.
[79] Franca, A.S., Oliveira, L.S. and Ferreira, M.E., Kinetics and equilibrium studies of methylene blue adsorption by spent coffee grounds. *Desalination,* 2009. 249(1): p. 267-272.
[80] Zou, W., Bai, H., Gao, S., Li, K., Characterization of modified sawdust, kinetic and equilibrium study about methylene blue adsorption in batch mode. *Korean Journal of Chemical Engineering,* 2013. 30(1): p. 111-122.
[81] Hameed, B., Equilibrium and kinetic studies of methyl violet sorption by agricultural waste. *Journal of Hazardous Materials,* 2008. 154(1): p. 204-212.
[82] Senthil Kumar, P., Palaniyappan, M., Priyadharshini, M., Vignesh, A.M., Thanjiappan, A., Fernando, P.S.A., Ahmed, R.T. and Srinath, R., Adsorption of basic dye onto raw and surface-modified agricultural waste. *Environmental Progress & Sustainable Energy,* 2014. 33(1): p. 87-98.
[83] Low, L.W., Teng, T.T., Rafatullah, M., Morad, N., Azahari, B., Adsorption studies of methylene blue and malachite green from aqueous solutions by pretreated lignocellulosic materials. *Separation Science and Technology*, 2013. 48(11): p. 1688-1698.
[84] Zhang, J., Zhou, Q. and Ou, L., Kinetic, Isotherm, and Thermodynamic Studies of the Adsorption of Methyl Orange from Aqueous Solution by Chitosan/Alumina Composite. *Journal of Chemical & Engineering Data,* 2012. 67: p. 412-419.

[85] Maurya, R., Ghosh, T., Paliwal, C., Shrivastav, A., Chokshi, K., Pancha, I., Ghosh, A. and Mishra, S., Biosorption of Methylene Blue by De-Oiled Algal Biomass: Equilibrium, Kinetics and Artificial Neural Network Modelling. *PLOS one,* 2014. 9(10): p. e109545.

[86] Belay, K. and Abebe, M., Removal of Malachite Green from Aqueous Solutions by Adsorption Using Low Cost Boisorbent Neem Leaf (Azadirachta Indica). 2014.

[87] Ong, S.-A., Ho, L.-N., Wong, Y.-S. and Zainuddin, A., Adsorption behavior of cationic and anionic dyes onto acid treated coconut coir. *Separation Science and Technology,* 2013. 48(14): p. 2125-2131.

[88] Banerjee, S. and Chattopadhyaya, M., Adsorption characteristics for the removal of a toxic dye, tartrazine from aqueous solution by a low cost agricultural by-product. *Arabian Journal of Chemistry*, 2013.

[89] Zhang, Z., O'Hara, I.M., Kent, G.A. and Doherty, W.O., Comparative study on adsorption of two cationic dyes by milled sugarcane bagasse. *Industrial Crops and Products,* 2013. 42: p. 41-49.

[90] Ramaraju, B., Reddy, P.M.K. and Subrahmanyam, C., Low cost adsorbents from agricultural waste for removal of dyes. *Environmental Progress & Sustainable Energy,* 2013. 13: p. 1-9.

[91] Kahraman, S., Yalcin, P. and Kahraman, H., The evaluation of low-cost biosorbents for removal of an azo dye from aqueous solution. *Water and Environment Journal,* 2012. 26(3): p. 399-404.

[92] Hameed, B., Mahmoud, D. and Ahmad, A., Equilibrium modeling and kinetic studies on the adsorption of basic dye by a low-cost adsorbent: Coconut (Cocos nucifera) bunch waste. *Journal of Hazardous Materials,* 2008. 158(1): p. 65-72.

[93] Bouguettoucha, A., Chebli, D., Mekhalef, T., Noui, A. and Amrane, A., The use of a forest waste biomass, cone of Pinus brutia for the removal of an anionic azo dye Congo red from aqueous medium. Desalination and Water Treatment, 2014 (ahead-of-print): p. 1-10.

[94] Nawaz, S., Bhatti, H.N., Bokhari, T.H. and Sadaf, S., Removal of Novacron Golden Yellow dye from aqueous solutions by low-cost agricultural waste: Batch and fixed bed study. *Chemistry and Ecology,* 2014. 30(1): p. 52-65.

[95] Zhao, Y., Xia, Y., Yang, H., Wang, Y. and Zhao, M., Synthesis of glutamic acid-modified magnetic corn straw: equilibrium and kinetic studies on methylene blue adsorption. *Desalination and Water Treatment,* 2014. 52(1-3): p. 199-207.

[96] Lin, L., Zhai, S.-R., Xiao, Z.-Y., Song, Y., An, Q.-D. and Song, X.-W., Dye adsorption of mesoporous activated carbons produced from NaOH-pretreated rice husks. *Bioresource Technology,* 2013. 136: p. 437-443.

[97] Sadaf, S. and Bhatti, H.N., Equilibrium modeling for adsorptive removal of Indosol Black NF dye by low-cost agro-industrial waste: batch and continuous study. *Desalination and Water Treatment,* 2013 (ahead-of-print): p. 1-16.

[98] Ahmad, A., Idris, A. and Hameed, B., Organic dye adsorption on activated carbon derived from solid waste. *Desalination and Water Treatment,* 2013. 51(13-15): p. 2554-2563.

[99] Sen, T.K., Afroze, S. and Ang, H., Equilibrium, Kinetics and Mechanism of Removal of Methylene Blue from Aqueous Solution by Adsorption onto Pine Cone Biomass of Pinus radiata. *Water, Air, & Soil Pollution,* 2011. 218: p. 499-515.

[100] Feng, Y., Zhou, H., Liu, G., Qiao, J., Wang, J., Lu, H., Yang, L. and Wu, Y., Methylene blue adsorption onto swede rape straw (Brassica napus) modified by tartaric acid: equilibrium, kinetic and adsorption mechanisms. *Bioresource technology,* 2012. 125: p. 138-144.

[101] Ata, S., Imran, D. M., Rasool, A., Qasim, I. and Mohsin, U.I., Equilibrium, Thermodynamics, and Kinetic Sorption Studies for the Removal of Coomassie Brilliant Blue on Wheat Bran as a Low-Cost Adsorbent. *Journal of Analytical Methods in Chemistry,* 2012. 2012.

[102] Kini, M.S., Saidutta, M. and Murty, V.R., Studies on biosorption of methylene blue from aqueous solutions by powdered palm tree flower (Borassus flabellifer). *International Journal of Chemical Engineering,* 2014. 2014.

[103] Khosla, E., Kaur, S. and Dave, P.N., Tea waste as adsorbent for ionic dyes. *Desalination and Water Treatment,* 2013. 13 (ahead-of-print): p. 1-10.

[104] Handan, U., Equilibrium, thermodynamic and kinetics of reactive black 5 biosorption on loquat (Eriobotrya japonica) seed. *Scientific Research Essays*, 2011. 6(19): p. 4113-4124.

[105] Safa, Y. and Bhatti, H.N., Kinetic and thermodynamic modeling for the removal of Direct Red-31 and Direct Orange-26 dyes from aqueous solutions by rice husk. *Desalination,* 2011. 272(1): p. 313-322.

[106] Hameed, B., Spent tea leaves: A new non-conventional and low-cost adsorbent for removal of basic dye from aqueous solutions. *Journal of Hazardous Materials,* 2009. 161(2): p. 753-759.

[107] Hameed, B., Evaluation of papaya seeds as a novel non-conventional low-cost adsorbent for removal of methylene blue. Journal of Hazardous Materials, 2009. 162(2): p. 939-944.

[108] Senthil Kumar, P., Ramalingam, S., Senthamarai, C., Niranjanaa, M., Vijayalakshmi, P. and Sivanesan, S., Adsorption of dye from aqueous solution by cashew nut shell: Studies on equilibrium isotherm, kinetics and thermodynamics of interactions. *Desalination*, 2010. 261(1): p. 52-60.

[109] Mane, V.S. and Babu, P., Studies on the adsorption of Brilliant Green dye from aqueous solution onto low-cost NaOH treated saw dust. *Desalination,* 2011. 273(2): p. 321-329.

[110] Saad, S., Isa, K.M. and Bahari, R., Chemically modified sugarcane bagasse as a potentially low-cost biosorbent for dye removal. *Desalination,* 2010. 264(1): p. 123-128.

[111] Ong, S., Lee, C. and Zainal, Z., Removal of basic and reactive dyes using ethylenediamine modified rice hull. *Bioresource Technology*, 2007. 98(15): p. 2792-2799.

[112] Bharathi, K. and Ramesh, S., Removal of dyes using agricultural waste as low-cost adsorbents: a review. *Applied Water Science,* 2013. 3(4): p. 773-790.

[113] Guechi, E.-K. and Hamdaoui, O., Sorption of malachite green from aqueous solution by potato peel: kinetics and equilibrium modeling using non-linear analysis method. *Arabian Journal of Chemistry,* 2011.

[114] Ren, X., Xiao, W., Zhang, R., Shang, Y. and Han, R., Adsorption of crystal violet from aqueous solution by chemically modified phoenix tree leaves in batch mode. *Desalination and Water Treatment,* 2013 (ahead-of-print): p. 1-11.

[115] Kuo, C.-Y., Wu, C.-H. and Wu, J.-Y., Adsorption of direct dyes from aqueous solutions by carbon nanotubes: Determination of equilibrium, kinetics and thermodynamics parameters. *Journal of Colloid and Interface Science,* 2008. 327(2): p. 308-315.
[116] Fedoseeva, M., Fita, P., Punzi, A. and Vauthey, E., Salt effect on the formation of dye aggregates at liquid/liquid interfaces studied by time-resolved surface second harmonic generation. *The Journal of Physical Chemistry C,* 2010. 114(32): p. 13774-13781.
[117] Zaheer, S., Bhatti, H.N., Sadaf, S., Safa, Y. and Zia-ur-Rehman, M., Biosorption Characteristics of Sugarcane Bagasse for the Removal of Foron Blue E-Bl Dye from Aqueous Solutions. *Journal of Animal and Plant Sciences,* 2014. 24(1): p. 272-279.
[118] Dağdelen, S., Acemioğlu, B., Baran, E. and Koçer, O., Removal of Remazol Brilliant Blue R From Aqueous Solution by Pirina Pretreated with Nitric Acid and Commercial Activated Carbon. *Water, Air, & Soil Pollution,* 2014. 225(3): p. 1-15.
[119] Chen, Z., Deng, H., Chen, C., Yang, Y. and Xu, H., Biosorption of malachite green from aqueous solutions by Pleurotus ostreatus using Taguchi method. *Journal of Environmental Health Science and Engineering*, 2014. 12(1): p. 63.
[120] Chieng, H.I., Lim, L.B. and Priyantha, N., Enhancing adsorption capacity of toxic malachite green dye through chemically modified breadnut peel: equilibrium, thermodynamics, kinetics and regeneration studies. *Environmental Technology,* 2014 (ahead-of-print): p. 1-12.
[121] Kong, L., Gong, L. and Wang, J., Removal of methylene blue from wastewater using fallen leaves as an adsorbent. *Desalination and Water Treatment,* 2013 (ahead-of-print): p. 1-12.
[122] Han, X., Yuan, J. and Ma, X., Adsorption of malachite green from aqueous solutions onto lotus leaf: equilibrium, kinetic, and thermodynamic studies. *Desalination and Water Treatment,* 2013 (ahead-of-print): p. 1-12.
[123] Djelloul, C. and Hamdaoui, O., Removal of cationic dye from aqueous solution using melon peel as nonconventional low-cost sorbent. *Desalination and Water Treatment,* 2013 (ahead-of-print): p. 1-10.
[124] Lagergren, S., Zur theoric der sogenannten adsorption geloster stoffe. *Kungliga Svenska Vetenskapsakad Handlinger,* 1898. 24(4): p. 1-39.
[125] Ho, Y.-S. and McKay, G., Pseudo-second order model for sorption processes. *Process Biochemistry,* 1999. 34(5): p. 451-465.
[126] Oladoja, N., Aboluwoye, C.O., Oladimeji, Y.B., Ashogbon, A.O. and Otemuyiwa, I.O., Studies on castor seed shell as a sorbent in basic dye contaminated wastewater remediation. *Desalination,* 2008. 227(1-3): p. 190-203.
[127] Weber, W.J. and Morris, J.C., Kinetics of adsorption on carbon from solution. *J. Sanit. Eng. Div. Am. Soc. Civ. Eng,* 1963. 89(17): p. 31-60.
[128] Zolgharnein, J., Gholami, F., Asanjarani, N., Zolgharnein, P. and Azimi, G., Removal of Methyl Violet Dye from Aqueous Solution by Platanus Carpinifolia Tree Leaves as Highly Efficient Sorbent: Multivariate Optimization, Isotherm Modeling, and Kinetic Studies. *Separation Science and Technology,* 2014. 49(5): p. 752-762.
[129] Dursun, A.Y., Tepe, O., Uslu, G., Dursun, G. and Saatci, Y., Kinetics of Remazol Black B adsorption onto carbon prepared from sugar beet pulp. *Environmental Science and Pollution Research,* 2013. 20: p. 2472-2483.

[130] Rebitanim, N.Z., Wan Ab Karim Ghani, W.A., Mahmoud, D.K., Rebitanim, N.A. and Mohd Salleh, M.A., Adsorption Capacity of Raw Empty Fruit Bunch Biomass onto Methylene Blue Dye in Aqueous Solution. *Journal of Purity, Utility Reaction and Environment,* 2012. 1(1): p. 45-60.

[131] Zolgharnein, J. and Bagtash, M., Hybrid central composite design optimization for removal of Methylene blue by Acer tree leaves: characterization of adsorption. *Desalination and Water Treatment,* 2014 (ahead-of-print): p. 1-10.

[132] Boudrahem, F., Aissani-Benissad, F. and Soualah, A., Removal of basic yellow dye from aqueous solutions by sorption onto reed as an adsorbent. *Desalination and Water Treatment,* 2014(ahead-of-print): p. 1-8.

[133] Zhou, L., Huang, J., He, B., Zhang, F. and Li, H., Peach gum for efficient removal of methylene blue and methyl violet dyes from aqueous solution. *Carbohydrate polymers,* 2014. 101: p. 574-581.

[134] Krishna, L.S., Reddy, A.S., Zuhairi, W.Y., Taha, M.R. and Reddy, A.V., Indian Jujuba Seed Powder as an Eco-Friendly and a Low-Cost Biosorbent for Removal of Acid Blue 25 from Aqueous Solution. *The Scientific World Journal,* 2014. 2014.

[135] Yagub, M.T., Sen, T.K. and Ang, M., Removal of cationic dye methylene blue (MB) from aqueous solution by ground raw and base modified pine cone powder. *Environmental Earth Sciences,* 2013. 13: p. 1-13.

[136] Saka, C., Şahin, Ö. and Çelik, M., The Removal of Methylene Blue from Aqueous Solutions by Using Microwave Heating and Pre-boiling Treated Onion Skins as a New Adsorbent. *Energy Sources, Part A: Recovery, Utilization, and Environmental Effects,* 2012. 34(17): p. 1577-1590.

[137] Duran, C., Ozdes, D., Gundogdu, A. and Senturk, H.B., Kinetics and isotherm analysis of basic dyes adsorption onto almond shell (Prunus dulcis) as a low cost adsorbent. *Journal of Chemical & Engineering Data,* 2011. 56(5): p. 2136-2147.

[138] Dulman, V. and Cucu-Man, S.M., Sorption of some textile dyes by beech wood sawdust. *Journal of Hazardous Materials,* 2009. 162(2): p. 1457-1464.

[139] Allen, S., Mckay, G. and Porter, J., Adsorption isotherm models for basic dye adsorption by peat in single and binary component systems. *Journal of Colloid and Interface Science,* 2004. 280(2): p. 322-333.

[140] El-Halwany, M., Study of adsorption isotherms and kinetic models for Methylene Blue adsorption on activated carbon developed from Egyptian rice hull (Part II). *Desalination,* 2010. 250(1): p. 208-213.

[141] Langmuir, I., The adsorption of gases on plane surfaces of glass, mica and platinum. *Journal of the American Chemical Society,* 1918. 40(9): p. 1361-1403.

[142] Baskaralingam, P., Pulikesia, M., Elangob, D., Ramamurthia, V. and Sivanesana, S., Adsorption of acid dye onto organobentonite. *Journal of Hazardous Materials,* 2006. 128(2): p. 138-144.

[143] Freundlich, H., Over the adsorption in solution. *Journal of Physical Chemistry,* 1906. 57: p. 385-470.

[144] Nausheen, S., Bhatti, H.N., Furrukh, Z., Sadaf, S. and Noreen, S., Adsorptive removal of Drimarine Red HF-3D dye from aqueous solution using low-cost agricultural waste: batch and column study. *Chemistry and Ecology,* 2014. 30(4): p. 376-392.

[145] Armagan, B. and Toprak, F., Using pistachio shell for Remazol Red removal from aqueous solutions: equilibrium, kinetics and thermodynamics. *Desalination and Water Treatment,* 2014 (ahead-of-print): p. 1-10.

[146] Çelekli, A., Çelekli, F., Çiçek, E. and Bozkurt, H., Predictive modeling of sorption and desorption of a reactive azo dye by pumpkin husk. *Environmental Science and Pollution Research,* 2014. 21(7): p. 5086-5097.

[147] Sadaf, S., Bhatti, H.N., Ali, S. and Rehman, K., Removal of Indosol Turquoise FBL dye from aqueous solution by bagasse, a low cost agricultural waste: batch and column study. *Desalination and Water Treatment,* 2014. 52(1-3): p. 184-198.

[148] Sen, T.K., Agricultural by-product biomass for removal of pollutants from aqueous solution by adsorption. *Journal of Environmental Research And Development Vol,* 2012. 6(3): p. 523-533.

[149] Dod, R., Banerjee, G. and Saini, S., Adsorption of methylene blue using green pea peels (Pisum sativum): A cost-effective option for dye-based wastewater treatment. Biotechnology and Bioprocess Engineering, 2012. 17(4): p. 862-874.

[150] Deniz, F. and Karaman, S., Removal of Basic Red 46 dye from aqueous solution by pine tree leaves. *Chemical Engineering Journal,* 2011. 170(1): p. 67-74.

[151] Ibrahim, S., Fatimah, I., Ang, H.-M. and Wang, S., Adsorption of anionic dyes in aqueous solution using chemically modified barley straw. Water Science and Technology: A Journal of the International Association on Water Pollution Research, 2010. 62(5): p. 1177.

[152] Ponnusami, V., Gunasekar, V. and Srivastava, S., Kinetics of methylene blue removal from aqueous solution using gulmohar (Delonix regia) plant leaf powder: Multivariate regression analysis. *Journal of Hazardous Materials,* 2009. 169(1): p. 119-127.

[153] Bhatnagar, A. and Sillanpää, M., Utilization of agro-industrial and municipal waste materials as potential adsorbents for water treatment—a review. *Chemical Engineering Journal,* 2010. 157(2): p. 277-296.

[154] Ahmedna, M., Marshall, W. and Rao, R., Production of granular activated carbons from select agricultural by-products and evaluation of their physical, chemical and adsorption properties. *Bioresource Technology,* 2000. 71(2): p. 113-123.

[155] Rehman, M.S.U., Kim, I. and Han, J.I., Adsorption of methylene blue dye from aqueous solution by sugar extracted spent rice biomass. *Carbohydrate Polymers,* 2012. 90(3): p. 1314-1322.

[156] Nyazi, K., Yaacoubi, A., Baçaoui, A., Bennouna, C., Dahbi, A., Rivera-Utrilla, J. and Moreno-Castilla, C., Preparation and characterization of new adsorbent materials from the olive wastes. *Journal of Physics IV France,* 2005: p. 121-124.

[157] Hameed, B., Krishni, R. and Sata, S., A novel agricultural waste adsorbent for the removal of cationic dye from aqueous solutions. *Journal of Hazardous Materials,* 2009. 162(1): p. 305-311.

[158] Ferrero, F., Dye removal by low cost adsorbents: Hazelnut shells in comparison with wood sawdust. *Journal of Hazardous Materials*, 2007. 142(1): p. 144-152.

[159] Doğan, M., Abak, H. and Alkan, M., Adsorption of methylene blue onto hazelnut shell: Kinetics, mechanism and activation parameters. *Journal of Hazardous Materials,* 2009. 164(1): p. 172-181.

[160] Hameed, B., Mahmoud, D. and Ahmad, A., Sorption equilibrium and kinetics of basic dye from aqueous solution using banana stalk waste. *Journal of Hazardous Materials,* 2008. 158(2): p. 499-506.

[161] Annadurai, G., Juang, R.S. and Lee, D.J., Use of cellulose-based wastes for adsorption of dyes from aqueous solutions. *Journal of Hazardous Materials,* 2002. 92(3): p. 263-274.

[162] Banat, F., Al-Asheh, S. and Al-Makhadmeh, L., Evaluation of the use of raw and activated date pits as potential adsorbents for dye containing waters. *Process Biochemistry,* 2003. 39(2): p. 193-202.

[163] Kumar, K.V. and Kumaran, A., Removal of methylene blue by mango seed kernel powder. *Biochemical Engineering Journal*, 2005. 27(1): p. 83-93.

[164] Kumar, K.V. and Porkodi, K., Relation between some two-and three-parameter isotherm models for the sorption of methylene blue onto lemon peel. *Journal of Hazardous Materials,* 2006. 138(3): p. 633-635.

[165] Shukla, A., Zhang, Y.-H., Dubey, P., Margrave, J.L. and Shukla, S.S., The role of sawdust in the removal of unwanted materials from water. *Journal of Hazardous Materials,* 2002. 95(1): p. 137-152.

[166] Ho, Y.-S., Chiu, W.-T. and Wang, C.-C., Regression analysis for the sorption isotherms of basic dyes on sugarcane dust. *Bioresource Technology,* 2005. 96(11): p. 1285-1291.

[167] Namasivayam, C. and Kanchana, N., Waste banana pith as adsorbent for color removal from wastewaters. *Chemosphere*, 1992. 25(11): p. 1691-1705.

[168] Hameed, B., Grass waste: A novel sorbent for the removal of basic dye from aqueous solution. *Journal of Hazardous Materials,* 2009. 166(1): p. 233-238.

[169] Kumar, K.V. and Porkodi, K., Mass transfer, kinetics and equilibrium studies for the biosorption of methylene blue using Paspalum notatum. *Journal of Hazardous Materials,* 2007. 146(1): p. 214-226.

[170] Ponnusami, V., Vikram, S. and Srivastava, S., Guava (Psidium guajava) leaf powder: Novel adsorbent for removal of methylene blue from aqueous solutions. *Journal of Hazardous Materials,* 2008. 152(1): p. 276-286.

[171] Singh, D. and Srivastava, B., Removal of basic dyes from aqueous solutions by chemically treated Psidium guyava leaves. *Indian Journal of Environmental Health,* 1999. 41(4): p. 333-345.

[172] McKay, G., Ramprasad, G. and Pratapa, M. P., Equilibrium studies for the adsorption of dyestuffs from aqueous solutions by low-cost materials. *Water, Air, & Soil Pollution,* 1986. 29(3): p. 273-283.

[173] McKay, G., Porter, J. and Prasad, G., The removal of dye colours from aqueous solutions by adsorption on low-cost materials. *Water, Air, & Soil Pollution,* 1999. 114(3): p. 423-438.

[174] Hameed, B. and El-Khaiary, M., Sorption kinetics and isotherm studies of a cationic dye using agricultural waste: broad bean peels. *Journal of Hazardous Materials*, 2008. 154(1): p. 639-648.

[175] Hameed, B. and El-Khaiary, M., Removal of basic dye from aqueous medium using a novel agricultural waste material: Pumpkin seed hull. *Journal of hazardous materials,* 2008. 155(3): p. 601-609.

[176] Özer, D., Dursun, G. and Özer, A., Methylene blue adsorption from aqueous solution by dehydrated peanut hull. *Journal of Hazardous Materials,* 2007. 144(1): p. 171-179.

[177] Low, K. and Lee, C., The removal of cationic dyes using coconut husk as an adsorbent. *Pertanika,* 1990. 13(2): p. 221-228.

[178] Oliveira, L.S., Franca, A.S., Alves, T.M. and Rocha, S.D.F., Evaluation of untreated coffee husks as potential biosorbents for treatment of dye contaminated waters. *Journal of Hazardous Materials,* 2008. 155(3): p. 507-512.

[179] Gong, R., Ding, Y., Li, M., Yang, C., Liu, H. and Sun, Y., Utilization of powdered peanut hull as biosorbent for removal of anionic dyes from aqueous solution. *Dyes and Pigments,* 2005. 64(3): p. 187-192.

[180] Yagub, M.T., Sen, T.K. and Ang, M., Removal of cationic dye methylene blue (MB) from aqueous solution by ground raw and base modified pine cone powder. *Environmental Earth Sciences,* 2014. 71(4): p. 1507-1519.

[181] Wang, X.S., Zhou, Y., Jiang, Y. and Sun, C., The removal of basic dyes from aqueous solutions using agricultural by-products. *Journal of Hazardous Materials,* 2008. 157(2): p. 374-385.

[182] Gong, R., Jin, Y., Chen, J. Hu, Y. and Sun, J., Removal of basic dyes from aqueous solution by sorption on phosphoric acid modified rice straw. *Dyes and Pigments,* 2007. 73(3): p. 332-337.

[183] Bulut, Y. and Aydın, H., A kinetics and thermodynamics study of methylene blue adsorption on wheat shells. *Desalination, 2006.* 194(1): p. 259-267.

[184] Batzias, F., Sidiras, D., Schroeder, E. and Weber, C., Simulation of dye adsorption on hydrolyzed wheat straw in batch and fixed-bed systems. *Chemical Engineering Journal,* 2009. 148(2): p. 459-472.

[185] Williams, P.T. and Nugranad, N. Comparison of products from the pyrolysis and catalytic pyrolysis of rice husks. *Energy,* 2000. 25(6): p. 493-513.

[186] Lakshmi, U.R., Srivastava, V.C., Mall, I.D. and Lataye, D.H., Rice husk ash as an effective adsorbent: evaluation of adsorptive characteristics for Indigo Carmine dye. *Journal of Environmental Management,* 2009. 90(2): p. 710-720.

[187] Chandrasekhar, S. and Pramada, P., Rice husk ash as an adsorbent for methylene blue—effect of ashing temperature. *Adsorption,* 2006. 12(1): p. 27-43.

[188] Kaur, S. and Prasad, N., Adsorption of dyes on rice husk ash. *Indian Journal of Chemistry,* 2001. 40: p. 388-391.

[189] Khan, T.A., Sharma, S., Khan, E.A. and Mukhlif, A.A., Removal of congo red and basic violet 1 by chir pine (Pinus roxburghii) sawdust, a saw mill waste: Batch and column studies. *Toxicological & Environmental Chemistry,* 2014 (ahead-of-print): p. 1-14.

[190] Khattri, S. and Singh, M., Colour removal from synthetic dye wastewater using a bioadsorbent. *Water, Air, and Soil Pollution,* 2000. 120(3-4): p. 283-294.

[191] Ho, Y.-S. and McKay, G., Kinetic models for the sorption of dye from aqueous solution by wood. *Process Safety and Environmental Protection,* 1998. 76(2): p. 183-191.

[192] Azhar, S.S., Liew, A.G., Suhardy, D., Hafiz, K.F. and Hatim, M.I., Dye removal from aqueous solution by using adsorption on treated sugarcane bagasse. *American Journal of Applied Sciences,* 2005. 2(11): p. 1499.

[193] Rachakornkij, M., Ruangchuay, S. and Teachakulwiroj, S., Removal of reactive dyes from aqueous solution using bagasse fly ash. *Songklanakarin Journal of Science and Technology,* 2004. 26(Suppl 1): p. 13-24.

[194] Asgher, M. and Bhatti, H.N., Evaluation of thermodynamics and effect of chemical treatments on sorption potential of (Citrus) waste biomass for removal of anionic dyes from aqueous solutions. *Ecological Engineering,* 2012. 38(1): p. 79-85.

[195] Tunali Akar, S., Gorgulu, A., Akar, T. and Ceik, S., Decolorization of Reactive Blue 49 contaminated solutions by Capsicum annuum seeds: Batch and continuous mode biosorption applications. *Chemical Engineering Journal,* 2011. 168(1): p. 125-133.

[196] Akar, T., Tosun, I., Kaynak, Z., Ozkara, E., Yeni, O., Sahin, E.N. and Akar, S.T., An attractive agro-industrial by-product in environmental cleanup: Dye biosorption potential of untreated olive pomace. *Journal of Hazardous Materials,* 2009. 166(2): p. 1217-1225.

[197] Hameed, B., Removal of cationic dye from aqueous solution using jackfruit peel as non-conventional low-cost adsorbent. *Journal of Hazardous Materials*, 2009. 162(1): p. 344-350.

[198] Pavan, F.A., Lima, E.C., Dias, S.L.P. and Mazzocato, A.C., Methylene blue biosorption from aqueous solutions by yellow passion fruit waste. *Journal of Hazardous Materials,* 2008. 150(3): p. 703-712.

[199] Sivaraj, R., Namasivayam, C. and Kadirvelu, K., Orange peel as an adsorbent in the removal of acid violet 17 (acid dye) from aqueous solutions. *Waste Management,* 2001. 21(1): p. 105-110.

[200] Weng, C.-H., Lin, Y.-T. and Tzeng, T.-W., Removal of methylene blue from aqueous solution by adsorption onto pineapple leaf powder. *Journal of Hazardous Materials,* 2009. 170(1): p. 417-424.

[201] Han, R., Zou, W., Yu, W., Cheng, S., Wang, Y. and Shi, J., Biosorption of methylene blue from aqueous solution by fallen phoenix tree's leaves. *Journal of Hazardous Materials,* 2007. 141(1): p. 156-162.

[202] Ofomaja, A.E., Sorptive removal of methylene blue from aqueous solution using palm kernel fibre: effect of fibre dose. *Biochemical Engineering Journal,* 2008. 40(1): p. 8-18.

[203] Elizalde-González, M.P., Mattusch, J., Peláez-Cid, A.A. and Wennrich, R., Characterization of adsorbent materials prepared from avocado kernel seeds: Natural, activated and carbonized forms. *Journal of Analytical and Applied Pyrolysis,* 2007. 78(1): p. 185-193.

[204] Hamdaoui, O., Saoudi, F., Chiha, M. and Naffrechoux, E., Sorption of malachite green by a novel sorbent, dead leaves of plane tree: equilibrium and kinetic modeling. *Chemical Engineering Journal,* 2008. 143(1): p. 73-84.

[205] McKay, G., Allen, S.J., McConvey, I.F. and Otterburn, M.S., Transport processes in the sorption of colored ions by peat particles. *Journal of Colloid and Interface Science,* 1981. 80(2): p. 323-339.

[206] Han, R., Wang, Y., Han, P., Shi, J., Yang, J. and Lu, Y., Removal of methylene blue from aqueous solution by chaff in batch mode. *Journal of Hazardous Materials,* 2006. 137(1): p. 550-557.

[207] Ofomaja, A.E., Sorption dynamics and isotherm studies of methylene blue uptake on to palm kernel fibre. *Chemical Engineering Journal,* 2007. 126(1): p. 35-43.

[208] Ofomaja, A.E. and Ho, Y.-S., Effect of pH on cadmium biosorption by coconut copra meal. *Journal of Hazardous Materials,* 2007. 139(2): p. 356-362.

[209] DIM, P.E., Adsorption of Methyl Red and Methyl Orange using different tree bark powder. *Adsorption,* 2013. 4(1).

[210] Dave, P.N., Kaur, S. and Khosla, E., Removal of Eriochrome black-T by adsorption on to eucalyptus bark using green technology. *Indian Journal of Chemical Technology,* 2011. 18(1): p. 53-60.

[211] Srivastava, R. and Rupainwar, D., A comparative evaluation for adsorption of dye on Neem bark and Mango bark powder. *Indian Journal of Technology,* 2011. 18: p. 67-75.

[212] Srivastava, R. and Rupainwar, D., Eucalyptus bark powder as an effective adsorbent: Evaluation of adsorptive characteristics for various dyes. *Desalination and Water Treatment,* 2009. 11(1-3): p. 302-313.

[213] Oei, B.C., Ibrahim, S., Wang, S., Ang, H.M., Surfactant modified barley straw for removal of acid and reactive dyes from aqueous solution. *Bioresource Technology,* 2009. 100(18): p. 4292-4295.

[214] Ozacar, M. and Sengil, I.A., Adsorption of metal complex dyes from aqueous solutions by pine sawdust. *Bioresource Technology,* 2005. 96(7): p. 791-795.

[215] Namasivayam, C., Kumar, D.M., Selvi, K., Begum, R.A., Vanathi, T. and Yamuna, R.T., 'Waste'coir pith—a potential biomass for the treatment of dyeing wastewaters. *Biomass and Bioenergy,* 2001. 21(6): p. 477-483.

[216] Chen, B., Hui, C.W. and McKay, G., Film-pore diffusion modeling and contact time optimization for the adsorption of dyestuffs on pith. *Chemical Engineering Journal,* 2001. 84(2): p. 77-94.

[217] Ho, Y.S. and McKay, G., Sorption of dyes and copper ions onto biosorbents. *Process Biochemistry,* 2003. 38(7): p. 1047-1061.

[218] Mittal, A., Malviya, A., Kaur, D., Mittal, J. and Kurup, L., Studies on the adsorption kinetics and isotherms for the removal and recovery of Methyl Orange from wastewaters using waste materials. *Journal of Hazardous Materials,* 2007. 148(1): p. 229-240.

[219] El-Geundi, M.S., Colour removal from textile effluents by adsorption techniques. *Water Research,* 1991. 25(3): p. 271-273.

[220] Ho, Y. and McKay, G., Kinetic models for the sorption of dye from aqueous solution by wood. *Trans. IChemE,* 1998. 76: p. 183-191.

[221] Sharma, N., Tiwari, D. and Singh, S., The Efficiency Appraisal for Removal of Malachite Green by Potato peel and Neem Bark: Isotherm and Kinetic Studies. *International Journal of Chemical & Environmental Engineering,* 2014. 5(2).

[222] Ho, Y.S., Chiang, T.H. and Hsueh, Y.M., Removal of basic dye from aqueous solution using tree fern as a biosorbent. *Process Biochemistry,* 2005. 40(1): p. 119-124.

[223] Kyzas, G.Z., Lazaridis, N.K. and Mitropoulos, A.C., Removal of dyes from aqueous solutions with untreated coffee residues as potential low-cost adsorbents: Equilibrium, reuse and thermodynamic approach. *Chemical Engineering Journal,* 2012.

[224] Foo, K. and Hameed, B., Adsorption characteristics of industrial solid waste derived activated carbon prepared by microwave heating for methylene blue. *Fuel Processing Technology,* 2012. 99: p. 103-109.

[225] Atmani, F., Bensmaili, A. and Mezenner, N., Synthetic textile effluent removal by skin almonds waste. *Environmental Science and Technology,* 2009. 2(4): p. 153 169.

[226] Wang, X.S., Liu, X., Wen, L., Zhou, Y., Jiang, Y. and Li, Z., Comparison of basic dye crystal violet removal from aqueous solution by low-cost biosorbents. *Separation Science and Technology,* 2008. 43(14): p. 3712-3731.

[227] Kannan, C., Buvaneswari, N. and Palvannan, T., Removal of plant poisoning dyes by adsorption on Tomato Plant Root and green carbon from aqueous solution and its recovery. *Desalination,* 2009. 249(3): p. 1132-1138.

[228] Ahmad, R., Studies on adsorption of crystal violet dye from aqueous solution onto coniferous pinus bark powder (CPBP). *Journal of Hazardous Materials,* 2009. 171(1): p. 767-773.

[229] Bhatti, H.N., Akhtar, N. and Saleem, N., Adsorptive Removal of Methylene Blue by Low-Cost Citrus sinensis Bagasse: Equilibrium, Kinetic and Thermodynamic Characterization. *Arabian Journal for Science and Engineering,* 2012: p. 1-10.

[230] Banat, F., Al-Asheh, S., Al-Ahmad, R. and Bni-Khalid, F., Bench-scale and packed bed sorption of methylene blue using treated olive pomace and charcoal. *Bioresource Technology,* 2007. 98(16): p. 3017-3025.

[231] Vadivelan, V. and Kumar, K.V., Equilibrium, kinetics, mechanism, and process design for the sorption of methylene blue onto rice husk. Journal of Colloid and Interface *Science,* 2005. 286(1): p. 90-100.

[232] Sharma, Y.C. and Upadhyay, S.N., Removal of a cationic dye from wastewaters by adsorption on activated carbon developed from coconut coir. *Energy & Fuels,* 2009. 23(6): p. 2983-2988.

[233] Chakraborty, S., Chowdhury, S. and Saha, P.D., Insight into biosorption equilibrium, kinetics and thermodynamics of crystal violet onto Ananas comosus (pineapple) leaf powder. *Applied Water Science,* 2012: p. 1-7.

[234] Parab, H., Sudersanan, M., Shenoy, N., Pathare, T. and Vaze, B., Use of Agro-Industrial Wastes for Removal of Basic Dyes from Aqueous Solutions. *CLEAN–Soil, Air, Water,* 2009. 37(12): p. 963-969.

[235] Chakraborty, S., Chowdhury, S. and Saha, P.D., Adsorption of Crystal Violet from aqueous solution onto NaOH-modified rice husk. *Carbohydrate Polymers,* 2011. 86(4): p. 1533-1541.

[236] Banerjee, S. and Dastidar, M., Use of jute processing wastes for treatment of wastewater contaminated with dye and other organics. *Bioresource Technology*, 2005. 96(17): p. 1919-1928.

[237] Deniz, F., Karaman, S. and Saygideger, S.D., Biosorption of a model basic dye onto Pinus brutia Ten.: Evaluating of equilibrium, kinetic and thermodynamic data. *Desalination,* 2011. 270(1): p. 199-205.

[238] Mahmoodi, N.M., Arami, M., Bahrami, H. and Khorramfar, S., Novel biosorbent (Canola hull): Surface characterization and dye removal ability at different cationic dye concentrations. *Desalination,* 2010. 264(1): p. 134-142.

[239] Deniz, F. and Saygideger, S.D., Removal of a hazardous azo dye (Basic Red 46) from aqueous solution by princess tree leaf. *Desalination,* 2011. 268(1): p. 6-11.

[240] Zou, W., Li, K., Bai, H., Shi, X. and Han, R., Enhanced cationic dyes removal from aqueous solution by oxalic acid modified rice husk. *Journal of Chemical & Engineering Data,* 2011. 56(5): p. 1882-1891.

[241] Hiwarkar, A.D. and Srivastava, V.C., Comparative studies on adsorptive removal of indole by granular activated carbon and bagasse fly ash. *Environmental Progress & Sustainable Energy,* 2014.

[242] Mittal, A., Jhare, D. and Mittal, J., Adsorption of hazardous dye Eosin Yellow from aqueous solution onto waste material De-oiled Soya: Isotherm, kinetics and bulk removal. *Journal of Molecular Liquids,* 2013. 179: p. 133-140.

[243] Sadaf, S., et al., Removal of Indosol Turquoise FBL dye from aqueous solution by bagasse, a low cost agricultural waste: batch and column study. *Desalination and Water Treatment,* 2013 (ahead-of-print): p. 1-15.

[244] Tanyildizi, M.Ş., Modeling of adsorption isotherms and kinetics of reactive dye from aqueous solution by peanut hull. *Chemical Engineering Journal,* 2011. 168(3): p. 1234-1240.

[245] Arami, M., Limaee, N.Y., Mahmoodi, N.M. and Tabrizi, N.S., Removal of dyes from colored textile wastewater by orange peel adsorbent: equilibrium and kinetic studies. *Journal of Colloid and Interface Science,* 2005. 288(2): p. 371-376.

[246] Namasivayam, C. and Kavitha, D., Removal of Congo Red from water by adsorption onto activated carbon prepared from coir pith, an agricultural solid waste. *Dyes and Pigments,* 2002. 54(1): p. 47-58.

[247] Isa, M.H., Lang, L.S., Asaari, F.A.H., Aziz, H.A., Azam R.N. and Dhas, J.P., A Low cost removal of disperse dyes from aqueous solution using palm ash. *Dyes and Pigments,* 2007. 74(2): p. 446-453.

[248] Arami, M., Limaee, N.Y., Mahmoodi, N.M. and Tabrizi, N.S., Equilibrium and kinetics studies for the adsorption of direct and acid dyes from aqueous solution by soy meal hull. *Journal of Hazardous Materials,* 2006. 135(1): p. 171-179.

[249] Sun, G. and Xu, X., Sunflower stalks as adsorbents for color removal from textile wastewater. *Industrial & Engineering Chemistry Research,* 1997. 36(3): p. 808-812.

[250] Saha, P.D., Chowdhury, S., Mondal, M. and Sinha, K., Biosorption of Direct Red 28 (Congo Red) from aqueous solutions by eggshells: Batch and column studies. *Separation Science and Technology,* 2012. 47(1): p. 112-123.

[251] Tan, L., Jain, K. and Rozaini, C., Adsorption of textile dye from aqueous solution on pretreated mangrove bark, an agricultural waste: equilibrium and kinetic studies. *Journal of Applied Sciences in Environmental Sanitation,* 2010. 5(3): p. 283-294.

[252] Abdelwahab, O., El Nemr, A., El Sikaily, A. and Khaled, A., Use of Rice Husk for adsorption of direct dyes from aqueous solution: A Case Study Of Direct F. Scarlet. *Journal of Aquatic Research,* 2005. 31: p. 1-11.

In: Agricultural Wastes
Editor: Camille N. Foster
ISBN: 978-1-63482-359-3

Chapter 9

Coffee Wastes as Adsorbents

George Z. Kyzas[1,2,*]
[1]Department of Oenology and Beverage Technology,
Technological Educational Institute of Kavala, Kavala, Greece
[2]Division of Chemical Technology, Department of Chemistry,
Aristotle University of Thessaloniki, Thessaloniki, Greece

Abstract

One of the most recent trends in environmental technology is the research turn to green chemistry. It is general accepted that one of the most promising techniques for wastewaters treatment is adsorption. In this basis, numerous adsorbent materials have been synthesized up to now. However, there is a novel concept nowadays, which promotes the use of materials with the lowest possible cost. valuation of them for removing of different pollutants (dyes, cations, anions, etc). In the last years, the instant coffee industry has experienced a constant growth as instant coffee has become one of the most popular kinds of coffee drunk by millions of people around the world. As a consequence, large amounts of coffee grounds, which are the solid residues obtained during the processing of coffee powder with hot water or steam to prepare instant coffee, have been generated worldwide (in the order of 6 millions of tons per year). This work investigates the use of coffee wastes or coffee-based materials as adsorbents for the treatment of wastewaters.

Keywords: coffee wastes, effluents, adsorption, pollutants

[*] Corresponding author address: University Campus (P.O. Box 116), Division of Chemical Technology, Department of Chemistry, Aristotle University of Thessaloniki, Thessaloniki 54124, Greece. E-mail: georgekyzas@gmail.com.

History of Coffee

Coffee has been consumed for over 1,000 years and today it is the most consumed drink in the world (more than 400 billion cups yearly) [1]. Arabia was responsible for the coffee culture propagation. The most ancient manuscripts mentioning the culture of coffee date from 575 in Yemen, but only in the century XVI in Persia, the first coffee beans were toasted to be turned into the drink that we know today.

Coffee began to be savored in Europe in 1615, brought by travelers. Germans, Frenchmen, and Italians were looking for a way of developing the plantation of coffee in their colonies. But it was the Dutchmen who got the first seedlings and who cultivated them in the stoves of the botanical garden of Amsterdam, a fact that made the drink one of the most consumed in the old continent and becoming a definitive part of the habits of the Europeans. Next, the Frenchmen were given a plant of coffee by the major of Amsterdam, and they began to cultivate in the islands of Sandwich and Bourbon. With the Dutch and French experiences, the coffee cultivation was taken to other European colonies. The European market growth favored the expansion of the plantation of coffee in African countries and was also through the European colonists that coffee reached Puerto Rico, Cuba, Suriname, São Domingos, and Guianas. Through the Guianas, coffee arrived to the north of Brazil. Then, the secret of the Arabs was spread by the entire world.

The coffee tree or shrub belongs to the family Rubiaceae. Coffee beans are produced from the plant *Coffea L.*, of which there are more than 70 species. However, only two of these species are commercially explored worldwide: *Coffea arabica* (Arabica), considered as the noblest of all coffee plants and providing 75% of world's production; and *Coffea canephora* (Robusta), considered to be more acid but more resistant to plagues, and provides 25% of world's production [2, 3]. *C. arabica* is a bush originally from Ethiopia and develops well in high altitudes (600-2,000 m), while *C. canephora* plantations adapt well in altitudes below 600 m.

Wastes in During Coffee Synthesis

The generation of residues and by-products is inherent in any productive sector. The agro-industrial and the food sectors produce large quantities of waste, both liquid and solid. Coffee is the second largest traded commodity in the world, after petroleum, and therefore, the coffee industry is responsible for the generation of large amount of residues [4]. In the last decade, the use of such wastes has been subject of several studies, but this concern did not exist in past decades (1930 to 1943) when 77 million bags of green coffee were simply burned and released to the sea and in landfills [5]. However, this is an important topic explored nowadays.

In general, "coffee residues" are generally called the solid wastes discarded from the extraction process of instant coffee manufacturing, and the final residues originated from cafeterias. In the last years, the instant coffee industry has experienced a constant growth as instant coffee has become one of the most popular kinds of coffee drink by millions of people around the world. As a consequence, large amounts of coffee grounds, which are the solid residues obtained during the processing of coffee powder with hot water or steam to prepare instant coffee, have been generated worldwide (6,000,000 t/yr) [6].

Coffee silverskin (CS) and spent coffee grounds (SCG) are the main coffee industry residues. CS is a tegument of coffee beans obtained as a by-product of the roasting process. It is a residue with high concentration of soluble dietary fiber (86% of total dietary fiber) and high antioxidant capacity, probably due to the concentration of phenolic compounds in coffee beans, as well as to the presence of other compounds formed by the Maillard reaction during the roasting process, such as melanoidins [7]. Microscopic examination shows the presence of fibrous tissues from the surface layers of the CS. The main components of these fibrous tissues are cellulose and hemicellulose. Glucose, xylose, galactose, mannose, and arabinose are the monosaccharides present in CS; glucose being found in major amounts. Proteins and extractives are also fractions present in significant amounts in this coffee waste.

SCG is a residue with fine particle size, high humidity (in the range of 80% to 85%), organic load, and acidity, obtained during the treatment of raw coffee powder with hot water or steam for the instant coffee preparation. Almost 50% of the worldwide coffee production is processed for soluble coffee preparation [8]. Therefore, SCG is generated in large amounts, with a worldwide annual generation of 6 million tons [9]. Numerically, 1 ton of green coffee generates about 650 kg of SCG, and about 2 kg of wet SCG are obtained to each 1 kg of soluble coffee produced [10].

ADSORPTION APPLICATIONS

The key-point of each adsorbent material is its adsorption capacity. Three isotherm models are given in recent literature in order to fit the experimental equilibrium data: the Langmuir equation (Eq. (1)) [11], the Freundlich equation (Eq. (2)) [12] and the combinational Langmuir-Freundlich (L-F) equation (Eq. (3)) isotherm model [13]:

$$Q_e = \frac{Q_{max} K_L C_e}{1 + K_L C_e} \tag{1}$$

$$Q_e = K_F C_e^{1/n} \tag{2}$$

$$Q_e = \frac{Q_{max} K_{LF} (C_e)^{1/b}}{1 + K_{LF} (C_e)^{1/b}} \tag{3}$$

where Q_e (mg/g) is the equilibrium concentration of pollutant in the solid phase; Q_{max} (mg/g) is the maximum amount of adsorption; K_L (L/mg) is the Langmuir adsorption equilibrium constant; K_F ($mg^{1-1/n} L^{1/n}/g$) is the Freundlich constant representing the adsorption capacity; n (dimensionless) is the constant depicting the adsorption intensity; K_{LF} $(L/mg)^{1/b}$ is the Langmuir-Freundlich constant; b (dimensionless) is the Langmuir-Freundlich heterogeneity constant.

The amount of total uptake of pollutant in equilibrium (Q_e) was calculated using the mass balance equation (Eq. (4)):

$$Q_e = \frac{(C_0 - C_e)V}{m} \qquad (4)$$

where m (g) is the mass of adsorbent; V (L) the volume of adsorbate; C_0 and C_e (mg/L) are the initial and equilibrium concentrations of pollutant in the liquid phase, respectively.

Recently, many works have been studied for the development of effective and low-cost adsorbents. A published study of Crini [14] reported in details the main low-cost and non-conventional adsorbents for dyes removal (natural materials, biosorbents, waste materials from industry and agriculture). Some of the reported adsorbents include clay materials (bentonite, kaolinite), zeolites, siliceous material (silica beads, alunite, perlite), agricultural wastes (bagasse pith, maize cob, rice husk, coconut shell) [14-16], industrial waste products (waste carbon slurries, metal hydroxide sludge) [14, 17], biosorbents (chitosan, peat, biomass) and others (starch, cyclodextrin, cotton) [14, 18]. However, there is a lack of literature dealing with the possible application of coffee residues as adsorbents (i.e for metals, dyes, phenols, etc) [19-24].

Nowicki et al. studied the sorption properties of chars and activated carbons obtained from coffee industry waste materials toward hydrogen sulphide [25]. The effects of pyrolysis temperature and method of activation as well as porous structure, acid-base character of the surface and mineral matter content on the efficiency of H_2S removal are checked. Moreover, four different variants of adsorption test are applied, in order to estimate the optimal conditions of hydrogen sulphide capture. Depending on the method of activation, the adsorbents prepared are characterized by diverse textural parameters, strong basic or medium-acidic character of the surface and various mineral matter content, varying from 1.2 to 58.4 wt%. The results obtained in our study have proved that through an appropriate choice of pyrolysis conditions and activation procedure for coffee industry waste materials it is possible to obtain adsorbents with high capacity of hydrogen sulphide, reaching to 281.5 mg H_2S/gads. The results of our study have also shown that the adsorption ability of activated carbons prepared depends first of all on the conditions of adsorption test, mineral matter content and basicity of the surface and only to a small degree on porous structure development.

The removal of lead by untreated coffee grounds was also investigated in a packed bed up-flow column [26]. The adsorbent used in this study was coffee grounds coming from cafeterias and constitute a waste. It was used with no further treatment just only dried at the ambient air. The experiments were conducted to study the effect of important design parameter such as flow rate (5, 7 and 10 mL/min). Data confirmed that the breakthrough curves were dependent on flow rate. At a bed height of 7.5 cm and flow rate of 10 mL/min, the metal-uptake capacity of coffee grounds for lead was found to be 78.95 mg/g. The breakthrough time increased and the saturation time decreased with the increase of flow rate. The Adams-Bohart, Thomas and BDST models were applied to the adsorption under varying experimental conditions to predict the breakthrough curves and to evaluate the model parameters of the fixed-bed column that are useful for process design. The Adams-Bohart model was in good agreement with the experimental data. The untreated coffee grounds column study states the value of the excellent adsorption capacity for the removal of Pb (II) from aqueous solution.

Yeung et al. prepared activated charcoal by the pyrolysis of coffee ground waste residues impregnated by phosphoric acid and potassium hydroxide at 600 °C [27]. In our study, spent coffee was collected from the university canteen and the prepared activated charcoal impregnated with potassium hydroxide gave the highest surface area of 708 m^2/g, determined by using the methylene blue adsorption method. The activated charcoal was characterized by using scanning electron microscopy and Fourier transform infrared spectroscopy. In order to ensure that the activated charcoal produced is bio-safe, all samples were tested for possible toxic substances, including toxic heavy metals by using ICP-OES, and toxic organics such as acrylamide by using chromatography methods. In addition, all samples showed no ecotoxicity towards Escherichia coli and no cytotoxicity toward human HaCaT skin cells. The phosphoric acid activated charcoal demonstrated a high Q_{max} of 95.2 and 38.2 mg/g, respectively for lead and copper adsorption. For potassium hydroxide activated charcoal, the values of Q_{max} were found to be 45.4 and 21.2 mg/g, respectively for lead and copper adsorption.

A study of Kumaraswamy presented presents the results of studies carried out on sorption of zinc ions from aqueous solutions by a coffee industry waste s a low-cost sorbent [28]. It was found that crushed coffee industry waste possess relatively high sorption capacity as 46.05 mg/g, when comparing with other sorbents. The biosorption studies were determined as a function of initial metal ion concentrations, pH, biosorbent dosage and biosorbent particle size. About 0.1g of coffee industry waste was found to be enough to remove 71.66% of zinc for 20 mg/L of metal ion concentration from 30 mL aqueous solution. The optimum pH value was found to be 7. The experimental equilibrium data were tested with the biosorption isotherms like Langmuir, Freundlich and their equilibrium parameters were determined. The best-adjusted model to the experimental equilibrium data for coffee industry waste was the Freundlich model.

Another work is concerned with a comparative study to evaluate the adsorptive properties of activated carbons produced from coffee grounds [29]. Activation was done using different chemical activation agents such as H_3PO_4 and $ZnCl_2$ either separately or mixed together. Characterization of these prepared samples was carried out by determining their physicochemical properties such as specific surface area, porosity, and surface acidity. The results indicated that the produced carbonaceous materials generally developed different and interesting porous structures. Indeed, H_3PO_4-activated carbon (PPAC-P) develops a porous volume of 0.98 cm^3/g and a surface acidity of 317.6 meq/100 g (with a high mesoporous proportion). By contrast, $ZnCl_2$-activated carbon (PPAC-Z) seems to be the most microporous material with a rather limited surface acidity. Adsorption trials with Malathion, a commercial pesticide widely used in Algeria, were carried out onto produced samples of carbonaceous materials in batch experiments at 30°C. Various parameters such as pH, contact time, nature, and amount of adsorbent were investigated for removal efficiency under different operational conditions. Results herein showed that an interesting removal efficiency of 96% was achieved under the following conditions: pH 6, adsorbent amount of 1 g/L, and an equilibrium time of 60 min.

Shen and Gondal used exhausted coffee ground powder (CGP) which proved to be efficient adsorbent for the removal of Rhodamine dyes (i.e., Rhodamine B and Rhodamine 6G) from aqueous solutions by batch adsorption experiments [30]. The morphology, chemical structure as well as the surface property of the as-prepared CGP adsorbent were investigated by using SEM, FTIR and contact angle meter analytical techniques. The adsorption kinetics and isotherm behaviors of Rhodamine molecules onto CGP were studied and compared using pseudo-1st, pseudo-2nd and Langmuir/Freundlich models, respectively. The maximum adsorption capacities of Rh B and Rh 6G were calculated at 5.255 and 17.369 mmol/g by Langmuir model fitting. The effects of temperature, ionic strength, solution volume and the co-existing anions on the sorption behavior were also investigated. Furthermore, the adsorption mechanism responsible for the efficient removal of dyes is discussed in terms of adsorption process caused by electrostatic and intermolecular forces.

Zuorro and Lavecchia investigated another type of coffee (low-grade coffee beans (LCBs)) as a potential low-cost adsorbent for the removal of methylene blue (MB) from wastewater [31]. The waste was characterized by SEM analysis and FTIR spectroscopy. Equilibrium and kinetic experiments were performed to study the adsorption process. The equilibrium data were found to be well described by the Langmuir model, from which a maximum adsorption capacity of 476.2 mg/g was derived. A half-adsorption time ranging from 12.5 to 96.2 min was estimated by fitting the experimental kinetic data to the pseudo-second-order model.

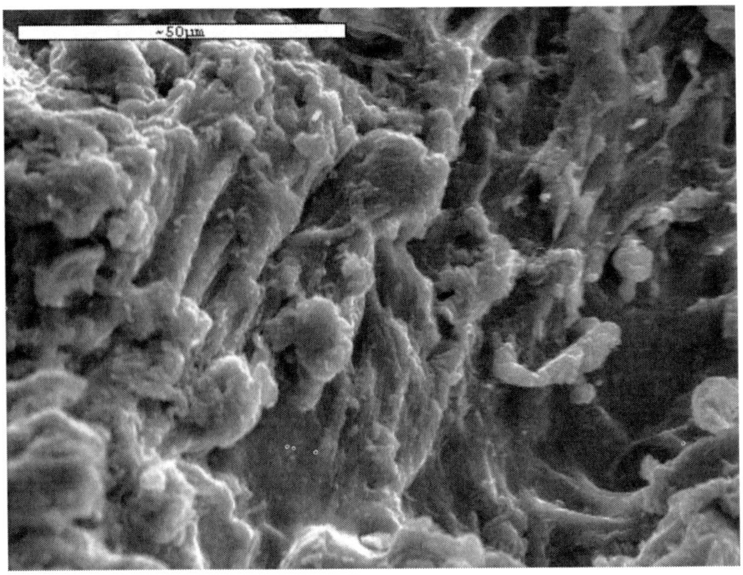

Figure 1. Scanning Electron Micrograph of the untreated coffee residues. Reprinted with permission by Elsevier [33].

Kyzas et al. studied the decolorization of industrial textile wastewaters firstly in batch mode and after the optimization of these conditions, in fixed-bed columns using Greek coffee wastes (COF) as low-cost adsorbents [32]. In this attempt, there is a cost potential given that there was no further modification of COF (just only washed with distilled water to remove dirt and colour, and dried in oven). Also, tests were realized both in synthetic and real textile wastewaters for comparative reasons. The optimum pH of adsorption was acidic (pH 2) for synthetic effluents, while experiments in free pH (non-adjusted) were carried out for real effluents. Equilibrium data were fitted to the Langmuir, Freundlich and Langmuir-Freundlich (L-F) model. The calculated maximum adsorption capacities (Q_{max}) for total dye (reactive) removal at 25 °C was 241 mg/g (pH 2) and 179 mg/g (pH 10). Kinetic data were fitted to the pseudo-first, -second and -third order model. The optimum agitation rate and the optimum pH for desorption were determined. The reuse potential showed 10 cycles of adsorptiondesorption, in which the reduction in adsorption percentages from the 1st to 10th cycle was 4% for COF adsorbents. Furthermore, experiments dealing the increase of mass of adsorbent showed a strong increase in total dye removal. To direct apply the optimized batch conditions, the same adsorbents were used in fixed-bed columns, performing a complete continuous adsorption/reuse system with low-cost.

A similar study of the above research team used industrial wastes of coffee (untreated coffee residues, UCR) were used as low-cost adsorbents for the removal of dyes (reactive and basic) from single-component aqueous solutions [33]. The cost potential is high given for the non-further treatment of the coffee residues (just only washing with distilled water to remove dirt/color, and drying in oven). The characterization of adsorbents was carried out with SEM micrographs (Figure 1), titrations for point of zero charge (PZC), BET analysis for surface area, titrations (Boehm method) for surface functional groups, and energy dispersive X-ray microanalysis (EDX) for elemental analysis/composition. The optimum pH found after adsorption experiments was pH 2 for reactive and pH 10 for basic dyes. The most inetersting of this study is the explanation of the pH-adsorption mechanism. To explain the possible pH-mechanism, the determination of PZC played an important role.

According to PZC, which is determined to be 3.2-3.4, for pH values above of 3.5, a predominant negatively charged surface of UCR is occurred. At lower pH values, the surface charge may get mainly positively charged. In the case of reactive dye (RB), the molecule was instantly dissociated as follows:

$$D-SO_3Na \xrightarrow{H_2O} D-SO_3^- + Na^+ \quad (5)$$

Furthermore, the coffee adsorbent was further protonated in acidic pH values (given the already positive charge found from PZC experiments):

$$UCR \xrightarrow{pH<3.5} UCR^+ \quad (6)$$

As a result, the adsorption process mainly proceeds through electrostatic interaction between the two counterions (UCR^+ and $D-SO_3^-$):

$$UCR^+ + D-SO_3^- \rightleftarrows UCR^{+\,-}O_3S-D \quad (7)$$

Increasing the pH of the solution (pH>3.5), due to the findings from PZC experiments, the surface of UCR is charged negatively:

$$UCR \xrightarrow{pH>3.5} UCR^- \quad (8)$$

However, the RB still remains in negatively charged form ($D-SO_3^-$). So, the interactions decrease, illustrating the reduction of dye (RB) removal. Also, UCR still adsorb dye molecules at intermediate values of pH region (6-8), but in lower percentages (31-22%).

This fact can be explained with a combination of other interactions, as Van der Waals forces and hydrogen bonding [34].

That consideration was strengthened from the surface functional groups of UCR, which are mainly found to be equally carboxylic and basic. In the highly alkaline region (pH 10-12), the strong deprotonation of UCR with simultaneous existence of phenolic groups, in line with the existence of the atom of chloride in the monochlorotriazinyl anchor of dye molecule (functional group) helped to still adsorb in extremely alkaline conditions. The resulting dissociated groups can substitute the chloride atom from the dye molecule in a similar manner as in the dyeing process of textiles, by covalent bonding [34]:

$$UCR-OH\ (\text{phenolic groups}) \rightarrow UCR-O^{-} \tag{9}$$

$$UCR-O^{-}+D-Cl \rightarrow UCR-O-D+Cl^{-} \tag{10}$$

In the case of basic dye (BB), the molecule presented constant positive charge due to the shift of electrons from oxygen to chloride and the already occurrence of amino groups. As described above, the coffee adsorbent was further protonated in acidic pH values (given the already positive charge found from PZC experiments). As a result, the adsorption process is very limited (14%). However, increasing the pH of the solution (pH>3.5), due to the findings from PZC experiments, the surface of UCR is charged negatively. In this way, the deprotonation of the surface of UCR is taken place, transformed to the negatively charged form. So, an interaction between the negative UCR and positive dye begins:

$$UCR^{-}+D^{+} \rightleftarrows UCR^{-}D^{+} \tag{11}$$

In alkaline conditions, where the full deprotonation dominates, the interaction between adsorbent and dye is completely controlled by electrostatic and coulombic strong forces, between the negatively charged functional groups of UCR (especially the carboxylic groups) and the constant localized positive charge of the cationic dye:

$$UCR-COO^{-}+D^{+} \rightleftarrows UCR-COO^{-}D^{+} \tag{12}$$

The equilibrium experimental data were fitted to Langmuir, Freundlich, and Langmuir-Freundlich model (best correlation: R^2>0.997). The calculated maximum adsorption capacities (Q_{max}) for the reactive dye at 25 °C were 179 mg/g (pH 2) and 295 mg/g for the basic one (pH 10) (Figure 2).

Pseudo-first, -second, and -third order kinetic equations were used to fit the kinetic data (pseudo-second order equation presented the most sufficient correlation, R^2>0.992). Some other adsorption parameters, as agitation rate, initial dye concentration and temperature (25, 45 and 65°C) were also determined.

The desorption was evaluated with experiments for the optimum desorption pH and desorption kinetics, while the ability of reuse was determined with 10 cycles of adsorptiondesorption (the reduction in adsorption percentages from the 1st to 10th cycle was approximately 7% for both dyes). Additionally, experiments in dyeing mixtures were realized in dyeing mixtures composed of (i) separately reactive or basic dyes, and (ii) simultaneously reactive and basic dyes. After the dosage of 3 g/L of adsorbent, a very slight change was observed in equilibrium for all types of dyeing mixtures studied.

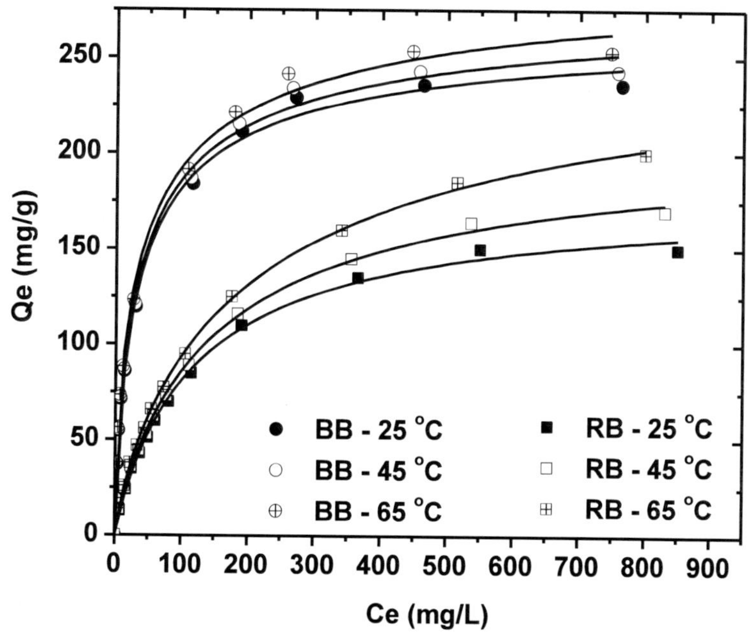

Figure 2. Effect of initial dye concentration and temperature (isotherms) on adsorption of RB and BB onto untreated coffee residues (pH=2, RB; pH=10, BB; 0-1000 mg/L dye concentration; 1 g/L adsorbent; T=25, 45, 65 °C; 140 rpm; 24 h contact). Reprinted with permission by Elsevier [33].

Roh et al. used some waste coffee-grounds (CG) with micro- and macropores as potential biosorbents for the removal of organics or heavy metal ions from aqueous solutions [35]. In several studies, CG was used as adsorbent for removal of heavy metal ions and organics (phenolic compounds). We investigated the potential application of CG as biosorbents for the removal of acid dye (Acid Red 44). To evaluate objectively the adsorption performance of the CG, conventional adsorbent (DA, Degussa alumina) was also tested and our previous reported data for mesoporous materials compared. In adsorption kinetics, experimental data followed the pseudo-second-order kinetic model and intraparticle diffusion was rate-controlled. The maximum uptake (Q_{max}) capacity of CG proved half of DA, but its adsorption rate was fast (less than 1 h). Namely, Q_{max} of CG is 27.8 mg/g, and smaller than that of mesoporous adsorbents. However, coffee-ground biosorbent still possesses economical advantages compared to inorganic adsorbents.

Zahoor prepared a type of carbon from Turkish coffee residue and used it to remove crystal violet dye from aqueous solutions [36]. The adsorbent was characterized by surface analyzer, FTIR spectroscopy, Boehm titration and mass titration. Adsorption was studied as a function of pH, dye concentration and contact time. Pseudo-first-order, pseudo-second-order and intraparticle diffusion models were used to describe the kinetic data. The adsorption of crystal violet dye on prepared adsorbent followed pseudo second order kinetic model rather than pseudo first order kinetic model. The adsorption equilibrium data were analyzed using the Langmuir, Freundlich and Temkin isotherms model. The pH of the solution affects the adsorption capacity of the adsorbent. An increase in the adsorption capacity of the dye was observed with increase in pH. However, this was up to pH 6.5. The increase was insignificant from pH 6.5 to 10. The desorption experiments showed the possibility of regeneration of the adsorbent. Equilibrium was attained within 4 hour for 30 mg/L of dye concentration, while for 25 mg/L and 15 mg/L it was less than 4 h.

Another study studied the use of coffee residues as biosorbents for xenobiotics removal [37]. Spent coffee grounds were magnetically modified by contact with water-based magnetic fluid. This new type of magnetically responsive biocomposite materials can be easily separated by means of commercially available magnetic separators or strong permanent magnets. Magnetic coffee grounds can be used as an inexpensive magnetic adsorbent for the removal of water-soluble dyes. Seven dyes (Crystal violet, Malachite green, Amido black 10B, Congo red, Bismarck brown Y, Acridine orange and Safranin O) were used to study the adsorption process. The dyes adsorption could be described with the Langmuir isotherm. The maximum adsorption capacities reached the value 73.4 mg of dye per g of dried magnetically modified coffee grounds for acridine orange; it corresponds to 276.6 mmol/g. This adsorbent can also be used for magnetic solid-phase extraction of crystal violet from extremely diluted solutions. To conclude, magnetic modification of spent coffee grounds resulted in the formation of a new, promising adsorbent for selected xenobiotics removal.

Another type of study was published by Khenniche et al., in which carbons were prepared from coffee residues using chemical activation with $ZnCl_2$ [38]. Among five carbons prepared by varying the activating agent ratio (mass of $ZnCl_2$/mass of coffee residue) from 0 to 100 %, the one with an activation ratio equal to 25 % (AC 25%) was the most effective sorbent showing the maximum phenol uptake (68%). Consequently, all the adsorption experiments were achieved with the carbon having an activation ratio equal to 25%. A comparative study of prepared activated carbon (AC 25%) and a commercial activated carbon (CAC) was undertaken to determine their capacities for phenol removal. For each adsorbent-phenol system, a pseudo-second order kinetic model described the adsorption kinetics accurately at all concentrations and temperatures for the two systems. The thermodynamics of the phenol-CAC and phenol-AC 25% systems indicate an exothermic process. Phenol adsorption isotherms onto the prepared and commercial activated carbons have been studied. They display two plateaus. The degree of coverage of the surface of two carbons by the phenol molecules was calculated, and it was revealed that the second plateau, appearing at high concentrations, is assigned to the desorption of water molecules fixed on the surface oxygen groups of the activated carbons and the occupation of these sites by phenol molecules in excess in the treated solutions. The Langmuir, Freundlich, and Elovich models were tested.

The same team used coffee wastes as source material to prepare activated carbons by chemical activation with $ZnCl_2$ [39]. The influence of impregnation ratio ($ZnCl_2$/coffee residue) on the physical and chemical properties of prepared carbons is studied in order to optimize this parameter. Texture properties of these carbons were determined by measuring the adsorption of nitrogen at 77 K. The nitrogen adsorption isotherms were interpreted by BET and Dubinin-Radushkevick (D-R) equations. The nature of carbon surface functionalities was studied by Boehm titration method. Phenol and salicylic acid removal from aqueous solutions by adsorption onto the prepared activated carbons was investigated.

The effect of parameters such as pH, agitation time, initial phenol and salicylic acid concentrations, temperature, adsorbent dosage and particle size on phenol and salicylic acid removal were observed. In addition, adsorption kinetics and adsorption isotherms study were realized. Maximum phenol removal was obtained at pH 3 and 20 °C, while for salicylic acid it was obtained at pH 3 and 25 °C. In the isotherm studies, Langmuir and Freundlich isotherm models were applied and it was observed that the phenol experimental data were perfectly described by the Langmuir model while the salicylic acid experimental data were correctly fitted by both Langmuir and Freundlich equations. Batch adsorbent capacity (Q_{max}) was calculated as 55 mg/g for phenol and 128 mg/g for salicylic acid. The rates of adsorption were found to conform to pseudo-second order kinetics with good correlation.

The potential to remove chromium(VI) from aqueous solutions through biosorption using coffee husk was also investigated [40]. The effects of pH, contact time, initial concentration and adsorbent dosage on the adsorption of Cr(VI) were studied. The data obeyed Langmuir and Freundlich adsorption isotherms. The Langmuir adsorption capacity was found to be 44.95 mg/g. Desorption studies indicated the removal of 60% of the Cr(VI). Infrared spectral studies revealed the presence of functional groups, such as hydroxyl and carboxyl groups, on the surface of the biomass, which facilitates biosorption of Cr(VI).

Boudrahem et al. showed that lignocellulosic materials are good precursors for the production of activated carbon. In this work, coffee residue has been used as raw material in the preparation of powder activated carbon by the method of chemical activation with zinc chloride for the sorption of Pb(II) from dilute aqueous solutions [41]. The influence of impregnation ratio ($ZnCl_2$/coffee residue) on the physical and chemical properties of the prepared carbons was studied in order to optimize this parameter. The optimum experimental condition for preparing predominantly microporous activated carbons with high pore surface area (890 m^2/g) and micropore volume (0.772 cm^3/g) is an impregnation ratio of 100%. The developed activated carbon shows substantial capability to adsorb Pb(II) ions from aqueous solutions and for relative impregnation ratios of 75 and 100%, the maximum uptake is practically the same. Thus, 75% represents the optimal impregnation ratio. Batch experiments were conducted to study the effects of the main parameters such as contact time, initial concentration of Pb(II), solution pH, ionic strength and temperature. The maximum uptake of lead(II) at 25 °C was about 63 mg/g of adsorbent at pH 5.8, initial Pb(II) concentration of 10 mg/L, agitation speed of 200 rpm and ionic strength of 0.005 mol/L. The kinetic data were fitted to the models of pseudo-first order and pseudo-second order, and follow closely the pseudo-second order model. Equilibrium sorption isotherms of Pb(II) were analyzed by the Langmuir, Freundlich and Temkin isotherm models. The Freundlich model gives a better fit than the others. Results from this study suggest that activated carbon produced from coffee residue is an effective adsorbent for the removal of lead from aqueous solutions and that $ZnCl_2$ is a suitable activating agent for the preparation of high-porosity carbons.

The potential use of spent coffee dusts was also investigated for the removal of Cr(VI) from aqueous solution by Prabhakaran et al. [42]. The removal mechanism was identified as the reduction reaction of Cr(VI) to Cr(III), followed by Cr(III) sorption to the biomass. The phenolic compounds in coffee dusts serve as electron-donor groups for rapid reduction of Cr(VI). The pH edge experiments revealed that Cr(VI) reduction by coffee dusts was independent of pH whereas reduced Cr(III) adsorption onto biomass was strongly dependent on pH. Isotherm experiments revealed that coffee dusts possess maximum chromium uptakes of 39.0 mg/g (pH 4). Among the two isotherm models (Langmuir and Toth), the Toth model better described the chromium biosorption isotherms with high correlation coefficients and low percent error values. A kinetic model based on the redox reaction between Cr(VI) and biomass successfully described the kinetic data.

Oliveira et al. proposes an alternative use for coffee husks (CH), a coffee processing residue, as untreated sorbents for the removal of heavy metal ions from aqueous solutions [43]. Biosorption studies were conducted in a batch system as a function of contact time, initial metal ion concentration, biosorbent concentration and pH of the solution. A contact time of 72 h assured attainment of equilibrium for Cu(II), Cd(II) and Zn(II). The sorption efficiency after equilibrium was higher for Cu(II) (89-98% adsorption), followed by Cd(II) (65-85%) and Zn(II) (48-79%). Even though equilibrium was not attained in the case of Cr(VI) ions, sorption efficiency ranged from 79 to 86%. Sorption performance improved as metal ions concentrations were lowered. The experimental sorption equilibrium data were fitted by both Langmuir and Freundlich sorption models, with Langmuir providing the best fit (R^2>0.95). The biosorption kinetics was determined by fitting first and second-order kinetic models to the experimental data, being better described by the pseudo-second-order model (R^2>0.99). The amount of metal ions sorbed increased with the biosorbent concentration in the case of Cu(II) and Cr(VI) and did not present significant variations for the other metal ions. The effect of the initial pH in the biosorption efficiency was verified in the pH range of 4-7, and the results showed that the highest adsorption capacity occurred at distinct pH values for each metal ion. A comparison of the maximum sorption capacity of several untreated biomaterial-based residues showed that coffee husks are suitable candidates for use as biosorbents in the removal of heavy metals from aqueous solutions.

Kaikake et al. studied the feasibility of using coffee beans after being dripped and degreased (DCB) as an adsorbent for base metals such as Cu(II), Zn(II), Pb(II), Fe(III) and Cd(II) were examined [44]. The plant cell wall in DCB has the porous structure from the scanning electron microscopy (SEM) analysis, and the specific surface area was determined to be 1.2 m^2/g using the specific surface area analyzer. Batch adsorption experiments on DCB were carried out at various pHs in order to elucidate the selectivity of metal ions. All metals were adsorbed at low pH region (3.0-5.0). Of particular interest was the adsorption characteristics of Cd(II) on DCB. The adsorption isotherm for Cd(II) at pH 8 fitted with a Langmuir equation to yield an adsorption equilibrium constant of 55.2 $mmol/dm^3$ and an adsorption capacity of 5.98×10^{-2} mmol/g. The desorption of Cd(II) was easily achieved over 90% by a single batchwise treatment with an aqueous solution of hydrochloric acid or nitric acid at more than 0.01 mol dm-3. These results suggested that DCB behaves as a cation exchanger.

The adsorption behavior of heavy metals on arabica and robusta roasted coffee beans was investigated by Minamisawa et al. [45]. To adsorb heavy metals, the coffee beans residues were suspended in aqueous solutions containing Cu(II) or Cd(II). Then the amount of heavy metal remaining in the solution was measured by atomic absorption spectrometry. The results show that the adsorption percentage of the heavy metal ions were above 90% for all coffee beans examined. Further, the adsorption capacities of Cu(II) and Cd(II) ions onto blend coffee were about 2.0 mg/g. This adsorption capacity is similar to that of zeolite, activated carbon and chitosan; and is higher than that of chitin and cerite. Blend coffee was thus found to be a good adsorbent for the removal of heavy metals from wastewater.

REFERENCES

[1] Sobesa, C. (2008). http://www.sobesa.com.br.

[2] Belitz, H.D., Grosch, W., Schieberle, P. (2009). Coffee, tea, cocoa. *Food Chem.*, 938-951.

[3] Etienne, H. (2005). Somatic embryogenesis protocol: Coffee (*Coffea arabica L.* and *C. canephora P.*). *Protocol for Somatic Embryogenesis in Woody Plants* 77, 167-179.

[4] Nabais, J.M.V., Nunes, P., Carrott, P.J.M., Ribeiro Carrott, M.M.L., García, A.M., Díaz-Díez, M.A. (2008). Production of activated carbons from coffee endocarp by CO2 and steam activation. *Fuel Processing Technology* 89, 262-268.

[5] Cunha, M.R. (1992). Apêndice estatístico. *150 Anos De café*, 286-388.

[6] Mussatto, S.I., Machado, E.M.S., Martins, S., Teixeira, J.A. (2011). Production, Composition, and Application of Coffee and Its Industrial Residues. *Food and Bioprocess Technology* 4, 661-672.

[7] Borrelli, R.C., Esposito, F., Napolitano, A., Ritieni, A., Fogliano, V. (2004). Characterization of a New Potential Functional Ingredient: Coffee Silverskin. *Journal of Agricultural and Food Chemistry* 52, 1338-1343.

[8] Ramalakshmi, K., Rao, L.J.M., Takano-Ishikawa, Y., Goto, M. (2009). Bioactivities of low-grade green coffee and spent coffee in different in vitro model systems. *Food Chem.* 115, 79-85.

[9] Tokimoto, T., Kawasaki, N., Nakamura, T., Akutagawa, J., Tanada, S. (2005). Removal of lead ions in drinking water by coffee grounds as vegetable biomass. *J. Colloid Interface Sci.* 281, 56-61.

[10] Pfluger, R.A. (1975). Soluble coffee processing. *Solid Wastes: Origin, Collection, Processing, and Disposal.*

[11] Langmuir, I. (1918). The adsorption of gases on plane surfaces of glass, mica and platinum. *J. Am. Chem. Soc.* 40, 1361-1403.

[12] Freundlich, H. (1906). Over the adsorption in solution. *Z. Phys. Chem.* 57, 385-470.

[13] Tien, C. (1994). Adsorption Calculations and Modeling. Butterworth-Heinemann, Boston, U.S.A., pp.

[14] Crini, G. (2006). Non-conventional low-cost adsorbents for dye removal: A review. *Bioresour. Technol.* 97, 1061-1085.

[15] Fiorentin, L.D., Trigueros, D.E.G., Módenes, A.N., Espinoza-Quiñones, F.R., Pereira, N.C., Barros, S.T.D., Santos, O.A.A. (2010). Biosorption of reactive blue 5G dye onto drying orange bagasse in batch system: Kinetic and equilibrium modeling. *Chem. Eng. J.* 163, 68-77.

[16] Kaushik, C.P., Tuteja, R., Kaushik, N., Sharma, J.K. (2009). Minimization of organic chemical load in direct dyes effluent using low cost adsorbents. *Chem. Eng. J.* 155, 234-240.

[17] Konaganti, V.K., Kota, R., Patil, S., Madras, G. (2010). Adsorption of anionic dyes on chitosan grafted poly(alkyl methacrylate)s. *Chem. Eng. J.* 158, 393-401.

[18] Kyzas, G.Z., Kostoglou, M., Vassiliou, A.A., Lazaridis, N.K. (2011). Treatment of real effluents from dyeing reactor: Experimental and modeling approach by adsorption onto chitosan. *Chem. Eng. J.* 168, 577-585.

[19] Azouaou, N., Sadaoui, Z., Djaafri, A., Mokaddem, H. (2010). Adsorption of cadmium from aqueous solution onto untreated coffee grounds: Equilibrium, kinetics and thermodynamics. *J. Hazard. Mater.* 184, 126-134.

[20] Boonamnuayvitaya, V., Chaiya, C., Tanthapanichakoon, W., Jarudilokkul, S. (2004). Removal of heavy metals by adsorbent prepared from pyrolyzed coffee residues and clay. *Sep. Purif. Technol.* 35, 11-22.

[21] Ahmad, M.A., Rahman, N.K. (2011). Equilibrium, kinetics and thermodynamic of Remazol Brilliant Orange 3R dye adsorption on coffee husk-based activated carbon. *Chem. Eng. J.* 170, 154-161.

[22] Baek, M.H., Ijagbemi, C.O., O, S.J., Kim, D.S. (2010). Removal of Malachite Green from aqueous solution using degreased coffee bean. *J. Hazard. Mater.* 176, 820-828.

[23] Franca, A.S., Oliveira, L.S., Ferreira, M.E. (2009). Kinetics and equilibrium studies of methylene blue adsorption by spent coffee grounds. *Desalination* 249, 267-272.

[24] Oliveira, L.S., Franca, A.S., Alves, T.M., Rocha, S.D.F. (2008). Evaluation of untreated coffee husks as potential biosorbents for treatment of dye contaminated waters. *J. Hazard. Mater.* 155, 507-512.

[25] Nowicki, P., Skibiszewska, P., Pietrzak, R. (2014). Hydrogen sulphide removal on carbonaceous adsorbents prepared from coffee industry waste materials. *Chem. Eng. J.* 248, 208-215.

[26] Azouaou, N., Sadaoui, Z., Mokaddem, H.: 2014, 'Removal of lead from aqueous solution onto untreated coffee grounds: A fixed-bed column study', Chemical Engineering Transactions, pp. 151-156.

[27] Yeung, P.T., Chung, P.Y., Tsang, H.C., Cheuk-On Tang, J., Yin-Ming Cheng, G., Gambari, R., Chui, C.H., Lam, K.H. (2014). Preparation and characterization of bio-safe activated charcoal derived from coffee waste residue and its application for removal of lead and copper ions. *RSC Advances* 4, 38839-38847.

[28] Kumaraswamy, K., Dhananjaneyulu, B.V., Vijetha, P., Siva Jagadish Kumar, M. (2014). Removal of zinc from aqueous solution using coffee industry waste. *International Journal of Pharma and Bio Sciences* 5, B201-B209.

[29] Bouchenafa-Saïb, N., Mekarzia, A., Bouzid, B., Mohammedi, O., Khelifa, A., Benrachedi, K., Belhaneche, N. (2014). Removal of malathion from polluted water by adsorption onto chemically activated carbons produced from coffee grounds. *Desalination and Water Treatment* 52, 4920-4927.

[30] Shen, K., Gondal, M.A. (2013). Removal of hazardous Rhodamine dye from water by adsorption onto exhausted coffee ground. *Journal of Saudi Chemical Society*.

[31] Zuorro, A., Lavecchia, R.: 2013, 'Removal of methylene blue from aqueous solution by adsorption on low-grade green coffee beans', Advanced Materials Research, pp. 72-76.

[32] Kyzas, G.Z., Lazaridis, N.K., Mitropoulos, A.C. (2012). Optimization of batch conditions and application to fixed-bed columns for a sequential technique of total color removal using "greek coffee" residues as materials for real dyeing effluents. *Journal of Engineering Science and Technology Review* 5, 66-75.

[33] Kyzas, G.Z., Lazaridis, N.K., Mitropoulos, A.C. (2012). Removal of dyes from aqueous solutions with untreated coffee residues as potential low-cost adsorbents: Equilibrium, reuse and thermodynamic approach. *Chem. Eng. J.* 189-190, 148-159.

[34] Zollinger, H. (1987). *Color Chemistry: Synthesis, Properties and Applications of Organic Dyes and Pigments.*

[35] Roh, J., Umh, H.N., Yoo, C.M., Rengaraj, S., Lee, B., Kim, Y. (2012). Waste coffee-grounds as potential biosorbents for removal of acid dye 44 from aqueous solution. *Korean J. Chem. Eng.* 29, 903-907.

[36] Zahoor, M. (2012). Removal of crystal violet from water by adsorbent prepared from turkish coffee residue. *Tenside, Surfactants, Deterg.* 49, 107-113.

[37] Safarik, I., Horska, K., Svobodova, B., Safarikova, M. (2012). Magnetically modified spent coffee grounds for dyes removal. *Eur. Food Res. Technol.* 234, 345-350.

[38] Khenniche, L., Benissad-Aissani, F. (2010). Adsorptive removal of phenol by coffee residue activated carbon and commercial activated carbon: Equilibrium, kinetics, and thermodynamics. *Journal of Chemical and Engineering Data* 55, 4677-4686.

[39] Khenniche, L., Aissani, F. (2009). Characterization and utilization of activated carbons prepared from coffee residue for adsorptive removal of salicylic acid and phenol: Kinetic and isotherm study. *Desalination and Water Treatment* 11, 192-203.

[40] Ahalya, N., Kanamadi, R.D., Ramachandra, T.V. (2010). Removal of hexavalent chromium using coffee husk. *International Journal of Environment and Pollution* 43, 106-116.

[41] Boudrahem, F., Aissani-Benissad, F., Aït-Amar, H. (2009). Batch sorption dynamics and equilibrium for the removal of lead ions from aqueous phase using activated carbon developed from coffee residue activated with zinc chloride. *J. Environ. Manage.* 90, 3031-3039.

[42] Prabhakaran, S.K., Vijayaraghavan, K., Balasubramanian, R. (2009). Removal of Cr(VI) ions by spent tea and coffee dusts: Reduction to Cr(III) and biosorption. *Ind. Eng. Chem. Res.* 48, 2113-2117.

[43] Oliveira, W.E., Franca, A.S., Oliveira, L.S., Rocha, S.D. (2008). Untreated coffee husks as biosorbents for the removal of heavy metals from aqueous solutions. *J. Hazard. Mater.* 152, 1073-1081.

[44] Kaikake, K., Hoaki, K., Sunada, H., Dhakal, R.P., Baba, Y. (2007). Removal characteristics of metal ions using degreased coffee beans: Adsorption equilibrium of cadmium(II). *Bioresour. Technol.* 98, 2787-2791.

[45] Minamisawa, M., Nakajima, S., Minamisawa, H., Yoshida, S., Takai, N. (2005). Removal of copper(II) and cadmium(II) from water using roasted coffee beans. In: *Environmental Chemistry: Green Chemistry and Pollutants in Ecosystems*, pp. 259-265.

In: Agricultural Wastes
Editor: Camille N. Foster

ISBN: 978-1-63482-359-3

Chapter 10

APPLICATIONS OF AGRICULTURAL WASTES ON BIO-HYDROGEN PRODUCTION WITH BACTERIAL TREATMENTS

Shao-Yi Hsia[1,*], Shiuh-Kuang Yang[2] and Huan-Chen Kuo[2]

[1]Department of Mechanical & Automation Engineering, Kao-Yuan University, Kaohsiung City, Taiwan
[2]Department of Mechanical and Electro-Mechanical Engineering, National Sun Yat-sen University, Kaohsiung City, Taiwan

ABSTRACT

At the present time, the demand for energy, goods, and materials is surging because of advanced technology and population growth. However, earth's resources are limited. For this reason, the issues concerning using resources effectively and converting them into energy are important. Taiwan creates a vast amount of agricultural waste every year, which is traditionally burned and buried. We do not reuse and recycle agricultural waste, and air pollution is increased when wastes are burned. Therefore, it is necessary to create methods for recycling and reusing agricultural wastes and to transform them into an energy source. This chapter is separated into two parts. The first part will convert agricultural waste into sugar. Agricultural waste is replete with wood fiber that can be reduced into sugar by a microbial method. The second part will use the biological hydrogen production capability of *Clostridium acetobutylicum* ATCC824, with sugar being added to the process. Also, this chapter used ultrasonic treatment for the production of biological hydrogen and calculated the natural frequency of ATCC824. The experiment was designed using the Taguchi method for increasing hydrogen production by using an ultrasonic treatment. Our results showed that the best combination is a temperature of 37 °C, 0.5 MHz ultrasonic frequency, 136 mW/cm^2 ultrasonic intensity, 10 s exposure time, pH 7.5, and a bacterial concentration of 20%. The outcome of our research can be applied to the production of biomass energy and the research and

* Corresponding author: syhsia@cc.kyu.edu.tw.

development of ripening techniques for accelerating fermentable food with biomechatronics.

Keywords: agricultural waste, bacterial, biological hydrogen production, ultrasonic, Taguchi method

1. INTRODUCTION

With the development and demands on present technology, energy development is urgently required. Fossil fuels have played a critical role since the Industrial Revolution. Present technology, industry, and transportation consume large amounts of fossil fuels, resulting in their short supply. As a result, with the urgent demand for clean and pollution-free energy, international researchers have been devoted to developing clean, efficient, and renewable environmental protection energy sources. Among the various emerging energy sources, hydrogen is the only one that does not emit pollutants when burned because it combines with oxygen to become water, which is not a greenhouse gas nor does it otherwise damage the environment. Hydrogen is regarded as the most secure and clean energy source on earth. In comparison with other fuels like petroleum, coal gas, and ethanol, hydrogen is harmless to human beings and the environment. Hydrogen, as the optimal choice for life and the environment at present, can therefore be used to replace fossil fuels. Hydrogen production is presently accomplished by thermochemical [1], electrochemical, and Biological methods. Since thermochemical and electrochemical methods require thermal and electric energy for reducing water to hydrogen and oxygen, so much energy would be consumed in those processes that they would not be environmentally friendly. Biological methods that apply microorganisms to produce hydrogen are hence the focus of potential hydrogen production methods. Substrates are required when hydrogen is produced by microorganisms, and glucose at high concentration and purity is used as a substrate to achieve high efficiency and large-scale production. Glucose is mainly acquired from agricultural processes; however, so much thermal and electric energy is consumed and agricultural waste produced that this is inconsistent with environmental protection goals. The technical literature reveals that the sugar required for the microbial decomposition of plant cellulose for hydrogen production can be utilized. Hence, this chapter aimed to collect and effectively decompose the cellulose in agricultural waste to extract sugar that can be used for biological hydrogen production. The latter included the biomechatronic effects of the hydrogen production process and changes in temperature, pH, and bacterial concentration, and applied ultrasound toward enhancing the growth of hydrogen-producing bacteria.

Several countries transform cellulose into carbohydrates for the subsequent production of biofuels like biodiesel and bioethanol as additives to fossil fuels [2]. The agricultural technology of Taiwan is ahead of other countries and yields diverse and abundant crops, including sugarcane, rice, bananas, pineapples, and various vegetables, in which sugarcane comprises a large proportion of crops. After squeezing sugarcane, the remaining plant fiber (bagasse) is regarded as an agricultural waste. However, bagasse contains a lot of high-glucose fiber, and how to renew and reuse its cellulose is an important issue.

Lin (2008) [3] shattered pineapple leaves with homogenizers and treated its cellulose with high-pressure steam containing various chemicals, and included enzymatic hydrolysis to investigate the possibility of sugar production with cellulose. He found that pineapple leaf cellulose pretreated with sodium hydroxide (NaOH) increased the sugar yield, comparing it to nongliusuan pretreatment, and that the increased temperature affected the concentration of sugar produced with enzymatic hydrolysis. Tai (2004) [4] separated thermostable cellulose bacteria T4 from compost from the Taiwan Sugar Corporation and compared it with other cellulose bacteria, finding that domestic cellulose bacteria T4 effectively decomposes cellulose at 60 °C and transforms it into glucose. This high-temperature characteristic could be applied to enhancing the mass-production of sugar with cellulose hydrolysis.

With the advance of technology and a deeper exploration of life, we are able to use ultrasound to affect the structure, state, and function of organism tissues or change their biological effects. The applications of ultrasound to organisms, the food industry, and medicine used to be limited to elimination and destruction, as it shattered and killed cells or other tissues with high-intensity sound and changed their chemical composition, or destroyed pathological tissues with physical approaches such as shattering stones or tumors, ultrasonic cleaning, and ultrasonic sterilization. Nevertheless, the possible applications of ultrasound to the constructive biological effects of activation, fostering, and treatment have seldom been discussed. Reviewing the previous development of destructive ultrasound biological effects, Ahmed and Russell [5] explored the death of microorganisms with ultrasound and hydrogen peroxide in 1975. Dooley et al. [6] irradiated mouse thymocyte suspensions with continuous and pulsed ultrasound at frequencies of 0.5 and 1 MHz and argued that medical ultrasound transducers did not appear to have an absolute correlation with damage to suspension cells. Van [7] discussed the flow of cysts and oocytes in protozoa by irradiating with ultraviolet and ultrasound in 2002. Broda [8] reduced the Bacillaceae attached in test tubes or on resin by increasing the temperature and adding high-acid disinfectants as well as increasing the temperature and adding ultrasound irradiation to study and observe the survival rate of bacteria. In terms of constructive ultrasound biological effects, Chiu [9] observed the biological reaction of *Paramecium* with ultrasound irradiation and estimated the quantity of protein. Chang [10] applied the mechanical effects of ultrasound in brewing rice wine in order to shorten the period of fermentation without destroying the quality of the wine. Guo et al. (2010) [11] emulsified silt with ultrasonic irradiation in water treatment plants to investigate hydrogen production from the bioreaction of anaerobic fermentation at particular acoustic intensities and exposure times and showed that low-energy ultrasound irradiation enhanced the hydrogen production rate and yield. Hsia et al. (2012) [12] utilized ultrasound irradiation during the dark anaerobic fermentation of silt and starch and discovered that ultrasound irradiation enhanced hydrogen yield and production rate and reduced the starch residual. Guo and Fei enhanced hydrogen production efficiency by irradiating silt with different approaches; however, there were various strains of bacteria in the silt that also generate methane gas during fermentation, which reduces the concentration of hydrogen. Also, bacteria strains that do contribute to hydrogen production were not selected, bacteria strains having favorable hydrogen production could not be further analyzed, and the natural frequency could not be calculated, but simply proceeding ultrasound irradiation with natural frequency. Furthermore, pH, temperature, and the quantity of bacteria in the silt affect their growth environment, activity, and hydrogen production efficiency, but were not discussed.

Consequently, dark fermentation was used for producing hydrogen in this chapter. *Clostridium acetobutylicum* ATCC 824 from the Food Industry Research and Development Institute of Taiwan was selected for our dark fermentation hydrogen production experiments wherein glucose from agricultural wastes was used as the growth medium to produce hydrogen in an anaerobic bioreactor. The mechanical effects of ultrasound irradiation at natural and non-natural frequencies were utilized for hydrogen production and acoustic intensity and exposure time were adjusted to enhance hydrogen production. Meanwhile, temperature, bacteria concentration, and pH were also adjusted. The Taguchi method was utilized for planning the experiment and recording hydrogen production. The integration of ultrasound and biohydrogen techniques was used to determine optimal hydrogen production conditions that would result in increased environmental protection.

2. Fundamental Theory

2.1. Extracting Sugar from Agricultural Wastes

The goal of this chapter was to effectively decompose waste bagasse cellulose with a biological method for reducing the amount of sugar used as a nutrient in hydrogen-producing bacteria. Figure 1 is the experiment flow chart for this chapter. This research tried to develop a method to decompose cellulose in agricultural waste (Figure 2) to extract sugar as the substrate for the bio-hydrogen production system.

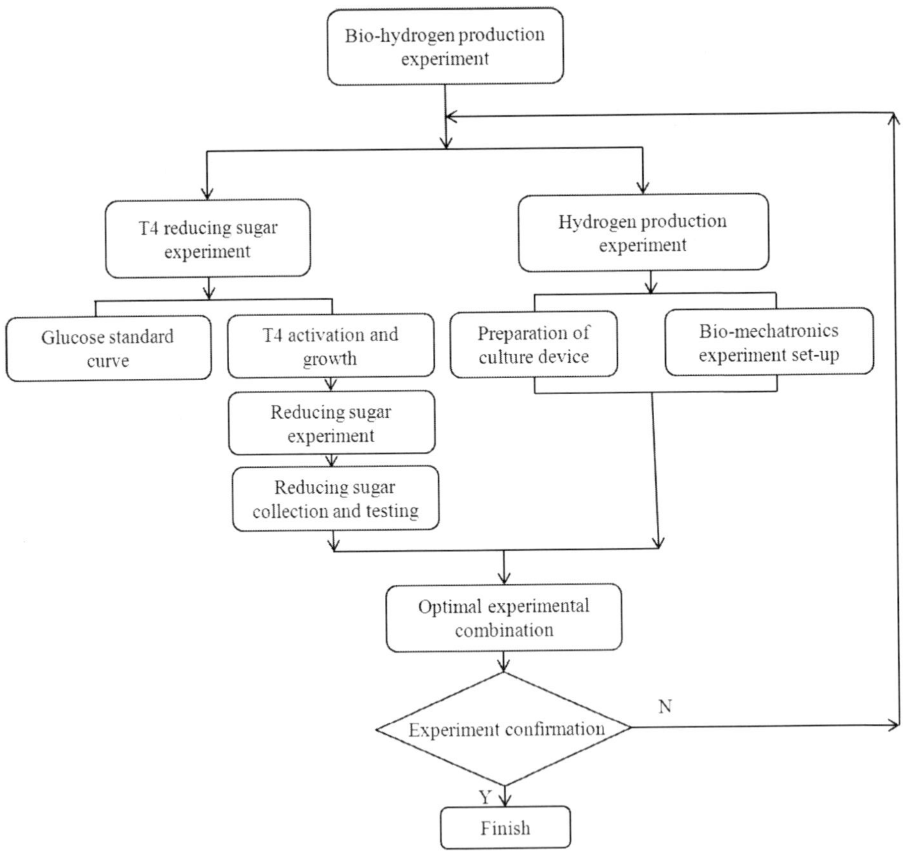

Figure 1. Experiment flow chart of bio-hydrogen production experiment.

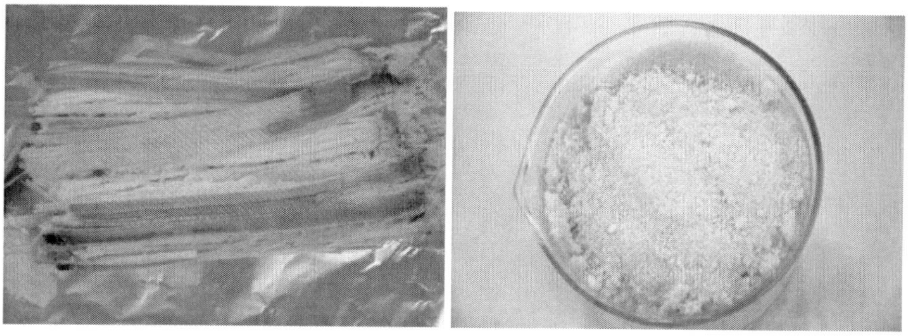

Figure 2. Cane fiber without treatment and post-treatment.

The composition and structure of lignocelluloses is discussed in the first part. The resources and characteristics of bacteria and required growth elements for cellulose bacteria cultivation and generation are explained in the second and third parts.

2.1.1. Lignocellulose

Lignocelluloses is regarded as the most abundant biomaterial, as it comes from many sources including agricultural residuals and wastes (bagasse, hull, stalk, and vegetable residuals), forest wastes (branches and wooden meal), and solid wastes from cities and towns. Cheap and sufficient materials are therefore available [13]. Lignocelluloses are composed of cellulose, hemicellulose, and lignin, which are also the major components of recycling resources [14]. They can be combined with other polysaccharides, such as xylan, mannan, polyuronides, and some galactans [15]. We acquired bagasse from the Lioho Tourist Night Market, the skin was removed to obtain soft sugarcane fiber, cut it into 5 cm sections, and reduced to powder with a homogenizer. After passing it through a No. 10 sieve, powder was baked at 80 °C for 10 hr.

2.1.2. Thermostable Cellulolytic Bacteria

Tai (2004) [4] separated thermostable cellulose bacterium T4 from compost from Taiwan Sugar Corporation and compared it with other cellulose bacteria. He found that domestic cellulose bacteria T4 effectively decomposed cellulose at 60 °C and transformed it into sugar (Figure 3). The high temperature could be utilized by industries to enhance the mass production of sugar with cellulose hydrolysis. This experiment therefore utilized bacteria strain T4 to transform cellulose into sugar for further hydrogen production. The 60 °C growth temperature is suitable for the high-temperature environments of industry. Bacteria strains for the present study were purchased from Bioresource Collection and Research Center (BCRC), numbered BCRC 17200.

2.1.3. Growth Medium

Nutrient agar broth and bagasse broth were also introduced into the growth medium. The major difference is that nutrient agar broth was used for initial T4 activation and growth. The growth medium contained rich nutrition and was suitable for the growth and activation of bacteria strains. On the other hand, bagasse broth was used for effectively transforming bagasse cellulose, the major nutrition source of T4, into sugar.

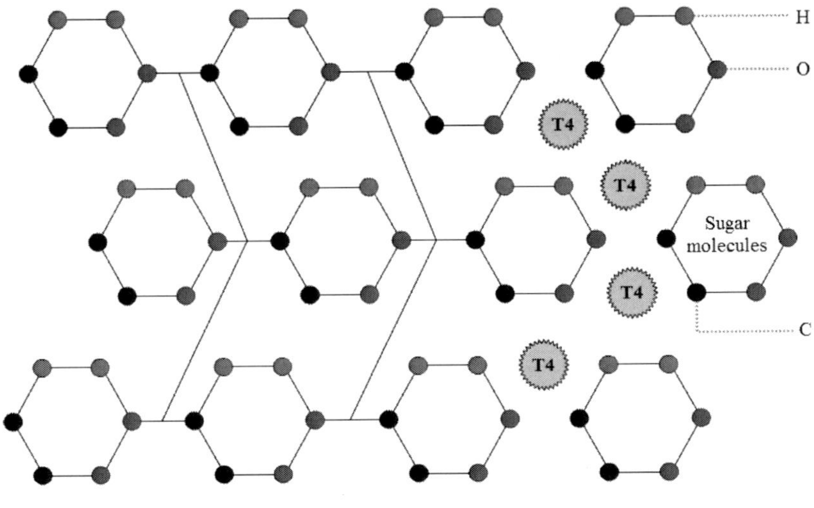

Figure 3. G. Thermostable Cellulolytic Bacteria T4.

2.1.4. Reduced Sugar in the Sample

To appropriately dilute the sample solution, 0.5 ml of the solution and 0.5 ml of DNS indicator were heated in a water tank at 100 °C for 5 min until the color of the indicator changed. The solution was then cooled in a cold-water bath, 3 ml of distilled water added, and a spectrophotometer measuring a wavelength of 540 nm utilized to measure absorbency. The amount of reduced sugar in the sample was calculated by comparison to the absorbency-to-glucose curve.

2.2. Approach and Introduction to Hydrogen Production

There are various options for hydrogen production, which are classified into thermochemistry, electrochemistry, and biological processing. The dark fermentation biological process, which effectively transforms organisms into energy through decomposition and biotransformation, was utilized for this experiment because it was regarded as the most suitable way to achieve environmental protection along with economic efficiency and resource recycling.

Anaerobic bacteria were utilized for dark fermentation hydrogen production in this experiment, as they effectively decompose organisms and produce hydrogen. Using glucose as the matrix, the reaction formula of hydrogen production with anaerobic bacteria is

$$C_6H_{12}O_6 + 2\,H_2O \rightarrow 2\,CH_3COOH + 2\,CO_2 + 4\,H_2 \quad (1)$$

$$C_6H_{12}O_6 \rightarrow C_3H_7COOH + 2\,CO_2 + 2\,H_2 \quad (2)$$

where CH_3COOH is acetic acid and C_3H_7COOH is butyric acid.

2.2.1. Biological Effects of Ultrasound [8]

Sonic waves transmit energy through a flexible and elastic medium by the propagation of waves generated from molecular movement. Sonic waves cannot be transmitted in a vacuum.

The movement of sonic waves is a physical phenomenon, with its properties being described in terms of wave frequency, wavelength, number of waves, and wave speed. Ultrasound is generally defined as high-frequency sound that cannot be heard by human ears. People normally hear sonic waves in the range of 20 Hz to 20 kHz; therefore, ultrasounds are sonic waves > 20 kHz.

Ultrasound transmits energy through particle vibration in a propagation medium and forms an ultrasound field. When ultrasound irradiates biological media with distinct frequencies and intensities, various physical effects appear between ultrasound energy and particles of matter. These effects are divided into thermal and non-thermal effects, and the latter is further divided into mechanical and cavitation effects. Cavitation effects are regarded as non-thermal and have their greatest impacts on biological tissues, as ultrasound transmits with waves of condensation and rarefaction in liquid that result in cavitation effects. Cavitation appears in liquids as tiny bubbles. Tiny bubbles in the liquid experience the processes of ultrasound vibration, growth, contraction, and crashing. When cells are subjected to high shear waves generated by vibrated bubbles or bubble crashing, a series of biological reactions appear.

Rayleigh-Plesset provided the mathematical movement model for inner cavitation vibration in an incompressible liquid [16]. Applying that theory to calculate the appearance of dark fermentation rod hydrogen-producing bacteria, we then can observe the activation effects by setting different pulse intensity for the vibration of the natural frequency. The following equation was used to calculate natural frequency:

$$(\omega_r')^2 = \frac{1}{\rho R_0^2}[3\gamma(P_0 + \frac{2\sigma}{R_0}) - \frac{2\sigma}{R_0}] - (\frac{2\eta}{\rho R_0^2})^2 \tag{3}$$

where R_0 is the radius, σ is surface tension, γ is the heat capacity ratio, ρ is density, η is the viscosity coefficient, and P_0 is the pressure of the bacteria.

2.2.2. Taguchi Quality Engineering [16, 17]

In the Taguchi experimental design, the quantified experimental results are called quality characteristics, which achieve an ideal target by determining the control factors in the parameter design. In order to solve a problem, an engineer needs to understand the quality characteristics of a product and the problem, and list the factor levels of quality characteristics with a fishbone diagram or equivalent. To conduct the experiment at the most efficient cost, the orthogonal array was selected based on control factors and levels, attempting to achieve the required quality characteristics with the most precise experiments. Experimental data are further analyzed with factor response analysis and analysis of variance to adjust the control factors, allowing quality characteristics to approach the optimal design. An orthogonal array presents the same frequency at all level combinations between two rows in the experimental design. Traditionally, the orthogonal array is named $L_a(b^c)$, representing c factors with b levels in each factor for a experiments. According to the objectives of quality characteristics, different S/N ratio equations are proposed, which are divided into Nominally the Best, Smaller is Better, and Larger is Better. Larger is Better was applied to this experiment using the following equation:

$$S/N_{LB} = -10\log\left[\frac{\sum_{i=1}^{n}\frac{1}{(y_i)^2}}{n}\right] \tag{4}$$

where y_i is the measured value and n the number of repeated measurements.

The factor response table and response graph were constructed after the experiment in order to understand the effects of the factors on the objective function of the experimental results. The mean was calculated according to the S/N ratio of the factors at the same level to construct the table with effects of the factors and levels on the result, which is further transformed into the response graph.

$$M_{ij} = \frac{\sum_{k=1}^{N} S/N_{ijk}}{N} \tag{5}$$

M_{ij} is the mean of S/N ratios containing i factors and j levels, k is the k-th S/N ratio with i factors and j levels, and N is the number of experiments with i factors and j levels.

3. Experiment

This chapter aimed to produce environmentally friendly reduced sugar from agricultural waste. Bagasse was therefore added to thermostable cellulose bacteria strain T4 to decompose cellulose with microorganisms and produce reduced sugar.

3.1. Activation and Growth of Bacteria Strain T4

Following the standard procedure for activating bacteria strains from the Food Industry Research and Development Institute, this experiment aimed to activate dried bacteria strains as well as ensure activation by growing it twice and mass producing it for the experiments. First, a bottle containing a double-layered tube with bacteria strains was held under laminar flow in order to open the bottle under aseptic conditions.

A Bunsen burner was then used to heat the top of the outer tube for 30 sec. In this process, bacteria in the tube cannot be heated. Aseptic water was added after 30 sec and the outer tube broken with tweezers. Meanwhile, the sheathing paper, inner tube, and cotton were taken out. Then, 0.3～0.5 ml of sterilized nutrient agar solution was dropped into the inner tube to dissolve the bacteria into a suspended bacterial solution. Finally, 0.1 ml of suspended bacterial solution was grown on a solid plate nutrient agar broth under four zone streak plate cultivations.

3.2. Organism Growth Curve Test

The test aimed to observe the growth cycle of bacteria strains in order to draw a growth curve, which was used to determine the growth peak of the strains for further experiments. The experiment was done on an aseptic plate. A 1 ml bacterial solution was inoculated into a 100 ml aseptic nutrient agar solution in an Erlenmeyer flask and cultivated in a water tank at constant temperature.

The equipment was set at 60 °C while rotating at 120 rpm. Within a 9 hr growth period, a spectrophotometer measuring a wavelength of 600 nm was used to measure absorbency every hour to draw the growth curve of the bacteria strains.

3.3. Production of Reduced Sugar

The cultivated T4 bacterial solutions were added to aseptic Bagasse broth at ratios of 1:6, 2:6, and 3:6. The goal was to observe the effects of varying concentrations on the production of reduced sugar. Bagasse broth with bacteria strains was then cultivated in a water tank for 3 hr. Equipment was set to 60 °C at a rotation of 120 rpm. The nutrient solution was further filtered bagasse and T4 bacteria to acquire reduced sugar for further utilization.

3.4. Glucose Test of Reduced Sugar

Based on the glucose standard curve, this experiment proceeds the glucose test of reduced sugar. First, 0.5 ml reduced sugar and 0.5 ml DNS indicator were poured into a tube and placed in a water tank at a constant temperature of 100 °C for 5 min and then cooled in a cold water bath. To this was added 3 ml of distilled water, and a spectrophotometer measuring a wavelength of 540 nm was utilized to measure absorbency for comparison with the glucose standard curve. The glucose content of the reduced sugar was then calculated.

3.5. Mechanical and Biological Effects

This experiment investigated the mechanical and biological effects of ultrasound irradiation on hydrogen production (Figure 1). This included frequency, acoustic intensity, exposure time, and temperature. Natural frequencies at 0.5 and 1 MHz were calculated with the Rayleigh-Plesset equation and compared to the non-natural 0.25 MHz frequency of the transducer. Acoustic intensity was set at 25, 79, and 136 mW/cm^2 to explore the effects on the growth and hydrogen production of the bacteria under irradiation. Exposure times were 10 sec, 5 min, and 15 min at a temperature of 37 °C, which is the most suitable for biological growth, and 30 °C for comparison. Bacterial concentrations of 10, 20, and 30% were used for determining the effects of bacterial concentration on hydrogen production. The pH was set at 6.5, 7.0, and 7.5 to explore the effects of acid or alkali growth media on bacterial growth and hydrogen production (Table 1). A full factorial experiment was not suitable because there are too many parameters in the Taguchi method. The latter can be a future reference for biohydrogen by irradiating the hydrogen producing bacteria with ultrasound.

Table 1. Six factors and two/three levels of the orthogonal array $L_{12}(2^1 \times 3^5)$

Factors	Level 1	Level 2	Level 3
A. Temperature (°C)	37	30	
B. Ultrasound frequency (MHz)	0.23	0.5	1.0
C. Ultrasound intensity (mW/cm^2)	25	79	136
D. Ultrasound exposure time	10s	5 min	15 min
E. pH value	6.5	7	7.5
F. Bacterial concentration (%)	10	20	30

3.5.1. Cultivation of Hydrogen-Producing Bacteria

Clostridium acetobutylicum ATCC 824 was purchased from the Food Industry Research and Development Institute, numbered BCRC 10639. After World War I, *C. acetobutylicum* bacteria strains were widely applied to produce acetone, ethanol, and butyl alcohol at the optimal growth temperature of 37 °C. *C. acetobutylicum* ATCC 824 decomposes sugar into various products with having commercial applications. Most significantly, acetone, ethanol, and butyl alcohol release carbon dioxide and hydrogen during fermentation and metabolic processes. Accordingly, glucose was added to the growth medium to produce hydrogen after decomposition. Oxygen is poisonous to the anaerobic *C. acetobutylicum*, and it becomes dormant and forms spores for protection when oxygenated. These spores can survive for years and then become active and initiate growth when favorable anaerobic environments are reinstated.

The hydrogen-producing bacteria growth medium FTM (fluid thioglycollate medium) designated by the Food Industry Research and Development Institute was utilized for activation and reproduction of the experimental bacteria. FTM includes 5 g yeast extract, 15 g peptone, 0.5 g sodium thioglycollate, 2.5 g sodium chloride, 0.5 g L-cystine, 0.75 g agar, and glucose added to make a 20 g solution. This was dissolved in 1 L of water and sterilized for further use, with sodium thioglycollate and L-cystine being used to reduce the oxygen concentration in the growth medium.

3.5.2. Ultrasound Irradiation Experiment

The natural frequency of *C. acetobutylicum* ATCC 824 was first calculated for setting the transducer's ultrasound frequency so as to achieve a resonant response that would achieve optimal effects in the bacteria. Because bacteria and yeast are microorganisms, surface tension data from Chang [10] on the biological effects of ultrasound on yeast were used. In the present study, we used the following physical parameters in Equation (3) to derive the bacterium's natural frequencies of 0.5 and 0.96 MHz: covered surface tension σ=72.75 dyn/cm, heat capacity ratio γ=1.4, water density ρ=1 g/cm^3, viscosity coefficient η= 9.197×10^{-3} g/cm·sec, pressure P_0= 760 torr [16], and size l_0 =5-8 μm.

The natural frequency of the bacteria was in the range of 0.5-0.96 MHz using a single-crystal straight-beam longitudinal-wave immersion transducer, so 0.5 and 1.0 MHz were selected for the natural-frequency and 0.25 MHz as the non-natural frequency for comparison. The ultrasound irradiation experiment aimed to control the transducer's ultrasound intensity by amplifying the waveforms generated by a power amplifier, Figure 4. Figure 5 shows the induced ultrasound waveforms, which are combined with 10 kHz square waves and sine waves.

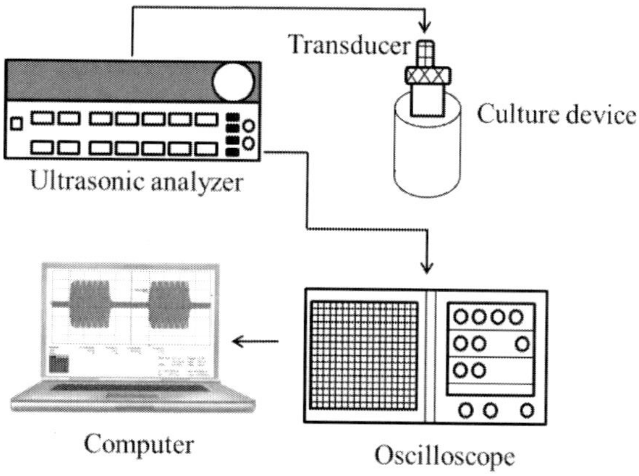

Figure 4. Layout of the ultrasonic experiment.

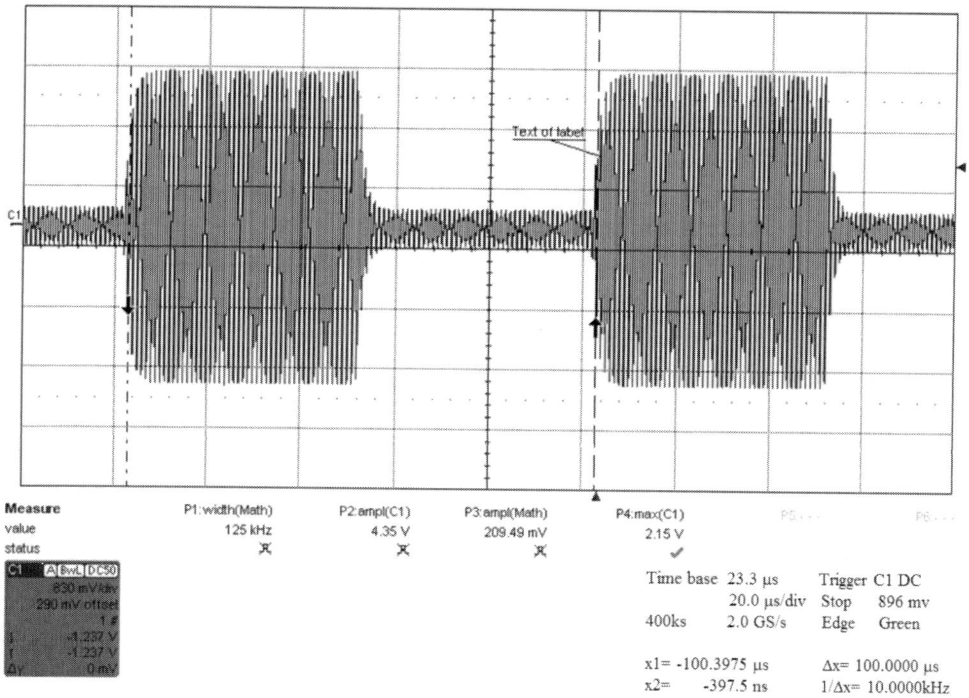

Figure 5. Ultrasound waveform induced by function generator.

Based on the transducer's irradiation frequency, sine wave frequency was set to 0.23, 0.5, and 1 MHz because the ultrasound transducer presented an unavoidable Gaussian distribution under different irradiation intensities. Nevertheless, in order to observe the changes in bacteria under different irradiation intensities, possible overlapping intensities were avoided when setting irradiation intensity.

There has been little relevant research on the effects of ultrasound exposure time on hydrogen-producing bacteria. Kao et al. utilized an ultrasonic emulsification instrument to irradiate silt and a growth medium for 10 sec. Hsia et al. [12] compared the optimal exposure time of irradiating anaerobic silt with ultrasound for 15 min and then stopping for another 15 min, and found that full-time irradiation resulted in lowered hydrogen production. However, Fei did not discuss the effects of less exposure time on hydrogen production, so a 5 min ultrasound irradiation group was added to our experiments. Exposure times were therefore set to 10 sec, 5 min, and 15 min, and irradiation was done every 8 hr.

Ultrasound irradiation parameters for hydrogen production by *C. acetobutylicum* ATCC 824 fermentation were controlled, and the control group (without ultrasound irradiation) and experimental group (with ultrasound irradiation) were separated in this chapter. Ultrasound irradiation experiments were further divided into natural and non-natural frequency test groups, both of which were also divided into three irradiation frequencies. Lastly, three exposure time periods were selected for irradiation.

3.5.3. Biological Effects

Temperature is a primary factor in the growth and fermentation of hydrogen-producing bacteria. Increasing temperatures accelerate bacteria growth and metabolism and hasten their reproduction. However, when the temperature is too high, bacterial growth is suppressed, bacteria are likely to age, fermentation products decrease in quantity, fermentation period and efficiency are largely reduced, and rapid death occurs. Under low temperatures, bacteria reduce nutrient absorption and decomposition, form spores, and enter dormancy. Temperature is, therefore, of primary concern in the fermentation process. The temperature of the growth medium in these experiments was 37 °C, which was one of the temperatures tested. The temperature was set to 30 °C in other experiments to determine the effects of ultrasound irradiation on hydrogen production when bacteria are held close to room temperature.

Mechanical and biological factors affect the growth and reproduction of microorganisms, including water quality, concentration of the growth medium, oxygen level, bacteria concentration, and pH. Sterilized distilled water was used for this experiment and the growth medium was described in Section 3.5.1. The hydrogen-producing bacteria that we used were anaerobes, so the experimental environment also had to be anaerobic. Bacterial concentration and pH affect delay time, and hydrogen production, so their impacts on hydrogen production were investigated in this experiment.

Cappelletti & Reginatto [18] utilized *C. acetobutylicum* ATCC 824 to produce hydrogen. They added a 20% bacteria liquid to a hydrogen-production fermentation system and found that higher bacterial concentrations decreased the delay time of hydrogen production to zero. For this reason, the effects of bacterial concentrations at 10, 20, and 30% on the delay time and changes in hydrogen production were explored in this experiment.

A pH of 5-9 appears to be suitable for microorganisms, although most prefer a neutral environment of pH 7. Only a few microorganisms survive in severe environments that are strongly acid (pH < 2) or strongly alkaline (pH > 10). The bacterium that we used prefers a pH $= 7$, so pH values of 6.5, 7.0, and 7.5 were used as the initial values for determining the effects on hydrogen yield, hydrogen production rate, and hydrogen production efficiency.

3.6. Taguchi Method Planning

Irradiation frequency, acoustic intensity, exposure time, temperature, pH, and bacterial concentration were integrated into the Taguchi method orthogonal array when planning our experiments. Data were analyzed with analysis of variance to determine the optimal combination for hydrogen production.

3.6.1. Factor Levels and Experimental Design

Mechanical and biological effects were considered before establishing factor levels (Table 1), which comprised two levels for temperature and three levels for the other parameters. The orthogonal array contains 12 experiments, one 2-level factor, and five 3-level factors, and can be written as $L_{12}(2^1 \times 3^5)$. Each experiment was repeated three times.

3.6.2. Taguchi Data Analysis

The Taguchi method looks for key factors and levels. Because a maximized hydrogen production was desired, the Larger is Better S/N ratio in Equation (4) was regarded as optimal. At the same control factor level, the mean S/N ratio was determined for each factor at each quality objective level. Based on this, the response table was constructed and transformed to the S/N ratio response graph.

4. Results and Discussion

To understand the effects of time on T4 bacterial growth and sugar production efficiency, the experiment was divided into two parts. In the first part, the growth curve of the T4 bacteria strain was acquired from the growth cycle and the growth peak from the microorganism's growth curve. Besides, the time was followed for acquiring the T4 bacterial solution with the best activation for the experiment of sugar production. In Figure 6, the horizontal axis represents time and the vertical axis the absorbency on the spectrophotometer. The activation of T4 was completed within 3~6 hr. Comparing the third and sixth hours, absorbency increased by a factor of 16.63. T4 apparently grew rapidly during that the time and under optimal conditions, with maximum absorbency appearing at the sixth hour.

The production of reduced sugar is shown in the second part. The previous standard curve for reduced sugar was followed before the production and analysis of reduced sugar. A DNS indicator was added to the glucose solution at different concentrations and heated until the solution's color was chemically changed. It was then diluted and a standard curve of reduced sugar to measured absorbency was drawn. The sugar production experiment was further run with ratios of bacterial-solution to bagasse-broth of 1:6 and 1:2. The concentration of reduced sugar was measured every 2 hr over an 8 hr period and graphed in Figure 7, and shows that an increasing concentration of bacteria results in insufficient nutrient availability after 2 hr. This is because the bacterial food (reduced sugar) was being consumed and not replaced, resulting in a decreasing production of sugar. In this case, the 8 h operation did not effectively produce reduced sugar. At ratios of 1:6 and 1:2, obvious impacts on sugar production appeared by hour two, whereas at the 1:6 ratio, sugar increased from 0.243 to 0.265 mg/ml by the second hour, a 0.022 mg/ml (9.05%) increase.

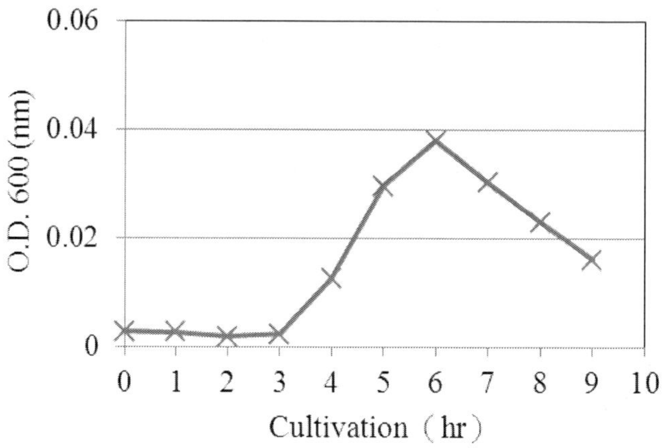

Figure 6. Growth Curve of Bacteria Strain T4.

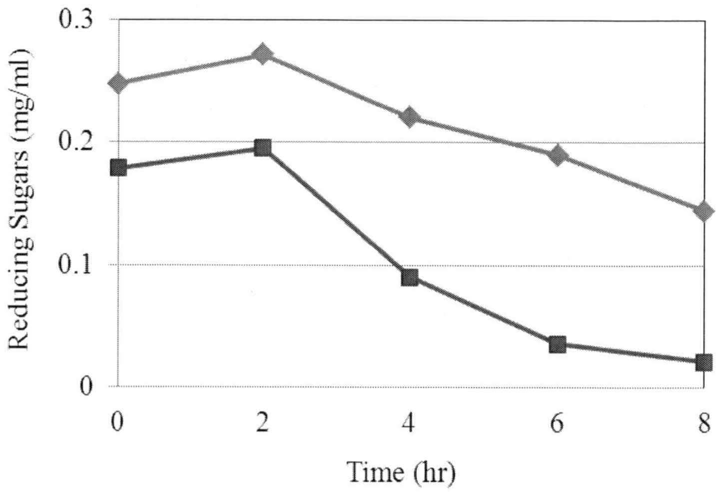

Figure 7. Reducing Sugars with ratios of bacterial-solution to bagasse-broth (◆=1 : 6, ■=1 : 2).

Both ratios appear to have better sugar reduction rates. Nevertheless, at a ratio of 1:3, sugar production was not observed by the third hour. The effects of different ratios of bacterial-solution to bagasse-broth on sugar reduction and changes in sugar production at hour two should be further explored. Experimental results are shown in Figure 8. When the ratio of bacterial-solution to bagasse-broth was 1:6, sugar production was stably increased compared to other groups and the reduced sugar level achieved its maximum by the third hour at a sugar reduction rate of 13.77% and increased sugar at 0.034 mg/ml. A 1:3 ratio did not show obvious effects other than a decrease in production by the first hour. The 1:6 ratio presents the optimal condition for reducing sugar when cultivated for 3 hr in this study (Figure 8).

Sugar cultivated under optimal conditions was utilized in the latter experiment to produce the growth medium for a hydrogen-producing bacterium, *C. acetobutylicum* ATCC 824. It was added to a growth medium (Figure 9, FTM) for hydrogen-production fermentation and ultrasound irradiation was done to foster growth. The hydrogen production experiment is depicted in a flowchart as Figure 1. The effects of control factors on hydrogen production were determined with a response table and response graph constructed in an orthogonal array according to the Taguchi method, and the optimal combination of treatments was determined. The bio-hydrogen liquid of mass culture used for the optimal experiments was appeared in Figure 10.

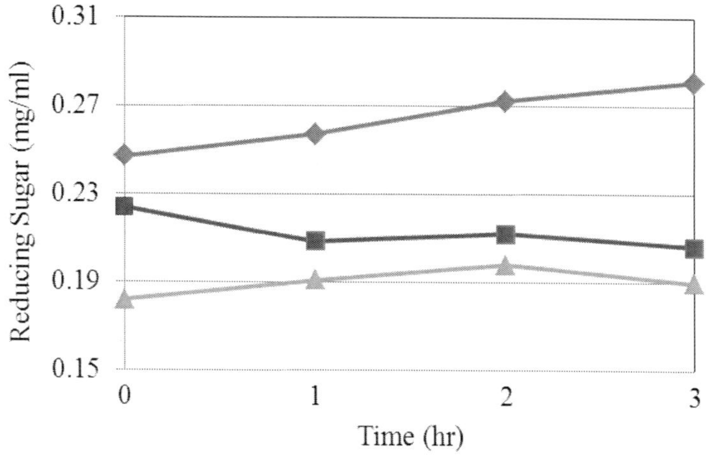

Figure 8. Reducing Sugars with ratios of bacterial-solution to bagasse-broth every 1 hr over a 3 hr period (◆=1 : 6, ■=1 : 3, ▲=1 : 2).

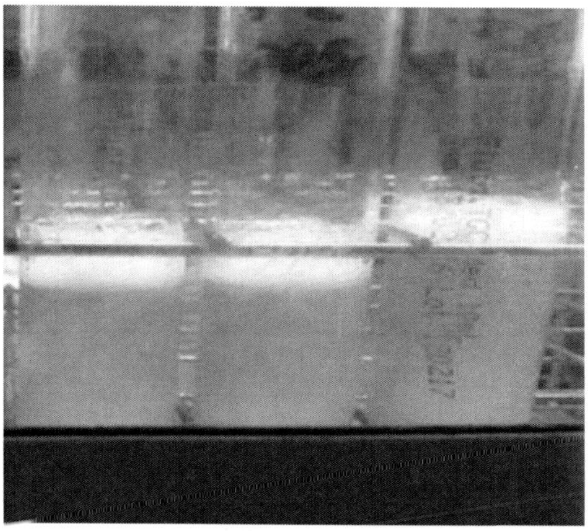

Figure 9. The hydrogen-producing bacteria growth medium FTM.

Our determination of optimized hydrogen production used the Larger is Better S/N ratio of Equation (4), which was regarded as the target of the experiment that was designed to reveal the maximum level of hydrogen production. Each experiment was repeated three times, and measured hydrogen production data are listed in Table 2 where the fifth and the sixth columns are hydrogen production and their corresponding S/N ratios, respectively, of the 12 experiments. Table 3 shows the responses of S/N ratios on the factors at the same level. For example, the factor response of Factor A at Level 1 is the mean S/N ratio of various hydrogen production experiments in Table 2, which also presents Factor C (acoustic intensity) as being able to reach 4.52 dB, which is the largest among all factors because of the level change.

Figure 10. Hydrogen liquid of mass culture used for the optimal experiments.

It is, therefore, an important factor, and is followed by B, E, D, A, and F. The response graph based on Table 3 is shown in Figure 11, where the horizontal axis represents control factors and levels and the vertical axis shows the S/N ratios for hydrogen production. Based on Figure 11, the optimal values for temperature, frequency, acoustic intensity, exposure time, pH, and bacterial concentration were 37 °C, 0.5 MHz, 136 mW/cm^2, 10 sec, 7.5, and 20%; *i.e.,* A1, B2, C3, D1, E3, and F2, respectively. This combination of control factors and levels maximizes hydrogen production.

Table 2. Experiments examining the effect of ultrasonic influences on hydrogen production

Exp.	y_1 (ml)	y_2 (ml)	y_3 (ml)	Average (ml)	S/N (dB)
1	1060	1150	1020	1076.67	60.61
2	1360	3360	2860	2526.67	66.01
3	960	1070	1100	1043.33	60.32
4	3520	3300	3420	3413.33	70.65
5	1000	1010	1000	1003.33	60.03
6	800	1080	1030	970.00	59.50
7	670	1050	1050	923.33	58.71
8	900	2170	520	1196.67	57.66
9	970	1130	1080	1060.00	60.45
10	1370	1350	1130	1283.33	62.07
11	2100	955	2740	1931.67	63.14
12	595	1270	2620	1495.00	59.22
			Ave.=	1493.61	61.53

Table 3. Response of hydrogen production based on S/N ratios

	A	B	C	D	E	F
Level 1	*62.86*	62.24	59.45	*63.28*	59.96	60.35
Level 2	60.21	*63.37*	61.17	61.04	60.99	*62.46*
Level 3		58.97	*63.97*	60.28	*63.65*	61.78
Rang	2.65	4.4	4.52	3	3.69	2.11
Rank	5	2	1	4	3	6

Values in italics represent the optimal values.

Finally, the optimal group was compared to the control group of the previous experiment without ultrasound for the hydrogen production, which was enhanced by the optimal combination of 37 °C, pH 7.0, and bacterial concentration of 20%. The comparison is shown in Table 4, in which the optimal combination presents a hydrogen production of 3413.33 ml while the control group was only 1213.33 ml.

Table 4. Comparison of experiments with control and optimal groups

	Hydrogen production
Control group (ml)	1213.33
Optimal group (ml)	3413.33
Improvement (%)	181.32

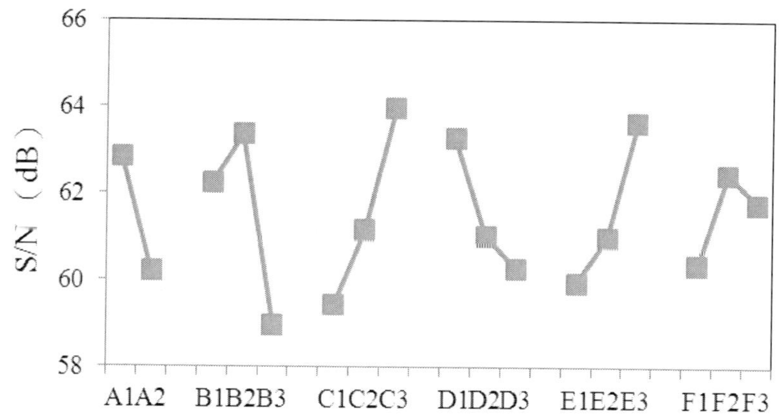

Figure 11. Response graph of S/N ratio in different levels.

Ultrasound at the calculated natural frequency, a higher acoustic intensity, a short-period irradiation, and increasing pH effectively enhance the hydrogen production by a factor of 1.81. The energy sources for ultrasound irradiation in the optimal combination were a power amplifier and signal generator, which consume 300 W (1080 kJ) per hour. Our 30 sec irradiation times consumed 9 kJ. One mole of hydrogen produces 25,447.81 ml, and the optimal group increased production by 2200 ml compared to the control group. In other words, there was an increase of 0.086 mole of hydrogen and the combustion of each mole of hydrogen released 286 kJ energy, so 0.086 mole of hydrogen produced 24.725 kJ of energy.

The ultrasound irradiation consuming 9 kJ results in a hydrogen production of 24.725 kJ of energy, which is economically effective. To evaluate the overall experiment from a similar energy perspective, the optimal group produced 3413.33 ml (0.134 mole) of hydrogen, which is equivalent to 38.361 kJ of energy. However, irradiation with ordinary 25 W light bulbs consumes 90 kJ energy, so the hydrogen produced would provide 25 W of light irradiation for 0.426 hr. Apparently, the reuse of agricultural wastes can produce environmentally clean energy. The experiments shown in this chapter are a preliminary study. More accurate results may depend on sophisticated equipments and more rigorous research methods.

CONCLUSION

This study aimed to include biomechatronics into the hydrogen production process, expecting to enhance the activity of hydrogen-producing bacteria to produce green energy hydrogen. The conclusions are listed below:

1. Bacterial solutions at higher concentrations cannot effectively enhance sugar production. Instead, lower ratios appear to have optimal sugar production efficiencies.
2. The sugar production period does not follow the organism growth curve, but shows a different time combination for T4 to bagasse-broth.
3. A 1:6 ratio for bacteria-solution to bagasse-broth and a cultivation period of 3 hr appears to yield the maximum sugar production rate.
4. The natural frequency for hydrogen-producing bacteria calculated by the Rayleigh-Plesset cavitation bubble equation was 0.5-0.96 MHz. The optimal hydrogen production occurred when irradiated with the natural frequency of 0.5 MHz.
5. Using the Taguchi method, the optimal combination for hydrogen production occurred with A1, B2, C3, D1, E3, and F2. The optimal combination is therefore suggested to be 37 °C, an irradiation frequency of 0.5MHz, acoustic intensity of 136 mW/cm^2, exposure time of 10 s, pH = 7.5, and bacterial concentration of 20%.
6. Compared to controls, the biomechatronic effects from higher acoustic intensities, shorter irradiation periods, and increasing pH levels effectively enhanced hydrogen production by a factor of 1.81.

ACKNOWLEDGMENTS

This work was supported in part by the National Science Council, Taiwan, Republic of China, under Grant NSC-100-2221-E-244 -010.

REFERENCES

[1] C. H. Wang, Fermentation Strategies for Biomass Hydrogen Production Using Anaerobic Bacterial Microflora, Ph.D. Dissertation, *Department of Chemical Engineering,* Feng Chia University, Taichung, Taiwan, 2006.

[2] W. H. Chen, K. L. Kuo, W. S. Huang and J. P. Wang, Development of Cellulosic Ethanol Technology, *Agricultural Biotechnology Industry Quarterly* 9, 62, 2007.

[3] R. B. Lin, Different Pretreatment of Agricultural Waste in the Hydrolsis of Cellulose Enzyms - An Example of Pinapple Leaf Cellulose, Master Dissertation, Department of Biomechatronics Engineering, *National Pingtung University of Science and Technology,* Pingtung, Taiwan, 2008.

[4] S. K. Tai, Characterizations and Phylogeny of Thermostable Cellulolytic Bacterial Isolates, Ph.D. Dissertation, Department of Biological Sciences, National Sun Yat-sen University, Kaohsiung, Taiwan, 2004.

[5] F. I. Ahmed and C. K. Russell, "Synergism Between Ultrasonic Waves and Hydrogen Peroxide in the Killing of Micro Organisms," *Journal of Applied Bacteriology*, Vol. 39 (1), pp. 31-40, 1975.

[6] D. A. Dooley, S. Z. Child, E. L. Cartensen and M. W. Miller, "The Effects of Continuous Wave and Pulsed Ultrasound on Rat Thymocytes *In Vitro,*" *Ultrasound in Med. & Biol.*, Vol. 9, pp. 379-384, 1983.

[7] D. W. Van, "The Effect of Ultraviolet Light, Cavitational Flow and Ultrasound on Protozoan Cysts and Oocysts, Bacteriophages and Clostridium," Water SA, Vol. 28, pp. 16-22, 2002.

[8] D. M. Broda, "The Effect of Peroxyacetic Acid-based Sanitizer, Heat and Ultrasonic Waves on the Survival of Clostridium Estertheticum Spores in Vitro," *Letters in Applied Microbiology*, Vol. 45(3), pp. 336-341, 2007.

[9] Wen-Kuei Chiu, Protein Variation of Paramecium in Ultrasonic Field, Master Dissertation, Department of Biomechatronics Engineering, National Pingtung University of Science and Technology, Pingtung, Taiwan, 2004.

[10] A. C. Chang, "Study of Ultrasonic Wave Treatments for Accelerating the Aging Process in a Rice Alcoholic Beverage," *Food Chemistry*, Vol. 92, pp. 337–342, 2005.

[11] Y. P. Guo, S. H. Kim, S. H. Sung and P. H. Lee, "Effect of Ultrasonic Treatment of Digestion Sludge on Bio-hydrogen Production from Sucrose by Anaerobic Fermentation," *International Journal of Hydrogen Energy*, Vol. 35, pp. 3450-3455, 2010.

[12] Shao-Yi Hsia, Ching-Hsiung Wang and Yu-Tuan Chou, "Study on Biohydrogen Production of Dark Fermentation with the Stimulation of Ultrasonic," 2012 International Conference on Information, *Communication and Engineering*, Dec. 15-20, Fuzhou, Fujian, P.R. China, No. 1391, 2012.

[13] Ch. E. Wyman, Twenty Years of Trials, Tribulations and Research Progress in Bioethanol Technology, *Appl. Biotech*, Vol. 91-93, p. 5021, 2001.

[14] S. Subramaniyan and P. Prema, Biotechnology of Microbial Xylanases: Enzymology, Molecular Biology, and Application, *Crit. Rev. Biotechnology*, Vol. 22, pp. 33-64, 2002.

[15] C. S. Gong, In: Tsao, G.T.(Ed.), Recen Advancens in D-xylose Conversion by Yeasts, Annual Reports of Fermentation Processes, Academic Press, Vol. 6, pp. 253-291, 1983.
[16] Shao-Yi Hsia and Shiuh-Kuang Yang, "Enhancing algal growth by stimulation with LED lighting and ultrasound," *Journal of Nanomaterials*, Vol. 2015, Article ID 531352, 2015.
[17] Shao-Yi Hsia, "Optimization of Microextrusion Preforming using Taguchi method," *Mathematical Problem in Engineering*, Vol. 2013, Article ID 305797, 2013.
[18] B. M. Cappelletti, V. Reginatto, "Fermentative Production of Hydrogen from Cassava Processing Wastewater by Clostridium Acetobutylicum," *Renewable Energy*, Vol. 36(12), pp. 3367-3372, 2011.

INDEX

#

A

B

C

D

E

F

G

H

I

N

O

P

Q

T

U

V

W

X

Y

Z